Monitoring Water Quality

Monitoring Water Quality

Pollution Assessment, Analysis, and Remediation

Satinder Ahuja

AMSTERDAM • BOSTON • HEIDELBERG • LONDON • NEW YORK • OXFORD
PARIS • SAN DIEGO • SAN FRANCISCO • SYDNEY • TOKYO

Elsevier
225, Wyman Street, Waltham, MA 02451, USA
The Boulevard, Langford Lane, Kidlington, Oxford OX5 1GB, UK
Radarweg 29, PO Box 211, 1000 AE Amsterdam, The Netherlands

British Library Cataloguing in Publication Data
A catalogue record for this book is available from the British Library

Library of Congress Cataloging-in-Publication Data
A catalog record for this book is available from the Library of Congress

ISBN: 978-0-444-59395-5

Printed and bound in Great Britain
13 14 15 16 17 10 9 8 7 6 5 4 3 2 1

Contents

Preface

Water supplies around the world have been polluted to the point where we have to purify water for human consumption. Even rainwater, nature's way of water purification, is not always pure. It is usually contaminated by various pollutants that have been added to our atmosphere. Although Earth is a water planet, fresh water comprises only 3% of the total water available to us. Of that, only 0.06% is easily accessible. This explains why over 80 countries now have a water deficit. The shortage of affordable clean water forces an estimated 1.2 billion people in the world to drink unclean water today. This causes water-related diseases that kill 5 million people a year, mostly children, around the world. The problem will not get any better—the UN estimates that 2.7 billion people will face water shortages by 2025. It is crystal clear that water is a scarce and valuable commodity and we need to sustain its quality to assure water sustainability (see Chapter 1).

Drinking water comes mainly from rivers, lakes, wells, and natural springs. These sources are exposed to a variety of conditions that can contaminate the water. The failure of safety measures relating to production, utilization, and disposal of thousands of inorganic and organic compounds can pollute our water supplies. The majority of water quality problems are now caused by diffuse nonpoint sources of pollution from agricultural land, urban development, forest harvesting, and the atmosphere. These nonpoint source contaminants are more difficult to effectively monitor, evaluate, and control than those from point sources (such as discharges of sewage and industrial waste). It should be noted that a number of water contaminants arise from the materials we use frequently to improve the quality of life (see Chapter 1).

To monitor contaminants in water, it is necessary to perform ultratrace analyses at or below parts-per-billion levels. Selective methodology that allows such separations and quantification is covered in this text. Exposures to environmental concentrations of endocrine-disrupting compounds (EDCs) are now a known threat to both human and ecological health (see Chapter 5). A large body of work has established that EDCs can agonize, antagonize, or synergize the effects of endogenous hormones, resulting in physiological and behavioral abnormalities in aquatic organisms. Among emerging contaminants of concern, the problem of pharmaceuticals and endocrine disruptors is gaining greater importance. Under the rules that the EPA finalized recently, seven sex hormones, six perfluorocarbons, and hexavalent chromium are among the 28 chemicals that utilities will have to test for in drinking water. Water quality monitoring of various contaminants is covered in most of the chapters of this book. Real-time monitoring can enable a quick response to water quality concerns that arise from natural or intentional contamination, and it allows the greatest protection of public health (see Chapter 8).

Contaminants may also come from Mother Nature, even where the soil has not been influenced by pollutants from human beings. For example, natural processes like erosion and weathering of crustal rocks can lead to the breakdown and translocation of arsenic from primary sulfide minerals (see Chapters 1, 11, 12). Other contaminants from nature include manganese, radionuclides, and various other chemicals.

Examples of water problems from point and nonpoint source pollution in less developed countries, as experienced by those in Asia, Africa, and Latin America, and in developed countries, as exemplified by the United States, are discussed in this book (see Chapters 1–4, 7, 10, 11, 13). Also covered at length are various approaches to monitor contaminants, risk assessment, analytical methods, and various remediation methods to achieve the desired water quality (see Chapters 1, 5–14). The active quantitative removal of trace levels of chemicals may require additional process treatments such as the use of in situ-generated radical species. These approaches are generally referred to as advanced oxidation/reduction processes. A variety of techniques for creating radicals in water are discussed in Chapter 9. Low-priced remediation approaches for removing arsenic are described in Chapters 10, 11, 13. Chapter 12 highlights the use of nanoparticles for removal of arsenic.

Continuous monitoring of the entire water supply network is a goal that has been achieved only recently. Historically, most monitoring outside of water treatment plants has been relegated to the occasional snapshot provided by grab sampling for a few limited parameters or the infrequent regulatory testing required by mandates such as the Total Coliform Rule. The development of water security monitoring in the years since September 11, 2001, has the potential to change this paradigm (see Chapter 14).

Many countries have been pushing simply for economic development, only to realize that the environmental costs (resource depletion, pollution, health problems, and frequent occurrences of calamities such as flooding) of such single-minded growth nullify all the gains. To understand the importance of water sustainability, it is important to realize that the three pillars of sustainability, viz., economic, social, and environmental, are intimately interwoven (see Chapters 1 and 15). Sustaining the water quality of our resources will advance only if we have data on contaminants, along with information on natural and human causative factors that affect water quality conditions. This suggests that water quality and sustainability should be given a very high priority in our society.

I believe water quality and sustainability can be achieved with the information and guidance provided in this book. It should be a useful text for academics, practitioners, regulators, and other interested individuals and groups.

Satinder Ahuja
August 16, 2012

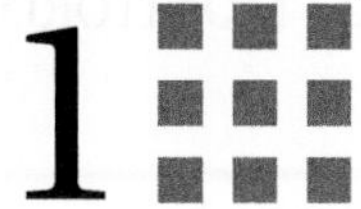

Monitoring Water Quality, Pollution Assessment, and Remediation to Assure Sustainability

Satinder (Sut) Ahuja
AHUJA CONSULTING, CALABASH, NC, USA

CHAPTER OUTLINE

Monitoring Water Quality, http://dx.doi.org/10.1016/B978-0-444-59395-5.00001-7

1.1 Introduction

Our civilization has managed to pollute our water supplies to the point where we have to purify water for drinking [1,2]. The expressions "clean as freshly driven snow" or "pure rainwater" are not true today. In the past, rain was nature's way of providing freshwater; however, rain is usually contaminated by various pollutants that we add to our atmosphere. The shortage of affordable pure water forces an estimated 1.2 billion people to drink unclean water. As a result, water-related diseases kill 5 million people a year, mostly children, around the world. The problem does not seem to be getting any better—the UN estimates that 2.7 billion people will face water shortages by 2025.

Although earth is composed largely of water, freshwater comprises only 3% of the total water available to us. Of that, only 0.06% is easily accessible. This is reflected by the fact over that 80 countries now have water deficits. It is patently clear that water is a scarce and valuable commodity and we need to sustain its quality and use it judiciously, i.e. assure water sustainability. To achieve sustainability, we must ensure that as we meet our needs, we do not compromise the requirements of future generations [3].

Drinking water comes largely from rivers, lakes, wells, and natural springs. These sources are exposed to a variety of conditions that can contaminate water. The failure of safety measures relating to the production, utilization, and disposal of thousands of inorganic and organic compounds causes pollution of our water supplies. The overwhelming majority of water-quality problems are now caused by diffuse nonpoint sources of pollution from agricultural land, urban development, forest harvesting, and the atmosphere. These nonpoint source contaminants are more difficult to effectively monitor, evaluate, and control than those from point sources, such as discharges of sewage and industrial waste. Many water contaminants arise from the materials we use frequently to improve the quality of life:

- Combustion of coal and oil
- Detergents
- Disinfectants
- Drugs (pharmaceuticals)
- Fertilizers
- Gasoline (combustion products) and additives
- Herbicides
- Insecticides
- Pesticides.

The failure of safety measures relating to production, utilization, and disposal of a large number of inorganic/organic compounds encompassing the entire range of the alphabet, from arsenic to zinc, can cause contamination of our water supplies [4]. For example, whereas zinc in small amounts is desirable, arsenic at concentrations as low as 10 parts per billion (ppb) is quite harmful.

1.1.1 What Is Potable Water?

Expressed simply, potable water is any water suitable for human consumption. National Primary Drinking Water Regulations control water quality in the United States. Water-quality regulations vary in the different parts of the world. For instance, Table 1-1A in the Appendix shows what one municipality in the United States—Brunswick County in North Carolina—does to monitor water quality.

However, it should be noted that some of the contaminants of concern that are not monitored on a regular basis include:

- MTBE (methyl *tertiary* butyl ether)
- Herbicides
- Fertilizers
- Pharmaceuticals
- Perchlorate
- Mercury
- Arsenic

1.1.2 Monitoring Water Quality

Water quality and the monitoring of various contaminants, are discussed in this book. Among emerging contaminants of concern, the problem of pharmaceuticals and endocrine disruptors is gaining greater importance. It was recently reported that liquid formula is the biggest culprit in exposing infants to bisphenol A, a potential hormone-disrupting chemical extracted from plastic containers [5].

Hexavalent chromium, six perfluorocarbons, and seven sex hormones are among the 28 chemicals that utilities will have to test for in drinking water under the rules that the United States Environment Protection Agency (USEPA) finalized recently [5a].

Contaminants may also come from mother nature, even where the soil has not been degraded by pollutants from human beings. For example, natural processes like erosion and weathering of crustal rocks, can lead to the breakdown and translocation of arsenic from primary sulfide minerals. Contaminants from nature include arsenic, manganese, radionuclides, and a host of other chemicals. Though arsenic contamination of groundwater has now been reported in a large number of countries worldwide, Bangladesh has suffered the most from this contamination. Other countries affected by the arsenic problem include Argentina, Australia, Cambodia, Canada, Chile, China, Ghana, Hungary, India, Mexico, Nepal, Thailand, Taiwan, UK, the United States, and Vietnam.

Prolonged drinking of arsenic-contaminated water can lead to arsenicosis in a large number of people, eventually resulting in a slow and painful death. It is estimated that arsenic contamination of groundwater can seriously affect the health of more than 200 million people worldwide. Arsenic (As) contamination of groundwater can occur from a variety of anthropogenic sources, such as pesticides, wood preservatives, glass manufacture, and other diverse uses of arsenic. These sources can be monitored and controlled. However, this is not so easy with naturally occurring arsenic. The natural content of arsenic in soil is mostly in a range below 10 mg/kg; however, it can cause major crises when it gets into groundwater [2].

Arsenic contamination of groundwater is described below briefly, as it serves as an excellent example of how water purity and quality problems can occur if adequate attention is not paid to monitor all the potential contaminants. In Bangladesh, groundwater contamination was discovered in the 1980s [1,2]. A large number of shallow tube wells (10–40 m), installed with the help of the United Nations Children's Emergency Fund (UNICEF) in the 1970s to solve the problem of microbial contamination of drinking water, were found contaminated with arsenic. The crisis occurred because the main focus was on providing water free of microbial contamination, a problem that was commonly encountered in surface water. Apparently, the project did not include adequate testing to reveal the arsenic. This unfortunate calamity could have been avoided, as analytical methods that can test for arsenic down to the parts-per-billion (ppb) level have been available for many years [6]. At times, speciation of a contaminant is necessary. For example, trivalent arsenic is more toxic than the pentavalent species. This demands a more selective method, and high pressure liquid chromatography-inductively couple plasma -mass spectroscopy (HPLC-ICP-MS) can resolve trivalent and pentavalent arsenic compounds at parts-per-trillion levels [7].

1.1.3 Monitoring Contaminants at Ultratrace Levels

It should be recognized that even with well-thought-out purification and reprocessing systems, trace (at parts-per million level) or ultratrace amounts (below 1 part-per-million level) of every substance present in untreated water is likely to be found in drinking water. To monitor contaminants in water, it is necessary to perform analyses at ultratrace levels. An example of ultratrace-level contaminants in drinking water in Ottawa, Canada, by gas chromatography/mass spectroscopy (GC/MS) is shown in Table 1-1 [8].

At ultratrace levels, sampling and sample preparation should be given great attention.

1.1.3.1 Sampling and Sample Preparation

It is abundantly clear that the sample used for ultratrace analysis (analysis at ppb level) should be representative of the "bulk material." The major considerations are [9]:

(1) Determination of the population of the "whole" from which the sample is to be drawn.
(2) Procurement of a valid gross sample.
(3) Reduction of the gross sample to a sample suitable for analysis.

Table 1-1 Gas Chromatography/Mass Spectroscopy of Ottawa Tap Water

Compound	Concentration Detected in Ottawa Tap Water (ppt)
α-Benzene hexachloride	17
Lindane	1.3
Aldrin	0.70
Chlordane	0.0053
Dibutyl phthalate	29
Di(2-ethylhexyl)	78
Phthalate	

The analytical uncertainty should be reduced to a third or less of sampling uncertainty [10]. Poor analytical results can also be obtained because of reagent contamination, operator errors in procedure or data handling, biased methods, and so on. These errors can be controlled by the proper use of blanks, standards, and reference samples. It is also important to determine the extraction efficiency of the method.

Preconcentration of the analyte may frequently be necessary because the detector used for quantitation may not have the necessary detectability, selectivity, or freedom from matrix interferences [11]. Significant losses can occur during this step because of very small volume losses to glass walls of recovery flasks or disposable glass pipettes and other glassware. However, with suitable precautions, preconcentration of metals at concentrations down to 10^{-12} g/g for copper, lead, and zinc, and 10^{-13} g/g for cadmium have been successfully demonstrated in a typical polar snow matrix [12]. The reader may benefit from the sample-collection techniques used by Chakraborti to analyze thousands of samples in the investigation of the problems of arsenic contamination of groundwater in various regions of Bangladesh and India (see Chapter 5 in Ref. [1]).

Discussed below are examples of water problems from point and nonpoint source pollution in underdeveloped countries, as experienced in Asia, Africa, and Latin America; as well as in developed countries as exemplified by the United States. Also discussed at length are methods used to monitor the assessment of risk of contaminants and the various remediation approaches used to achieve the desired water quality.

1.2 Water-Quality Status and Trends in the United States

Information on water quality is critical to ensuring long-term availability and sustainability of water that is safe for drinking and recreation, and suitable for industry, irrigation, and fish and wildlife (Chapter 3). Water-quality challenges are increasingly complex. First, the majority of water-quality problems are now caused by diffuse nonpoint sources from agricultural land, urban development, forest harvesting, and the atmosphere. These nonpoint-source contaminants are more difficult to effectively monitor, evaluate, and control than contaminants from point sources, such as discharges of sewage and industrial waste. Concentrations can vary from hour to hour and season to season, making it difficult to monitor and quantify possible effects on the health of human and aquatic ecosystems. A second challenge facing us is emerging diversity in water-quality issues. The dominant concerns regarding water quality were focused largely on the sanitary quality of rivers and streams. For example, in the United States, we focused on bacteria counts, oxygen levels in the water, nutrients, and a few measurements like temperature and salinity. While these are still important, the issues have become increasingly complex. Hundreds of synthetic organic compounds, such as pesticides and volatile organics in solvents and gasoline, have been introduced into the environment. Improved laboratory techniques have led to the discovery of microbial and viral contaminants, pharmaceuticals, and endocrine disruptors in our waters that were not measured before. We are also finding that many contaminants (such as arsenic and radon) can originate from a wide range of natural sources, which are of potential concern for human health, even in relatively undeveloped settings that might not be perceived as vulnerable to contamination. These natural and organic contaminants often end up in our waters as complex mixtures of organic compounds; many of these can, even at very low concentrations, potentially affect the health of humans and/or the reproductive success of aquatic organisms.

Water-quality issues are not solely determined by our activities on the landscape, but are also governed by large-scale natural processes. Natural factors (such as hydrology, geology, soils, and changing climate) control the timing and amount of surface and groundwater flow and the transport of waterborne constituents and contaminants. Human actions, such as the use and disposal of chemicals, how we convert land over time, our use of water, and our practices of land management are carried out in this context. We can sustain the quality of our water resources only if we have data on the contaminants and information on natural and human causative factors that affect water-quality conditions. All of these must be part of our investment in water monitoring and science.

1.3 Rivers in Africa Are in Jeopardy

The challenges of water use in Africa come in three forms: Water is either too much, too little, or unclean (see Chapter 2). The rainfall pattern has changed because of climatic variability. The drought and flooding cycle has changed from a frequency of 7 years to 2.5–3 years. The water is so dirty because of many factors, which may include natural processes such as seasonal trends, underlying geology and hydrology, weather and climate, as well as human activities, including domestic, agricultural, industrial, and environmental engineering activities. Some of these are analyzed below. Africa has 84 rivers, of which 17 are major ones, each with catchments of over 100,000 km^2. There are more than 160 lakes larger than 27 km^2. In addition, Africa has vast wetlands and widespread groundwater. It receives plentiful rainfall, with annual precipitation levels equal to that of Europe and North America in differentiated regions. Water withdrawal is mainly for three purposes—agriculture, domestic consumption, and industrial use at 3.8% of total annual renewals. There are large disparities in water availability among regions, with 50% availability of the surface water in the Congo basin and 75% of water found in eight other major basins.

The causes for Africa's water pollution may be found in high population growth (the population was 1 billion in 2009 and is growing at a rate of about 2.4% yearly); overstocking and overgrazing; poor agricultural practices; desertification; atmospheric deposition; industrialization; and lack of sustainable sanitary facilities. The high load of soil sediments deposited into the river systems was last estimated at 54 tons per hectare per year. Human and animal waste is deposited in water because most rural and suburban dwellings have no sewage treatment plants. Another cause is the lack of efficient municipal and industrial waste collection, treatment, and disposal. Other contributing causes arise from weak legal instruments coupled with weak enforcement of the existing laws. The lack of quality data on water to assist policy formulation, is another major cause of weak laws. The high load deposits of large quantities of nitrogen and phosphorus plant nutrients (possibly from biomass burning), sewage, industrial effluents, and dust contribute to the eutrophication of rivers and lakes.

Water- and sanitation-related diseases collectively account for 80% of sicknesses in developing countries. The waterborne diseases caused by pathogenic bacteria, viruses, and parasites (e.g. protozoa and helminths) are the most common and widespread health risks associated with drinking water. The public health burden is determined by the severity of the illnesses associated with pathogens, their infectivity, and the size of the population exposed. The coliform counts in most river waters are often very high. All human and animal wastes in the bush are carried by storm water into the rivers, causing high coliform loads.

1.4 Septic Systems in the Coastal Environment: Multiple Water-Quality Problems in Multiple Areas

The septic system is one of the most popular means of treating human sewage in coastal regions, especially in areas away from metropolitan locations, in developing and developed countries. Such systems can function well, given proper hydrological and soil conditions; however, numerous reports from coastal regions in the United States find that these systems discharge excessive levels of nutrients and fecal microbes into coastal waters, including those used for shell fishing (see Chapter 4). In many situations, coastal soils are sandy and porous, allowing rapid transmission of constituents. When this is coupled with a high groundwater table, treatment is insufficient and septic leachate plumes can extend over 100 miles into coastal waters. In other areas, such as the Florida Keys, carbonate karst soils allow similar rapid movement of pollutants from septic drain fields. Nutrient plumes contribute to coastal eutrophication, and microbial pollution is a human health issue in shellfish consumption and water recreation. Thus, alternative treatment options are required in such situations.

1.5 Assessment of Risk from Endocrine-Disrupting Compounds

Exposures to environmental concentrations of endocrine-disrupting compounds (EDCs) are now a known threat to both human and ecological health (see Chapter 6). A large body of work has established that EDCs can agonize, antagonize, or synergize the effects of endogenous hormones, resulting in physiological and behavioral abnormalities in aquatic organisms. Examples of disruption in fish include altered secondary sexual characteristics and male production of female reproductive proteins. The universe of potential EDCs is expanding as new pesticides and pharmaceuticals constantly enter the marketplace; and the monumental tasks of prioritizing the backlog of compounds to be assessed and reducing their release into the environment is essential. Until recently, the majority of EDC research has focused on reproductive impacts, particularly those caused by estrogenic compounds or, to a lesser extent, androgenic compounds. Attention is now being directed toward impacts inflicted via novel mechanisms and toward impacts on other aspects of the endocrine system. Examples of lesser-known impacts of EDCs on fish, include changes in somatic growth and modulation of the immune system. EDCs are known to disrupt pathways mediated by thyroid hormone, glucocorticoids, progestogens, and prostaglandins via receptor binding, interfering with cellular signaling cascades, or altering steroidogenesis. The challenge for ecotoxicologists is to determine the end points that should be measured in fish in order to most accurately predict the impact at the population level and even at the ecosystem level. Furthermore, in addition to assessing risk at multiple biological scales, the effects of complex environmental mixtures, differences in species sensitivity, adaptation to pollution, and the potential for epigenetic change must also be integrated into determinations of “safe” EDC concentrations. Considering the propensity of EDCs to exert effects at low doses and to exhibit unimodal or biphasic responses, this task will require increased collaboration and ingenuity among researchers in the field.

1.6 Water-Quality Monitoring and Environmental Risk Assessment

Environmental monitoring may have several goals (see Chapter 5): determining compliance with discharge and ambient standards, identifying areas with persistent problems for remediation efforts, establishing cause and effect relationships, and providing information for modeling and management tools, such as the total maximum daily loads (TMDLs). Monitoring programs need not be extensive in time and space to serve some of these goals, such as compliance determinations for point source discharges; but long-term, spatially comprehensive monitoring programs provide additional power to address larger-scale questions about environmental hazards, particularly with regard to the parameters that describe different aspects of those hazards. Such large-scale approaches lend themselves to more formal risk assessments, which can help focus attention of managers, policymakers, the regulated community, and the public at large on the most important problems.

Risk assessment protocols include four steps: identification of the hazard; identification of the population exposed to that hazard; establishment of dose response or other quantitative cause-and-effect relationships for the hazard and for the exposed population; and overall assessment of the risk posed by the specific hazard to the exposed population. Environmental hazards can be diverse, encompassing a wide array of pollutants, direct and indirect impacts, and interactive effects. In an environmental context, one must consider both risks to human users (who generally create, expose themselves to, and manage those risks), and risks to ecological integrity, the ability of an ecosystem to support its "normal" suite of ecosystem services from an anthropocentric perspective, and to support its inherent ecological properties, such as stability, resilience, and biodiversity, among others. Cause-and-effect relationships of environmental hazards are sometimes sufficiently well-established to justify regulations and standards, so that more informative risk assessments can be conducted. This chapter addresses risk assessments for four environmental parameters in aquatic ecosystems (dissolved oxygen, chlorophyll *a*, fecal coliform bacteria, and turbidity) for which regulatory standards have been established and that have been monitored extensively in time and space in southeastern North Carolina.

1.7 Analytical Measurements to Improve Nonpoint Pollution Assessments in Indiana's Lake Michigan Watershed

Lake Michigan's southern shoreline borders approximately 59 miles of northwest Indiana. A large industrial base has occupied the southern Lake Michigan shoreline, from the Illinois border of the lake to the Michigan border, since the turn of the twentieth century. This is due to the beneficial functions provided by the lake. When both Milwaukee, Wisconsin, and Chicago, Illinois, are considered, the southern Lake Michigan shoreline is home to nearly 8 million people and is heavily urbanized. Many municipal wastewater treatment plants discharge treated wastewater into the lake, and the combined sewer overflow and sanitary sewer overflow increase municipal waste pollution (Chapter 7). It is estimated that 24 billion gallons of untreated sewage and storm water flow into the Great Lakes each year. Municipal waste has increased in complexity while wastewater treatment plants have aged, thus loading freshwater bodies with chemicals of emerging concern—a problem recognized worldwide and highlighted by the USEPA.

The Salt Creek watershed in Porter County, Indiana, is a sub-watershed within the Little Calumet-Galien watershed, the only one in Indiana that discharges to Lake Michigan. The state-approved watershed management plan for Salt Creek requires monthly measurements of general chemistry parameters, discharge, turbidity, and *Escherichia coli* from May to October. This pilot study was designed to utilize the strengths of the watershed management component in combination with research science measurements and analyses, to more accurately assess the nonpoint sources of pollution. The outcomes of the first year of the study that focused on dissolved anions, fluorescent whitening agents, total suspended solids, and discharge loads from September 2010 through April 2011, demonstrate the substantial increase in comprehension gained from such management–research partnerships. While basic measurements of pH, dissolved oxygen, and other parameters offered an idea of the waterway's health, the chloride, nitrate, and total suspended solids showed significant seasonal stresses, and the chloride and nitrate concentrations provided insights into specific nonpoint pollution sources.

1.8 Real-Time and Near Real-Time Monitoring Options for Water Quality

Current water-quality issues are complex and involve a wide range of chemical and microbial contaminants that are of concern. Coupled with increasing urban populations and aging infrastructure, poor water quality can have a large impact on public health. Modern water management requires more reliable and quicker characterization of contaminants, to allow for a more timely response. Real-time monitoring can enable a quick response to water-quality concerns that arise from natural or intentional contamination and allows the greatest protection of public health (see Chapter 8).

Ideally, all water-quality monitoring should be performed in real time, allowing for the most accurate and precise view of water quality. However, there are relatively few technologies that are able to provide true real-time measurements. Traditional monitoring of water quality involves on-site sampling and the transportation of samples to testing facilities to determine physical, chemical, and biological characteristics. Laboratory methods are lengthy and expensive, and samples may be compromised during transportation. With ever-changing water characteristics, the results obtained from these trials do not necessarily reflect the current characteristics of water.

Recent advancements in both sensor technology and computer networks have brought about new technologies that can determine water characteristics in a dynamic way. Online real-time monitoring and screening methods can be used to provide up-to-date information on water systems and warn against contamination. This enables faster response times and quicker adjustments to treatment methods, reducing the risk to public health. Real-time information can be used to assess changes in water quality; identify trends and determine the state of water quality and ecosystem; identify emerging issues and contaminants; and achieve rapid screening of water for toxic substances and pathogens. Another important application is the monitoring and optimization of water and wastewater treatment processes to ensure compliance to water-quality standards. In spite of the need, there are limited options for the implementation of real-time or near real time (within 1–4 h) monitoring of water quality.

With limited funds and increasingly complex water-monitoring issues, municipalities and water utilities are looking to operate water quality monitoring strategically. Typically, a tiered approach to monitoring water quality is preferred. Such a system would check for general changes in water quality and screen for possible contaminants. Then, more specific testing would identify the type and extent of contamination. A strategic approach to water monitoring may also include the selection of specific contaminants to be tested and the frequency of testing. The quantity of testing methods selected may depend on financial restrictions as well as regulatory requirements. Prioritization should be given to tests that are easy to perform (especially if they can be done on-site), can detect a substance that presents a high risk to public health, or give a wide description of the quality of the water.

Real-time and near real-time detection methods are more limited than laboratory methods, but their variety and capabilities are continuously improving. The majority of the current real-time monitoring applications are based on one or more of the following water-quality parameters: turbidity, conductivity, temperature, dissolved oxygen, pH, and chlorophyll-a. Systems that can measure ultraviolet absorbance (UVA) at 254 nm and total organic carbon have been used for indirect measurements of dissolved organic matter (DOM). These parameters are relatively easy to measure and provide useful information on the daily, weekly, monthly, and seasonal changes in water quality. Sudden and uncharacteristic changes in these parameters also serve as early warnings against intentional contamination of water with toxic chemicals or biological agents. However, none of these parameters can identify a specific chemical or biological agent, and they merely serve as screening methods. Efforts continue to employ advanced spectroscopy and molecular methods for in situ detectors, but the majority are still in the research phase and not ready for field applications. Other methods include utilizing organisms such as fish, clams, mussels, daphnia, and algae as biomonitors, and employing bioluminescence-based assays for toxicity testing.

1.9 Advanced Oxidation and Reduction Process Radical Generation in the Laboratory and at Large Scale

The adverse ecological impacts of anthropologically generated chemicals in our waters, especially EDCs antibiotics, and pesticides/herbicides, are of major concern to regulatory groups and the public. Traditional water treatment relies primarily upon adsorptive, chemical-physical, and microbial-based processes to transform or remove these unwanted organic contaminants. However, standard large-scale water treatments are not always sufficient for quantitative removal of small (ng L^{-1}) levels of dissolved chemicals (Chapter 9). This is complicated by the presence of much higher levels of water constituents, such as carbonates and DOM.

The active quantitative removal of trace levels of these chemicals requires additional process treatments, and therefore the use of in situ generated radical species has been proposed. These approaches are generally referred to as advanced oxidation/reduction processes (AO/RPs). Radicals can be created in water using a variety of techniques (see Table 1-1 for a summary of some major AO/RP processes and the radicals they generate) requiring either external energy deposition into the water or directly into a deliberately-added chemical, such as persulfate. Other radical-producing processes, or combinations of these processes, have also been used for contaminant treatment. While direct destruction of chemical contaminants by UV light or pulsed UV light is also widely utilized, this approach does not generate oxidization or reduction of radicals in water, and hence, is not considered further here.

The most widespread AO/RPs are based on the use of the oxidizing a hydroxyl radical (•OH), which is a powerful oxidant ($E^o = 2.8$ V) that reacts with both organic and inorganic chemicals. At treatment scale, the hydroxyl radical is usually generated using some combination of O_3/H_2O_2, O_3/UV-C, or H_2O_2/UV-C. Additional techniques such as ionizing radiation (electron beam and gamma irradiation), Fenton's reaction, UV irradiation of titanium dioxide, and sonolysis also generate this radical, facilitating the study of AO/RP radical chemistry in the laboratory. Some of these AO/RPs can generate other radicals, such as reducing hydrated electrons (e^-_{aq}) and hydrogen atoms (H•). While these reducing radicals can also react to destroy chemical contaminants in water, their use in real world waters is problematic because of the presence of air, where the relatively high level of dissolved oxygen ($[O_2] \sim 2.5 \times 10^{-4}$ M) preferentially scavenges these radicals to create the inert superoxide radical, O_2^-•. Another approach was preferentially converting these two reducing radicals to another oxidizing radical, specifically the sulfate radical (SO_4^-•), through the deliberate addition of persulfate. The sulfate radical is also strongly oxidizing ($E^o = 2.3$ V), thus it can also react with most organic contaminants.

The optimal, quantitative removal of chemical contaminants from waters through the use of these processes requires a thorough understanding of the redox chemistry occurring between free radicals and the chemicals of concern. This can be accomplished if absolute kinetic rate constants and efficiencies are determined for all the reactions occurring in the AO/RP system. These kinetic and mechanistic data allow for quantitative computer modeling of AO/RP systems to establish the efficiency and large-scale applicability of using radicals for specific contaminant removal. One important aspect of this understanding is to ascertain how these systems generate their radicals. The generation and reactivity of AO/RP radicals is the focus of Chapter 9.

To augment traditional adsorptive and chemical-physical water treatments for the complete remediation of chemically contaminated waters, radical-based AO/RPs are being considered for large-scale usage. While most AO/RPs use the hydroxyl radical in treatment, the use of other oxidizing and reducing radicals is also of interest. The chemistry behind the AO/RP-based generation of these radicals, both in the laboratory and in large-scale operations, is presented along with the important reactions that highlight the reactivities of these radicals. Specific examples of radical reactions with contaminant chemicals are presented to highlight this chemistry.

1.10 Cactus Mucilage as an Emergency Response Biomaterial for Providing Clean Drinking Water

The development of effective and inexpensive water purification systems would be very desirable. Many communities could benefit from water-purification methods that use renewable, sustainable, low-toxicity-producing, low-cost technology. A study was conducted to investigate cactus-mucilage-based separation as an emergency response biomaterial to provide clean drinking water for communities affected by natural disasters (Chapter 10). The mucilage is an extract from the *Opuntia ficus-indica*, commonly known as nopal or prickly pear cactus. This readily available and inexpensive natural extract has been shown to remove sediments, bacteria, and arsenic from contaminated water in previous studies. Samples of tap water (from a treatment center), well water, surface water, and distributed water (bottled water and water from tanker trucks) were collected from 10 different locations in Port-au-Prince, Haiti. A wide-ranging analysis of elements was performed on acidified raw samples to establish water quality and type of contamination. High concentrations of Fe, B, Ba, and Se were found in some of the sites, with concentrations above the maximum contaminant level and provisional guidelines recommended by USEPA (2009). Simple batch experiments were performed with two mucilage extracts from *O. indica-ficus* gelling and nongelling extracts. Analysis of samples from the top of water columns treated with the mucilage, revealed that high metal-removal rates were achieved using both types of mucilage extracts. Cactus mucilage effectively decreased the hazardous metal content to levels accepted by water-quality agencies.

1.11 Potable Water Filter Development

Clean potable water can be a matter of life or death. The scarcity of potable water is becoming the single most technologically challenging problem of the twenty-first century because of human activities changing the environment and rendering significant amounts of water unfit for life (see Chapter 11). Whatever the origin of water, it is now clear that potable water can be obtained only through a process of filtration either at the source or at the sink. At present, there is no clean source of water in major population centers of the world. This overview deals with some of the basic principles of aquatic chemistry, such as speciation and surface complexation reactions through computational equilibrium models, a continuous proton binding model to calculate pK_a spectra for solid sorbents and DOMs octanol–water partitioning models, and a bio-concentration factor for non-electrolytes. The urgent need for water filter development is highlighted with some recent developments in the use of iron-based sorbents in removing arsenic species from groundwater. The utility of a multi-physics approach with finite element analysis software (COMSOL) is exemplified by a sorption column for arsenic removal. The role of analytical chemistry in the study of aquatic chemistry is highlighted by special reference to quality-control issues. It can be argued that understanding water filtration as a science is the key to the solution of the present water crisis.

1.12 Removal and Immobilization of Arsenic in Water and Soil Using Nanoparticles

Some of the latest developments in the preparation and applications of polysaccharide-bridged or polysaccharide-stabilized magnetite nanoparticles are summarized in Chapter 12. While both starch and sodium carboxymethyl cellulose (CMC) are able to prevent aggregation of the nanoparticles, starch-coated nanoparticles offer more favorable sites for arsenate than CMC-coated counterparts. Bridged nanoparticles are obtained at low concentrations of starch, where nanoparticles are inter-bridged by starch molecules and are present as gravity-settleable flocs. Fully stabilized nanoparticles are prepared at higher concentrations of starch or CMC. Starch-stabilized nanoparticles offer a high Langmuir capacity of 62.1 mg/g, which is 72% and 2.3 times greater than for CMC-stabilized and CMC nonstabilized counterparts, respectively. The high Langmuir capacity was maintained over a broad pH range, and salt at 10 wt% did not suppress it. Stabilized nanoparticles were deliverable to soil under pressure, yet remained immobile under typical groundwater flow conditions, holding great potential to facilitate in situ immobilization of arsenic in soil and groundwater.

1.13 Transforming an Arsenic Crisis into an Economic Enterprise

Arsenic contamination of the drinking water of millions of people living in different parts of the world has been the focus of attention of public health scientists and engineers around the world (Chapter 13). Drinking arsenic-contaminated water over a long period of time causes severe damage to the human body, culminating in various forms of cancers. Apart from being fatal, health-related impairments are also known to cause a wide range of socio-economic problems, especially in the developing countries. The arsenic crisis prevailing over a large area of Bangladesh and India is one of the worst water calamities of the world in recent times. The crisis is also slowly unfolding in Southeast Asia, affecting several other countries including Cambodia, Vietnam, Laos, and Myanmar.

The genesis of arsenic in groundwater is considered to be the geochemical leaching of arsenic from underground arsenic-bearing rocks, caused by excessive extraction of the groundwater for agricultural purposes. Surface water does not contain any arsenic but its use for drinking purposes in many developing nations is restricted by the wide range of microbial contamination caused by the absence of proper sanitation practices. The use of groundwater in these regions is favored by its easy availability, microbial safety, and the absence of proper infrastructure for treatment and distribution of surface water. While the best solution to the crisis is to switch over to treated surface water, the development and maintenance of surface-water-based drinking water systems is expensive, time-consuming, and investment-intensive. With all these difficulties, it is unlikely that a developing country will switch the source of water from groundwater to surface water within a short period of time. In order to save lives before such a changeover is possible, it is imperative to build sustainable arsenic-removal systems on an urgent basis. Several technologies have been developed for the removal of arsenic from contaminated drinking water. Many of them are capable of selective removal of arsenic from the contaminated water; however, only a few of them have gained wide-scale application in the field. An example of one such arsenic-removal system has been presented that, apart from finding wide acceptance in the field, resulted in transforming the crisis into an economic enterprise. Between 1997 and 2010, more than 150 community-scale, arsenic-removal units were installed in the villages of West Bengal, a state of India that neighbors Bangladesh. The treatment units, along with ancillaries and protocols, were developed by the Bengal Engineering and Science University, in India, jointly with Lehigh University, in the United States. The project has been implemented by the Bengal Engineering and Science University with financial help from Water for People, Denver, Colorado. For all the treatment units, the villagers themselves take care of the maintenance, upkeep, and management related to the units. An account of the evolution of the treatment units is presented along with their performance for arsenic removal. The chapter also describes the extent and effect of community participation in this project and records the resulting socio-economic changes.

1.14 Monitoring from Source to Tap: The New Paradigm for Ensuring Water Security and Quality

Drinking water is one of the United States' key infrastructural assets that have been deemed vulnerable to deliberate terrorist attacks; however, terrorism is not the only risk (see Chapter 14). Currently our water supplies are vulnerable to a wide variety of potential hazards. These hazards can be intentional or accidental in nature. They range from the threat of deterioration of water quality due to aging infrastructure, to accidental contamination of water sources and treated water from pollution, to the potential of degrading our water quality as we make treatment changes to comply with new regulatory regimes. The rapid detection and characterization of breaches of integrity in the water-supply network is crucial in initiating appropriate corrective action. The ability of a technology system to detect such incursions on a real-time basis and give indications of their cause can dramatically reduce the impact of any such scenario.

Continuous monitoring of the entire water-supply network is a goal that seemed unachievable. Historically, most monitoring outside of the water treatment plant has been relegated to the occasional snapshot provided by grab sampling for a few limited parameters or the infrequent regulatory testing required by mandates such as the Total Coliform Rule. The development of water-security monitoring since 9/11 has the potential to change this paradigm.

An attack on the drinking water supply system could take many and varied forms depending upon the components of the system that are vulnerable. The provision of drinking water to our homes is a complex process that involves many steps and components. All of these steps are to some degree vulnerable to compromise by terrorist acts.

1.15 Evaluation of Sustainability Strategies

Sustainability is a method and an approach to create and maintain conditions under which humans can exist in harmony with nature, while enabling the economic, social, and other requirements of present and future generations. It is important to ensure that we will continue to have all the resources required, especially water, to protect human existence and its future growth.

Sustainability has three realms or pillars: economic, social, and environmental (see Chapter 15). All the three realms are highly important in achieving sustainability, as we cannot have one without the others, especially in the long term. For example, many countries—before the emergence of the term sustainability—have been pushing simply for economic development, only to realize that the environmental costs (resource depletion, pollution, health problems, and frequent occurrences of calamities, such as flooding) of such single-minded growth nullify all the gains. To understand the importance of sustainability, it is important to realize that the three pillars of sustainability are intimately entwined. A vibrant community is absolutely essential for healthy business growth, and for the resources for concern about environmental problems. A healthy economy offers ample revenues and resources to minimize social and environmental problems. And finally, a healthy environment means a healthy supply of resources to sustain economic and social growth for a very long period of time. When we understand these interdependencies, we are in a much better position to make sound decisions and to set long-term goals. A lack of this understanding will surely result in poor decisions that may give short-term profits in any of the three realms, but will result in long-term damages that may not be easily repaired or recovered. Sustainability avoids the trade-off between any two of these three realms and aims to optimize all three.

1.16 Conclusions

(1) It is important to monitor water quality rigorously to find and assess point and nonpoint source pollution.
(2) Appropriate means of remediation need to be found to assure sustainability of water quality.

Appendix

Table 1-1A Water-Quality Results of 2011 for Brunswick County, NC

Substances	USEPA's MCL	Amount Detected	Source of Contaminant
Turbidity	Treatment technique	Average 0.057 Maximum 0.651 ntu	Soil runoff
Raw water TOC	Treatment technique	Average 6.3 ppm	Naturally present in environment
Finish water TOC	Treatment technique	Average 3.2 ppm	Naturally present in environment
Total Organic Carbon (TOC)	Treatment technique	Removal efficiency average 46.6%	Naturally present in environment
Inorganic chemicals			
Chlorite	1 ppm	Average 0.71 ppm	By-product of disinfection
Chlorine dioxide	0.8 ppm	Average 0.022 ppm	Water additive used to control microbes
Fluoride	4 ppm	Average 0.75 ppm	Water additive that promotes strong teeth
Nitrate	10 ppm	1.01 ppm	By-product of disinfection
Sulfate	250 ppm	21 ppm	Part of treatment process, erosion of natural deposits
Orthophosphate	17 ppm	Average 2.32 ppm	Water additive used to control corrosion
Lead and copper Action Level (AL)			
Copper 90th percentile January 6 to September 30, 2011	1.3 ppm	90% of samples are $\leq$0.95 ppm	Corrosion of household plumbing
Lead 90th percentile January 6 to September 30, 2011	0.015 ppm	90% of samples are $\geq$0.95 ppm	Corrosion of household plumbing

(Continued)

Table 1-1A Water-Quality Results of 2011 for Brunswick County, NC—*Cont'd*

Substances	USEPA's MCL	Amount Detected	Source of Contaminant
Organic chemicals USEPA's MCL			
Monochloramine disinfectant residual	4 ppm	Average minimum 3.03 ppm	Water additive used to control microbes
Total trihalomethanes	80 ppb	Average 21.5 ppb	By-product of disinfection
Total haloacetic acids	60 ppb	Average 13.2 ppb	By-product of disinfection
Pesticides, volatile pesticides, volatile and synthetic organic chemicals in Cape Fear River: None of these regulated compounds was detected.			
Radionuclides			
Beta 5/27/11	10 pCi/l	None detected	Erosion of natural deposits
Unregulated contaminants			
Sodium	Not regulated	21.9 ppm	Part of treatment process Erosion of natural deposits
Cryptosporidium	Not applicable	0.210 oocysts	

References

[1] Ahuja S. Handbook of water purity and quality. Amsterdam: Elsevier; 2009.

[2] Ahuja S. Arsenic contamination of groundwater; mechanism, analysis, and remediation. New York: Wiley; 2008.

[3] Brundtland GH. Our common future, the World Commission on Environment and Development, United Nations, 1987.

[4] Ahuja S. Assuring water purity by monitoring water contaminants from arsenic to zinc, American Chemical Society Meeting, Atlanta, March 26–30, 2006.

[5] Chemical and engineering news, p. 42, November 17, 2008. (a) Chemical and engineering news, p. 21, May 7, 2012.

[6] Ahuja S. Ultratrace analysis of pharmaceuticals and other compounds of interest. New York: Wiley; 1986.

[7] Ahuja S, Water sustainability and reclamation, American Chemical Society Meeting, San Diego, March 25–29, 2012.

[8] McNeil EE, Otson R, Miles WF, Rahabalee FJM. J Chromatogr 1977;132:277.

[9] Kratochvil B, Taylor JK. Anal Chem 1981;53:924 A.

[10] Youden WJ. J Assoc Off Anal Chem 1967;50:1007.

[11] Karasek FW, Clement RE, Sweetman JA. J Assoc Off Anal Chem 1981;53:1050.

[12] Wolff EW, Landy M, Peel DA. J Assoc Off Anal Chem 1981;53:1566.

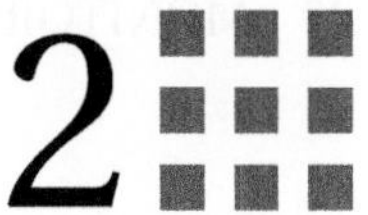

Water Quality Status and Trends in the United States

Matthew C. Larsen*, Pixie A. Hamilton, William H. Werkheiser

U.S. GEOLOGICAL SURVEY, RESTON, VA, USA

**CORRESPONDING AUTHOR*

CHAPTER OUTLINE

2.1 Introduction

National interest in water quality issues culminated in the 1972 enactment of the Clean Water Act (CWA) [1]. This law was passed in response to public concerns about burning rivers and dead lakes and a national consensus built over the previous 60 years that pollution of our rivers and lakes was unacceptable. Control of point-source

Monitoring Water Quality, http://dx.doi.org/10.1016/B978-0-444-59395-5.00002-9

contamination, traced to specific "end of pipe" points of discharge, or outfalls, such as factories and combined sewers, was the primary focus of the CWA. Significant progress toward cleaner water resulted through actions, such as implementing changes in manufacturing processes and wastewater treatment.

Water-quality challenges are now increasingly complex. The majority of water-quality problems are caused by diffuse nonpoint sources from agricultural land, urban development, forest harvesting, and the atmosphere (Table 2-1). These nonpoint-source contaminants are more difficult to effectively monitor, evaluate, and control than those from point sources (for example, discharges of sewage and industrial waste). We need improved quantification and understanding of human activities associated with nonpoint sources and how those human activities take place on the landscape—primarily information on how we use and dispose of chemicals, how we convert land over time, our use of water, and our land-management practices.

Several factors add to this complexity. First, the amount of pollution from nonpoint sources varies over short periods—hourly to seasonally—making it difficult to monitor and quantify the sources over time. Single or periodic measurements are not adequate to characterize water-quality conditions. Measurements are needed over seasons, hydrologic and meteorological events, and in real time.

We face large water-quality challenges because of the increasingly complex and emerging diversity of issues. When the CWA was passed, the dominant concern was the sanitary quality of rivers and streams. The focus was on temperature, salinity, bacteria counts, oxygen levels, and suspended solids, in large part, collected for day-to-day evaluations of compliance or permitting decisions.

While these remain important, there are now hundreds of synthetic organic compounds (such as pesticides and volatile organic compounds in solvents and gasoline) that are introduced into the environment every day. Improved laboratory techniques have led to the identification of microbial and viral contaminants, pharmaceutical compounds, and endocrine disruptors in our waters that were not previously measured. We are also finding that many contaminants, such as arsenic and radon, can originate

Table 2-1 The Changing National Focus on Water-Quality Challenges

Past Focus	Present and Future Focus
Point sources	Nonpoint sources
End-of-pipe approach	Watershed approach (landscape, human activities)
One-time, periodic reporting	Seasonal, hydrologic events, continuous, real time
Nutrients, dissolved oxygen, bacteria	Organic compounds
Single pollutants	Mixtures
Surface water	Total resource
Chemistry	Chemistry, biology, habitat, hydrology, landscape
Short-term monitoring	Long-term monitoring
Monitoring	Monitoring and prediction

from a wide range of natural sources and are of potential concern with respect to human health, even in relatively undeveloped settings that are perceived as less vulnerable to contamination. This is a critical concern with respect to the quality and safety of water from domestic or "private" wells, which are a source of drinking water for about 40 million people or 15% of the U.S. population, many of whom are based in rural and less-developed settings [2]. Domestic wells are not regulated under the federal Safe Drinking Water Act (SDWA) and are the responsibility of the homeowner. Natural and organic contaminants often end up in our waters as complex mixtures of organic compounds; many of these can, even at very low concentrations, potentially affect the health of humans and/or the reproductive success of aquatic organisms in our waters.

Our understanding of water-quality challenges has expanded with our understanding of the importance of the hydrologic cycle for water-quality conditions. Whereas our concerns were focused mainly on streams and rivers, we now recognize water-quality issues as part of an integrated hydrologic system. For example, groundwater and surface water are highly inter-related; reduced base flow from groundwater pumping often results in increased stream temperatures, drying wetlands, and habitats unsuitable for fish and other aquatic species [3]. The historic approach was to look at quality mostly in terms of concentrations independent of hydrology; however, concentrations and types of contaminants and their potential effects on ecosystems and drinking water supplies vary over time and depend largely on the amount of water flowing in streams and the amounts and directions of groundwater flow.

Other natural processes, including geology and geomorphology, also control the timing and amount of surface and groundwater flow and the transport of waterborne constituents and contaminants. Furthermore, natural complexity is increasing because of changes in climate, resulting in new patterns of seasonal precipitation, runoff, and the spatial and temporal distribution of snow versus rain [3–6]. Unfortunately, there is a continuing high degree of uncertainty in climate model predictions [7,8]. This set of challenges will continue and probably intensify as both nonclimatic and climatic factors, such as predicted rising temperature and associated changes in runoff, continue to develop [9–12].

Water-quality challenges extend beyond chemistry. We now realize that water quality, habitat disturbances, streamflow alterations, biological systems, and ultimately, ecosystem health are all closely interconnected. Meaningful water-quality assessments must therefore integrate biological monitoring and ecosystem health, such as inclusion of benthic invertebrates and other biological indicators as critical tools to understanding water quality [13–15].

Given our improved understanding of the spatial and temporal complexities in water quality and its numerous natural and human causes, the importance of long-term monitoring is increasingly clear. Comparable data must be collected over time if long-term trends are to be distinguished from short-term fluctuations and if natural fluctuations are to be distinguished from the effects of human activities. Long-term tracking is particularly critical for groundwater and sediment because slow flow paths and long residence times may not allow water quality issues to appear for years or even decades.

Monitoring alone does not provide understanding of the causes of water-quality conditions, given the complex interrelations among water quality, natural changes, and human actions over time and space. Furthermore, federal and state resources are increasingly limited, so we cannot expect to monitor water resources in all places and at all times. The value of data collected at individual sites is enhanced by applying assessment tools, including models that use monitoring data in conjunction with our understanding of the hydrologic and aquatic systems, the natural landscape, and human activities to develop more generalized knowledge of the status, trends and causes of these conditions for broader areas, including entire stream reaches and aquifers, large river basins, ecoregions, the states, and the nation as a whole. The integration of monitoring and assessment with modeling and predictive tools is the strategy needed to provide comprehensive statewide, regional, and national water-quality assessments. This strategy also will provide the needed national "water census" of water-quality status and trends and an increased ability to anticipate conditions in the future [16].

The brief overview below of water-quality status and trends in the United States is based largely on the past two decades of work accomplished by U.S. Geological Survey (USGS) scientists and is funded by USGS programs (principally the Cooperative Water- Groundwater Resources- Hydrologic Research and Development-, National Water Quality Assessment-, and Toxic Substances Hydrology Programs). These congressionally supported programs are a key part of the larger USGS portfolio of science and information aimed at informing decision makers at all levels in the United States [17].

2.2 Monitoring and Assessments of Complex Water Quality Problems

The USGS strives to continually improve and enhance science to provide unbiased information to decision makers and the public [18]. Goals of USGS programs have been adapted to the evolving complexity of water-quality challenges to track the status and trends of five priority water-quality issues across the nation: nutrients, pesticides, organic wastewater compounds, sediment-bound compounds, and mercury.

2.2.1 Nutrients

According to the U.S. Environmental Protection Agency (USEPA), nutrient pollution has for decades ranked as one of the top three causes of degradation in U.S. streams and rivers, and the concern is expected to continue for the foreseeable future. Since the 1990s the USGS has placed a major focus on studies of nutrients (Figure 2-1) [19] to provide the most comprehensive national-scale assessment of two key nutrients, nitrogen and phosphorus, in our streams and groundwater. In this study, water samples were collected

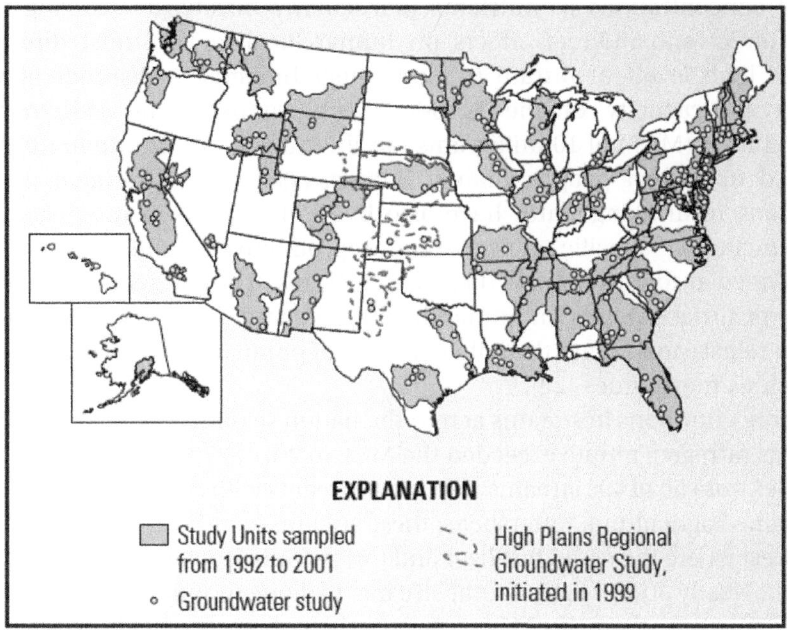

FIGURE 2-1 USGS assessments of nutrients followed a nationally consistent approach in 51 of the nation's major river basins and aquifer systems during 1992–2001. *From Ref. [20].* (For color version of this figure, the reader is referred to the online version of this book.)

from 499 stream sites monthly and during periods of high and low streamflow, usually for a minimum of 2 years. Biological communities were assessed at about 1400 stream sites. Groundwater samples were collected from 5101 wells, including domestic and public-supply wells. Trends in nutrient concentrations in streams were assessed at 171 and 137 stream sites, for phosphorus and nitrogen respectively, sampled from 1993 to 2003. Changes in nitrate concentrations in groundwater were assessed by measuring concentrations in 495 wells from 1988 to 1995, then again from 2001 to 2004. Most water samples were analyzed for five measures of nitrogen- and phosphorus-containing nutrients: total nitrogen, nitrate, ammonia, total phosphorus, and orthophosphate. The study provides improved science-based explanations of when, where, and how elevated concentrations reach streams, aquifers, and nearshore areas, and how they affect aquatic life and the quality of drinking water.

Findings [19] show that despite major federal, state, and local effort and expenditure to control the sources and movement of nutrients within our Nation's watersheds, national-scale progress is limited. For example, USGS findings show that widespread concentrations of nitrogen and phosphorus remain two to ten times greater than levels recommended by the USEPA to protect aquatic life. Most often, these elevated levels were found in agricultural and urban streams. These findings show that continued reductions in nutrient sources and implementation of land-management strategies for reducing nutrient delivery to streams may be needed to meet USEPA-recommended levels in most regions.

Elevated concentrations of nutrients, particularly nitrate, in drinking water may have both direct and indirect effects on human health. The most direct effect of ingestion of high levels of nitrate is methemoglobinemia, a disorder in which the oxygen-carrying capacity of the blood is compromised; the USEPA maximum contaminant level (MCL) of 10 milligrams per liter (mg/L) for nitrate in drinking water was adopted to protect people, mainly infants, against this problem. High nitrate concentrations in drinking water have also been implicated in other human health problems, including specific cancers and reproductive problems [21], but more research is needed to corroborate these associations. The indirect effects of nutrient enrichment of surface waters on human health are many and complex, including algal blooms that release toxins and the enhancement of populations of disease-transmitting insects, such as mosquitoes [22].

Nitrate concentrations in streams across the nation seldom exceeded the USEPA MCL of 10 mg/L as nitrogen; nitrate exceeded the MCL in 2% of 27,555 samples, and in one or more samples from 50 of 499 streams [19]. Most streams with concentrations greater than the MCL drained agricultural watersheds; these streams were particularly common in the upper Midwest where the use of fertilizer and (or) manure is relatively high and tile drains are common. Nearly 30% of agricultural streams had one or more samples with a nitrate concentration greater than the MCL, compared to about 5% of the streams draining urban land. None of the samples from streams draining undeveloped watersheds had a concentration greater than the MCL. For perspective on the relevance of these findings to surface water used for drinking water supplies, 12% of the Nation's 1679 public water-supply intakes withdraw water from streams that drain watersheds with predominantly agricultural land, whereas most water-supply intakes are in watersheds draining undeveloped land [19].

Nitrate concentrations greater than the MCL are more prevalent and widespread in groundwater than in streams. Concentrations exceeded the MCL in 7% of about 2400 private wells sampled by the USGS. Contamination by nitrate was particularly severe in shallow private wells in agricultural areas, with more than 20% of these wells exceeding the MCL. Elevated nitrate follows distinct geographic patterns—mostly related to land use. Elevated concentrations of nitrate were largely associated, for example, with intensively farmed land, such as in parts of the Midwest Corn Belt and the California Central Valley [2]. The quality and safety of water from private wells (which are a source of drinking water for about 40 million people or 15% of the U.S. population) are not regulated by the SDWA and are the responsibility of the homeowner [2,19].

Concentrations exceeding the MCL were less common in public-supply wells (about 3% of 384 wells) than in private wells. The lower percentage in public wells reflects a combination of factors, including (1) greater depths and hence age of the groundwater; (2) longer travel times from the surface to the well, allowing denitrification and/or attenuation during transport; and (3) the location of most public wells near urbanized areas where sources of nitrate generally are less prevalent than in agricultural areas [19,23].

A USGS national statistical model of the vulnerability of relatively deep groundwater (more than 164 feet below land surface) estimated that almost 500,000 people live in areas where nitrate concentrations are predicted to be greater than the MCL, and more than 1.2 million people live in areas predicted to have nitrate concentrations between 5- and 10 mg/L. A similar model suggests that the number of people exposed to nitrate concentrations greater than the MCL would be 14% greater if they obtained their water from shallow wells (33 feet or less) rather than deep wells (164 feet or greater) [19].

Nitrate concentrations are likely to increase in deep aquifers typically used for drinking water supplies despite nutrient-reduction strategies, as shallow groundwater with high nitrate concentrations moves downward to deeper aquifers. USGS findings show that the percentage of sampled wells with nitrate concentrations greater than the USEPA drinking water standard, increased from 16% to 21%, starting in the early 1990s [19,24]. Similarly, the probability of nitrate concentrations exceeding the MCL has increased from less than 1% in the 1940s to greater than 50% by 2000 for young groundwater in agricultural settings [19,24].

A 60-year record of nitrate concentrations at a public-supply well in Nebraska provides a local example of the long response time in groundwater (Figure 2-2). Nitrate concentration in shallow groundwater decreased after implementation of fertilizer-management strategies starting in the 1980s. This decrease was not noted in the deeper groundwater, where concentrations continued to increase for 25 years before beginning to decrease in 2005 (Figure 2-2). Consistent and systematic long-term monitoring is critical, especially for groundwater, when evaluating the effectiveness of environmental and land-management strategies, because of the slow response of groundwater to changes in chemical use or land-management practices [25].

The USGS generally does not assess nutrients in estuarine waters or the effects of nutrients on levels of dissolved oxygen and hypoxia; however, the USGS does assess the downstream transport of nutrients from major rivers to major receiving waters, such as

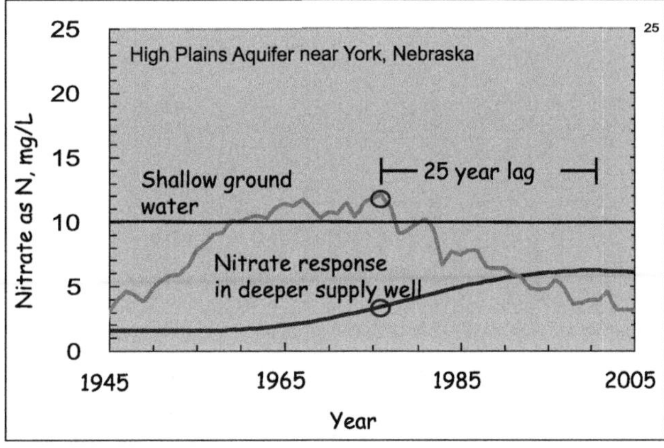

FIGURE 2-2 Nitrate concentration in a public-supply well. *From Ref. [25].* (For color version of this figure, the reader is referred to the online version of this book.)

the Gulf of Mexico and the Chesapeake Bay, as described below (Figure 2-3). Hypoxia, such as in the northern Gulf of Mexico, is caused by excess nutrients delivered from the Mississippi River in combination with seasonal stratification of Gulf waters [26]. Excess nutrients promote algal growth. When the algae die, they sink to the bottom and decompose, consuming available oxygen. Stratification of fresh and saline waters prevents mixing of oxygen-rich surface water with oxygen-depleted bottom water. Immobile species such as oysters and mussels are particularly vulnerable to hypoxia and become physiologically stressed and die, if exposure is prolonged or severe. Fish and other mobile species can avoid hypoxic areas, but these areas still impose ecological and economic costs, such as reduced growth in commercially harvested species and loss of biodiversity, habitat, and biomass [27]. Fish kills can result from hypoxia, especially when the concentration of dissolved oxygen drops rapidly.

A USGS hybrid statistical/mechanistic watershed model, known as Spatially Referenced Regression On Watershed attributes (SPARROW), is used to relate in-stream nutrient loads to upstream nutrient sources and record watershed characteristics affecting transport. SPARROW also provides information on the delivery of nitrogen and

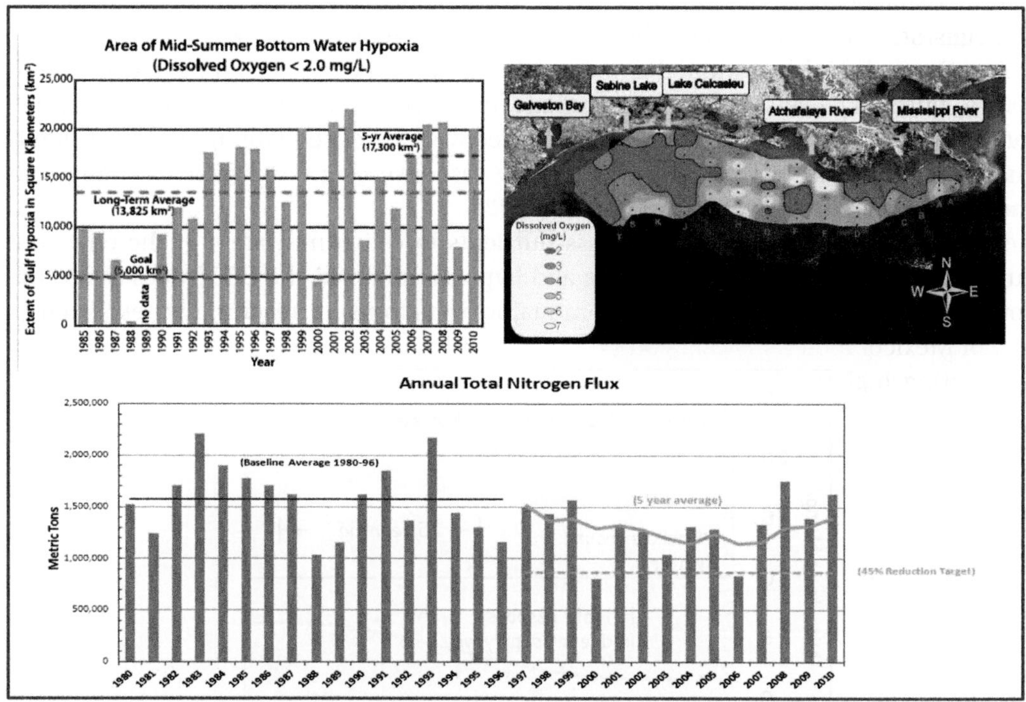

FIGURE 2-3 Areal extent of Gulf of Mexico hypoxic zone (top chart and map) and Mississippi River nitrogen flux (bottom chart). *Sources: Charts from Aulenbach and associates report [28] and Aulenbach written communication (2011); map of the 2010 Dead Zone from Ref. [29] with a larger image available at http://www.noaanews.noaa.gov/stories2010/images/dissolved_o2_day7.jpg.* (For color version of this figure, the reader is referred to the online version of this book.)

phosphorus from 62,000 stream reaches to the Nation's major rivers and estuaries [30]. Modeled findings show, for example, that the cultivation of corn and soybeans was the largest contributor of nitrogen to the Gulf of Mexico, whereas animal manure on pasture, rangelands, and corn and soybean cultivation were the largest contributors of phosphorus. In addition, modeled findings showed that large rivers contribute a larger percentage of their nitrogen to downstream receiving water bodies than small streams, in large part because nitrogen removal in streams rapidly declines as water depth and stream size increase [30].

Robertson et al. [31] used the SPARROW model to identify watersheds with the highest nutrient yields delivered to downstream waters. The watershed results provided information for management strategies to reduce the hypoxic zone and improve the water quality of rivers and streams. Additionally, the results were used to develop a statistically-reliable method for identifying high-priority areas for management, based on a probabilistic ranking of delivered nutrient yields from watersheds throughout a basin [31]. The method was designed to be used by managers to prioritize watersheds where additional stream monitoring and evaluations of nutrient-reduction strategies could be undertaken [31]. The study identified 150 watersheds having the highest delivered nutrient yields to the Gulf of Mexico; these were in the Central Mississippi, Ohio, and the Lower Mississippi River Basins.

USGS trend analyses suggest that despite major federal, state, and local nonpoint source, nutrient-control efforts for streams and watersheds across the nation, limited national progress has been made in reducing the impact of nonpoint sources of nutrients. Instead, concentrations have remained the same or increased in many streams and continue to pose risks to aquatic life and human health. For example, nitrate transport to the Gulf of Mexico during the spring, is one of the primary determinants of the size of the Gulf hypoxic zone. During high streamflow in spring in the period studied, the concentration of nitrate decreased at the study site near where the Mississippi River enters the Gulf of Mexico, indicating that some progress has been made in reducing nitrate transport during high flow conditions. However, during times of low to moderate spring streamflow, concentrations increased. The net effect of these changes is that nitrate transport to the Gulf was about 10% higher in 2008 than in 1980. This increase in nitrate transported to the Gulf can be attributed primarily to the upstream nitrate increases in the Mississippi River Basin above the Clinton (Iowa) monitoring site and in the Missouri River Basin [32]. There are some exceptions elsewhere in the nation to the findings in the Mississippi. For example, recent findings show decreased nutrient concentrations in the Susquehanna and Potomac rivers since 2000; but increasing concentrations in the Rappahannock and James rivers [33].

2.2.2 Pesticides

About 1×10^9 pounds of conventional pesticides are used annually in the United States to control weeds, insects, and other pests. The use of pesticides has a range of benefits, including increased food production and the reduction of insect-borne disease, but also has adverse effects on water quality. A USGS national assessment indicates that pesticides are widespread, albeit often at low concentrations, in river basins and aquifer systems across a wide range of landscapes and land uses [34]. Overall, at least one pesticide was found in about 95% of water samples and in 90% of fish samples from streams in agricultural and urban areas and in about 55% of shallow wells sampled in agricultural and urban areas [34].

Pesticide occurrence is closely linked to land use (Figure 2-4). For example, insecticides such as diazinon, carbaryl, chlorpyrifos, and malathion were detected more frequently and usually at higher concentrations in urban streams than in agricultural streams (Figure 2-5). Many herbicides (most commonly atrazine and its breakdown product deethylatrazine (DEA), metolachlor, alachlor, and cyanazine) were generally detected more frequently and usually at higher concentrations in streams and shallow groundwater in agricultural areas than in urban areas [35]. Some herbicides detected in urban areas, such as simazine and prometon, are those with substantial nonagricultural uses, such as weed control in commercial areas, golf courses, and along roadsides. The same relative patterns of different compounds were observed in groundwater, but at much lower detection frequencies [34].

These USGS findings provide a "signature" of what can be expected in streams and groundwater influenced by land use. This signature can be used by water managers to identify and anticipate types of contaminants, leading to better management of the use of chemicals.

It should be noted that detection does not necessarily translate to risk. The USGS intentionally analyzed for very low levels, sometimes 10–100 times lower than guidelines or standards established to protect drinking water. The intent of the low-level analysis is

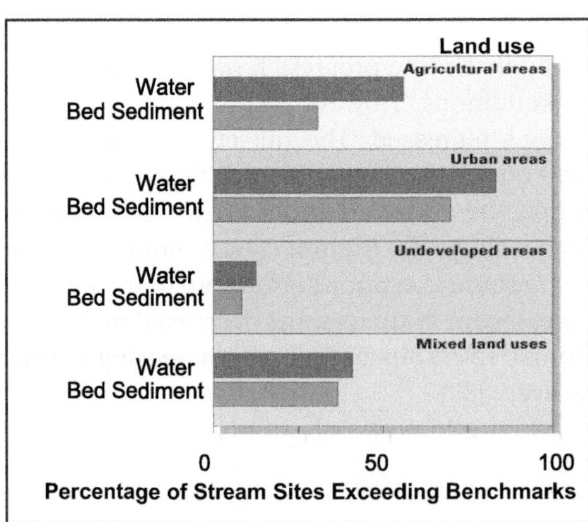

FIGURE 2-4 Pesticides and their significance to aquatic life. *From Ref. [34].* (For color version of this figure, the reader is referred to the online version of this book.)

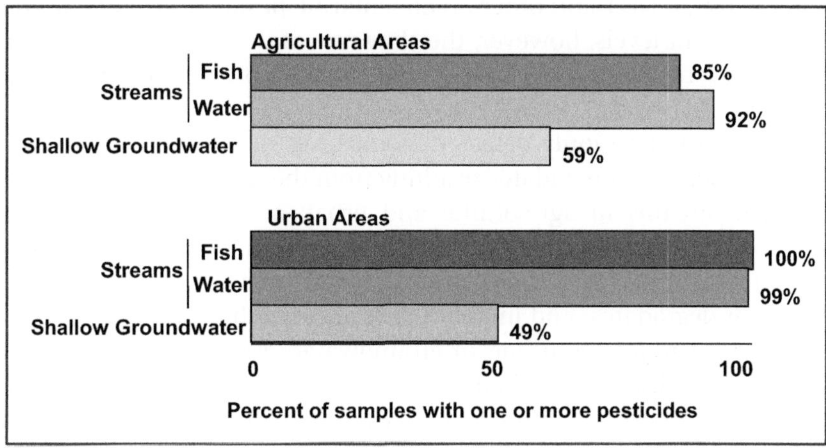

FIGURE 2-5 Pesticides in streams and shallow groundwater. In agricultural areas, these commonly include atrazine, [deethylatrazine], metolachlor, cyanazine, and alachlor. In urban areas, common pesticides found include diazinon, carbaryl, malathion, chlorpyrifos, atrazine, simazine, prometon, 2,4-D, and diuron. *From Ref. [34].* (For color version of this figure, the reader is referred to the online version of this book.)

to detect and evaluate emerging problems, as well as to track contaminant levels over time.

Nationally, the USGS findings show that pesticides are seldom present at concentrations likely to affect humans, but do occur in many streams, particularly those draining urban and agricultural areas, at concentrations that may affect aquatic life or fish-eating wildlife [34]. Concentrations of pesticides were greater than water-quality benchmarks for aquatic life and (or) fish-eating wildlife, in more than half of the streams with substantial agricultural and urban areas in their watersheds. Of the 178 streams sampled nationwide that have watersheds dominated by agricultural, urban, or mixed land uses, 56% had one or more pesticides in water that exceeded at least one aquatic-life benchmark. Urban streams had concentrations that exceeded one or more benchmarks at 83% of sites (mostly of the insecticides diazinon, chlorpyrifos, and malathion) although frequencies of exceedance declined during the study period. Concentrations exceeded benchmarks in 95% of urban streams sampled during the period 1993–1997 and in 64% of streams sampled during the years 1998–2000. Agricultural streams had concentrations that exceeded one or more benchmarks at 57% of sites—most frequently by chlorpyrifos, azinphos-methyl, atrazine, *p,p'*-DDE, and alachlor. As the use of alachlor declined through the study period, its benchmark exceedances also declined, with none during the last 3 years of study [34].

Aquatic-life benchmarks for organochlorine pesticide compounds in bed sediment also were frequently exceeded in urban areas (70% of urban stream sites). Most compounds that exceeded aquatic-life benchmarks for sediment were derived from organochlorine pesticides that had not been used since before the study began, such as DDT, chlordane, aldrin, and dieldrin. In agricultural streams, aquatic-life benchmarks were exceeded at 31% of sites (most often by DDT compounds and dieldrin) [34]. Federal regulation on the use of DDT and other organochlorine pesticides clearly has resulted in decreased contaminant levels; however, the slow rate of decreasing trends for DDT and the continuing concern for human exposure indicate that organochlorine pesticides will remain a concern.

Pesticides are most commonly detected as mixtures of multiple compounds, rather than individually, including degradates resulting from the transformation of pesticides in the environment. Streams in agricultural and urban areas almost always contained complex mixtures of pesticides and degradates. In more than 90% of cases, water samples from streams with agricultural-, urban-, or mixed land-use watersheds contained two or more pesticides or degradates, and in about 20% of cases they had 10 or more complex mixtures. Mixtures were less common in groundwater. Nevertheless, about half of the shallow wells in agricultural areas and about a third of shallow wells in urban areas contained two or more pesticides and degradates—less than 1% percent had 10 or more. The herbicides atrazine (and its degradate, deethylatrazine), simazine, metolachlor, and prometon were common in mixtures found in streams and groundwater in agricultural areas. The insecticides diazinon, chlorpyrifos, carbaryl, and malathion were common in mixtures found in urban streams [34].

The widespread abundance of pesticide mixtures, particularly in streams, means that the total combined toxicity of pesticides in water and other media often may be greater than that of any single pesticide compound that is present. Continued research is needed on the potential toxicity of pesticide mixtures, including degradates, to humans, aquatic life, and wildlife. USGS data on the presence and characteristics of mixtures and degradates is helping to target and prioritize toxicity assessments.

USGS studies continue to track pesticides over time. One such study in the Midwest Corn Belt, based on the analysis of 11 pesticides for 31 stream sites, showed that concentrations of several major pesticides mostly declined or stayed the same in rivers and streams from 1996 to 2006 [36]. Pesticides included in the trend analyses were the herbicides atrazine, acetochlor, metolachlor, alachlor, cyanazine, S-Ethyl dipropylthiocarbamate (EPTC), simazine, metribuzin and prometon, and the insecticides chlorpyrifos and diazinon. Declines in pesticide concentrations closely followed declines in their annual applications (generally within 1–2 years) indicating, not surprisingly, that reducing pesticide use is an effective and reliable strategy for reducing pesticide contamination in streams. Only one pesticide in the [36] study—simazine, (which is used for both agricultural and urban weed control) increased in the years from 1996 to 2006. Concentrations of simazine in some streams increased more sharply than its trend in agricultural use, suggesting that non-agricultural uses of this herbicide, such as for controlling weeds in residential areas and along roadsides, increased during the study period.

A key finding of the [36] study is that elevated concentrations can affect aquatic organisms in streams as well as the quality of drinking water in some high-use areas where surface water is used for municipal supply. Four of the 11 pesticides evaluated for trends were among those most often found in previous USGS studies at levels of potential

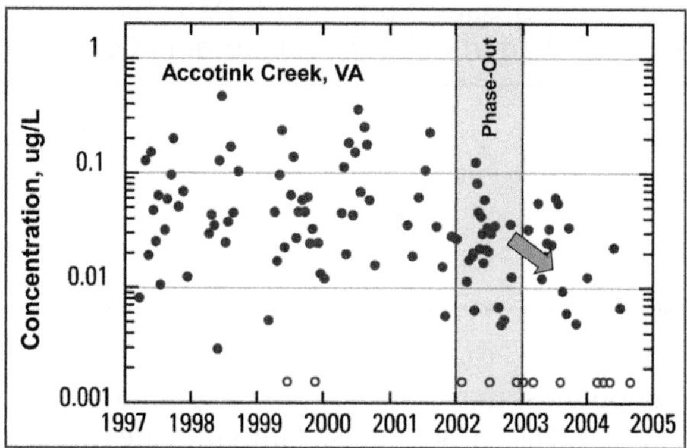

FIGURE 2-6 The concentration of diazinon at a stream in Virginia shows a downward trend following the phaseout of this chemical. *From Ref. [34].* (For color version of this figure, the reader is referred to the online version of this book.)

concern for healthy aquatic life. Atrazine, the most frequently detected, is also regulated in drinking water.

Trends in pesticide detection have been demonstrated in parts of the nation outside of the Midwest Corn Belt; these trends are also tied to use and regulations. Concentrations of diazinon, for example, show reductions beginning in 2002 in Accotink Creek, an urban stream in Virginia (Figure 2-6). Most urban uses of diazinon, such as on lawns and gardens, have been phasedout since 2001 [34].

Declines from 2000 to 2006 in concentrations of the insecticide diazinon correspond to the USEPA national phaseout of non-agricultural uses. Similarly, declines in concentrations of agricultural herbicides cyanazine, alachlor, and metolachlor show the effectiveness of USEPA regulatory actions, as well as the influence of new pesticide products. The USGS works closely with the USEPA, which uses USGS findings on pesticide trends to track the effectiveness of changes in pesticide regulations and use.

Although overall use is the most dominant factor driving changes in concentrations, other factors controlling declines may be related to improved management practices. This is suggested by rapid declines (even more rapid than their estimated use) in concentrations of atrazine and metolachlor in some streams that were sampled by the USGS in a Midwest Corn Belt study. The steeper declines in these instances may be caused by agricultural management practices that have reduced pesticide transport. However, data on management practices are not adequate to definitively draw this conclusion [36].

2.2.3 Mixtures of Organic Wastewater Compounds

Improved sampling and laboratory methods enable documentation of many compounds (known as emerging contaminants) in the Nation's waters (Table 2-2). Several national assessments have documented the presence of many commonly used substances, such as caffeine, personal use products, pharmaceuticals, and hormones (generally called

Table 2-2 Principal Organic Compounds Detectable in U.S. Wastewater (Described in Ref. [37] as Emerging Contaminants)

Antibiotics	Fragrances
Antioxidants	Fumigants
Detergents	Hormones
Disinfectants	Insecticides/repellants
Drugs	Plastics
Fire retardants	Steroids

organic wastewater compounds) in surface water and groundwater. In a landmark study, the USGS sampled 139 streams in 30 states; sampling sites represented "worst case" situations, purposely selected downstream from wastewater treatment plants, domestic septic systems, industrial discharge, animal feeding lots, and aquaculture [37]. The study showed that at least one organic wastewater compound was found in 80% of sampled streams and that 82 of the 95 targeted compounds were detected at least once. Fortunately, measured concentrations generally were low—only 5% were greater than 1 ppb (equivalent to a drop of water in an Olympic-sized swimming pool). However, detections of multiple compounds were common, with as many as 38 compounds in one sample. Nearly 35% of the samples contained more than 10 compounds. The mixtures resulted in high total concentrations in some samples, as high as 80 ppb.

Complex mixtures of organic wastewater compounds can, even at very low concentrations, adversely affect the health or reproductive success of aquatic organisms. Barber et al. [38] studied the North Shore Channel of the Chicago River (Chicago, Illinois) and determined that these waters contained mixtures of natural and anthropogenic chemicals that persisted through the water-treatment processes. They noted that aquatic organisms such as fish (largemouth bass and carp, in this study) are continuously exposed to biologically active chemicals throughout their life cycles. More than 100 organic chemicals were measured in the study, and 23 compounds were detected in all of the water samples analyzed. The majority of male fish exhibited vitellogenin induction, a physiological response consistent with exposure to estrogenic compounds.

2.2.4 Trends in Selected Sediment-Bound Compounds in Lakes and Reservoirs

Sediment cores from urban and agricultural reservoirs and lakes are used by USGS scientists to track changes in sediment-bound compounds over long periods [39,40]. The cores provide a measure of trends, make use of the watershed effect of integration of hydrologic inputs over space and time, and permit evaluation of many sediment-bound contaminants such as lead, chlordane, DDT, zinc, and polycyclic aromatic hydrocarbons (PAHs).

Findings showed a striking national reduction in concentrations of lead, which began to decrease after it was removed from gasoline in the 1970s (Figure 2-7). Specifically,

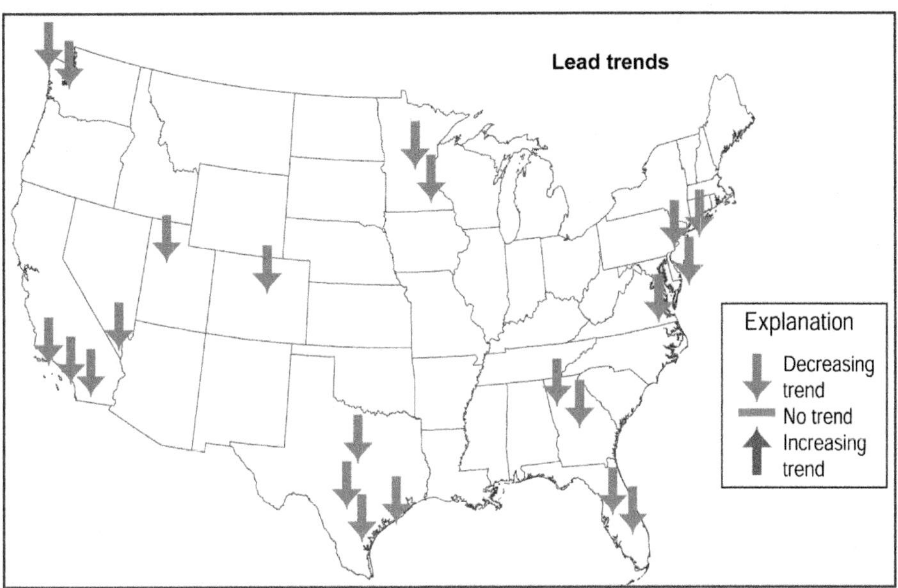

FIGURE 2-7 Samples taken from reservoir sediment cores show a decline from 1975 to 1997 in lead concentration. *From Ref. [42].* (For color version of this figure, the reader is referred to the online version of this book.)

trends in lead concentrations in sediment cores in nearly 85% of 35 sampled reservoirs and lakes in urban and reference areas, showed decreasing trends. This marks a positive environmental advancement resulting from the federal-led regulation, although concentrations are not yet back to background levels [41].

The USGS–reconstructed, water-quality histories for 38 urban and reference lakes across the United States for organochlorine compounds, showed significant trends in DDT, *p,p'*-DDE (a degradation, or breakdown product of DDT), and total polychlorinated biphenyls (PCBs). Similar to lead, the decreases reflect positive advancements resulting from federal regulations (Figure 2-8). Trends in chlordane were split evenly between upward and downward directions in concentration [43].

In contrast, trends were not so encouraging with PAHs, formed by the incomplete combustion of hydrocarbons (coal, oil, gasoline, and wood), resulting in many urban sources including industrial and power-plant emissions, home heating, car exhaust, tires, and asphalt in roads, roofs, and driveways (Figure 2-9). PAHs have increased in the majority of urban lakes sampled across the United States since about 1970 [43].

In more recent USGS work, coal-tar-based pavement sealant was determined to be the largest source of PAHs to 40 urban lakes [44,45]. On average, coal-tar-based sealcoat accounted for half of PAHs in the lakes, while vehicle-related sources accounted for about one-quarter. Elevated concentrations are generally associated with lakes receiving a large contribution of PAHs from sealcoat, in many cases at levels that can be harmful to aquatic life. Historical trends for a subset of studied lakes indicate that sealcoat use has been the

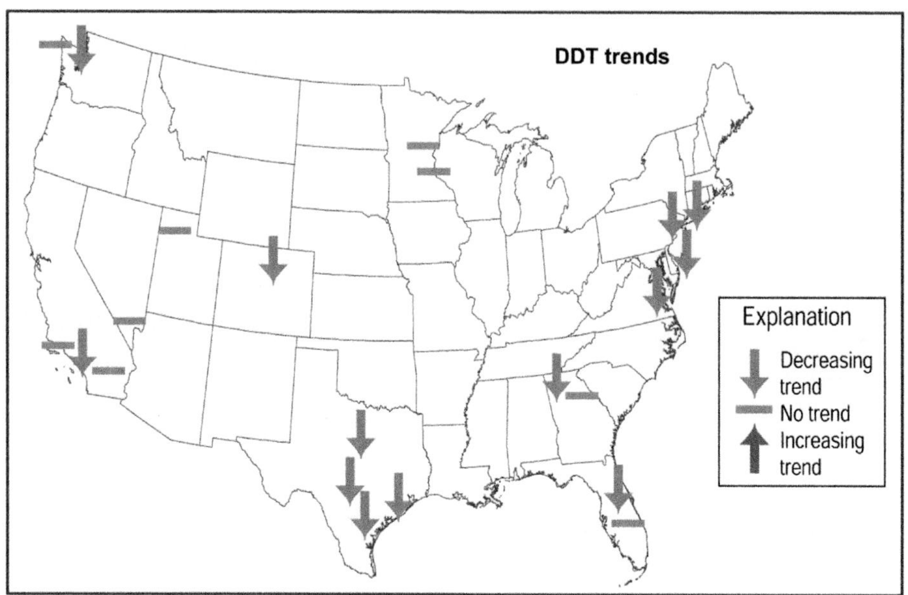

FIGURE 2-8 DDT trends from 1965 to 1997 are identified using sediment cores from reservoirs. *From Ref. [39].* (For color version of this figure, the reader is referred to the online version of this book.)

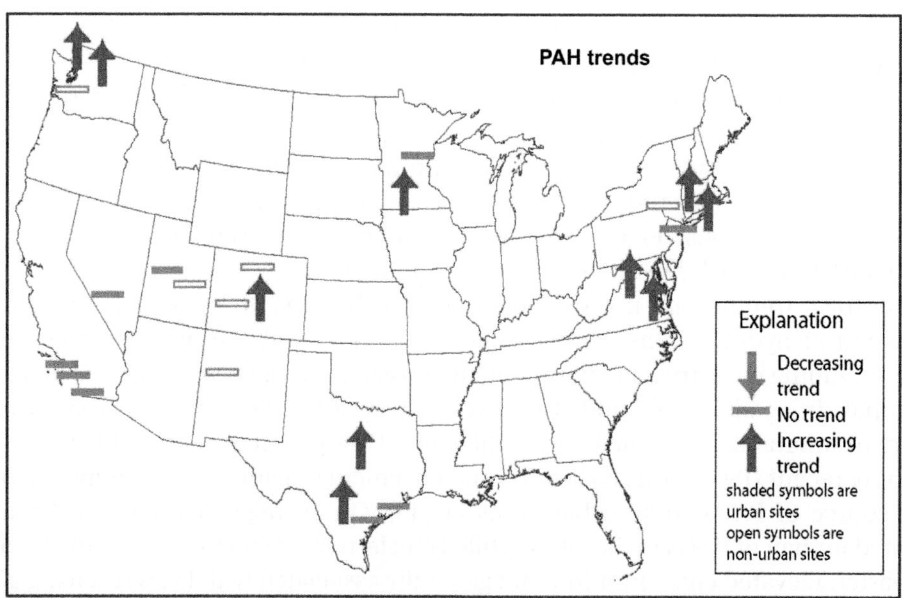

FIGURE 2-9 Polyaromatic hydrocarbon (PAH) trends from 1975 to 2000. *From Ref. [40].* (For color version of this figure, the reader is referred to the online version of this book.)

primary cause of increases in PAHs since the 1960s. These findings have led to local legislation, including a ban of coal-tar sealants, by the cities of Austin, (Texas) and Madison (Wisconsin), and in the state of Washington.

2.2.5 Mercury

Mercury, a neurotoxin, is one of the most hazardous contaminants to threaten the quality of our nation's waters. The USGS assessed mercury contamination in fish, bed sediment, and water in nearly 300 streams across the nation (Figure 2-10). The study revealed mercury contamination in fish sampled in the nearly 300 streams nationwide [46]. Sources of mercury (mostly from atmospheric deposition) are not uniform in their effects or distribution. Watershed characteristics make some streams more vulnerable to mercury deposition than others. In general, the presence of wetlands, organic material, and large amounts of dissolved organic carbon enhance the process of converting total mercury to methylmercury, the most toxic form of mercury and the form readily taken up by fish.

These factors are highlighted in a USGS study of total mercury in sediment in urban watersheds in the Boston metropolitan area, in which concentrations were highest in areas with many urban sources, including historical point-source discharges, nonpoint sources, and atmospheric deposition (Figure 2-11). Concentrations were lowest in sediment in watersheds with more forest cover in Maine and New Hampshire [47]. In

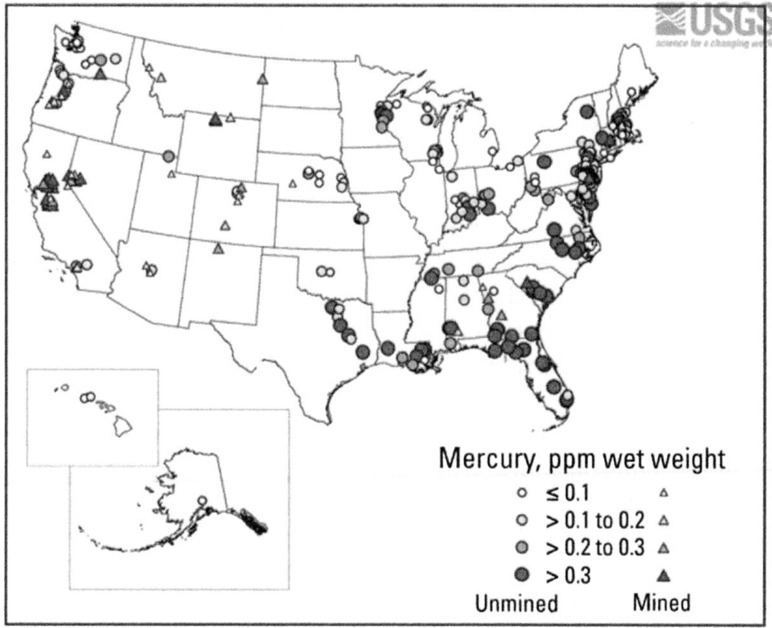

FIGURE 2-10 Spatial distribution of total mercury concentrations in game fish, 1998–2005. *From Ref. [46].* (For color version of this figure, the reader is referred to the online version of this book.)

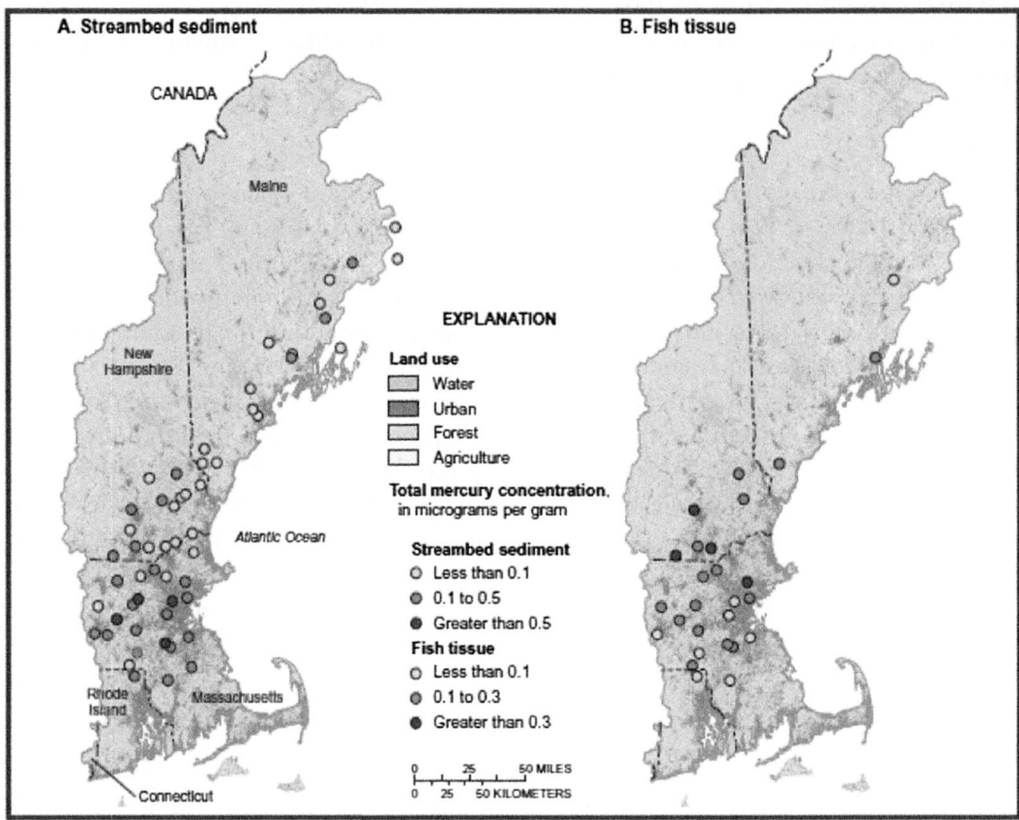

FIGURE 2-11 Mercury in fish varies by environmental setting. *From Ref. [47].* (For color version of this figure, the reader is referred to the online version of this book.)

contrast, concentrations of total mercury in fish (more than 95% of which is methylmercury) were higher in the forested watersheds near the Boston metropolitan area than in fish in the more urban watersheds. Elevated concentrations in fish in the forested watersheds result largely from natural factors, such as the presence of wetlands in these forested watersheds—an environment that enhances the process of converting total mercury to methylmercury [47].

In another study, the USGS assessed trends in mercury levels in fish, using a compilation of state and federal fish-monitoring data from 1969 to 2005 in U.S. rivers and lakes [48]. Findings showed significant decreases in fish mercury concentration at 22 of 50 sites sampled across the nation from 1969 to 1987, whereas only four sites showed increases. In those areas of decreases, mercury concentrations in fish decreased rapidly in the 1970s and more gradually or not at all during the 1980s. Most waters examined during this time period were medium to large rivers, draining areas of mixed land use. Trends were more variable from 1996 to 2005, during which time data were assessed for six states in the

Southeast and Midwest. More upward mercury trends in fish were documented in the Southeast compared to the Midwest. Upward mercury trends in fish in the Southeast were associated with increases in wet mercury deposition found at sites in the region that are part of the Mercury Deposition Network (a network of stations within the USGS National Atmospheric Deposition Program). Upward trends may, in part, be attributed to a greater influence of long-range global mercury emissions in the Southeast. In general, however, mercury concentrations in fish did not change in most aggregated state data from 1996 to 2005.

2.3 USGS Strategies to Assess Status and Trends

The USGS uses targeted designs and assessments to assess the status and trends of water quality and relations between water-quality conditions and natural and human factors that cause those conditions. These assessments consider the water-quality effects of human activities, such as agriculture and urban development, across different landscapes, geologic and hydrologic conditions, and during different seasons as well as over long periods of time.

This USGS approach is intended to answer questions such as: What are water-quality conditions? What are the contributing factors, and when do changes occur? Do certain natural features, land uses, or human activities and management actions affect the occurrence and movement of certain contaminants? Do water-quality conditions change over time?

USGS monitoring sites are not selected randomly, such as within a grid. Rather, sites are selected because they represent certain human activities, environmental settings, or hydrologic conditions during different seasons or times of year. For example, sites may be selected to assess the effects of agriculture and urban development on pesticide and nutrient contamination in streams.

2.3.1 Water Quality and the Natural Landscape

About 105 million people, or more than one-third of the Nation's population, receive their drinking water from one of the 140,000 public water systems across the United States that rely on groundwater pumped from public wells. Several USGS studies complement the extensive monitoring of public water systems that is routinely conducted for regulatory and compliance purposes by federal, state, and local drinking-water programs. USGS findings assist water utility managers and regulators in making decisions about future monitoring needs and drinking-water issues.

The natural landscape and geologic setting can be a source of natural contamination in our waters, such as with radon in groundwater (Figure 2-12). Radon is a naturally-occurring and soluble compound derived from uranium, and the USEPA classifies it as a human carcinogen. Uranium-bearing minerals are present in granite, pegmatite, and their derivative metamorphic rocks and sediments [2].

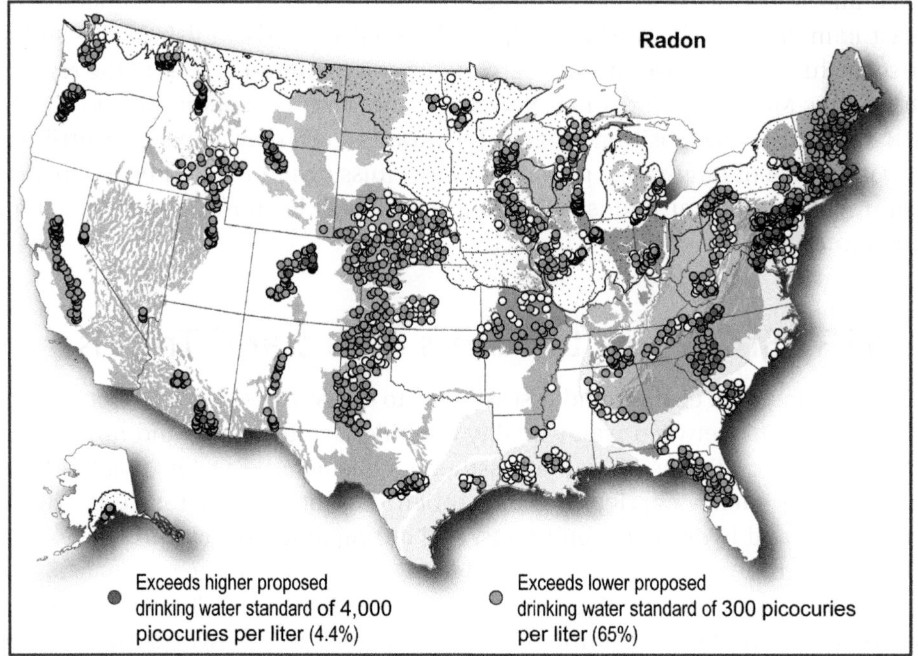

FIGURE 2-12 Radon concentrations in groundwater. Sampling sites reflect data from approximately 2000 private wells in 48 states. Regional coloring shows the extent of 30 principal aquifers sampled. Circles show radon concentrations; white circles show sites with concentrations below 300 picocuries per liter. *From Ref. [2].* (For interpretation of the references to color in this figure legend, the reader is referred to the online version of this book.)

Radon levels are elevated throughout much of the nation. A USGS national assessment of about 2100 domestic wells showed that concentrations of radon, greater than the lower USEPA-proposed MCL of 300 picocuries per liter (pCi/L), were found in 65% of the wells sampled across the nation, which were located in all 48 states and in parts of 30 regionally extensive principal aquifers [2]. Concentrations greater than the higher proposed MCL of 4000 pCi/L, were found in 4.4% of domestic wells. The standard is still under development by the USEPA (http://water.epa.gov/lawsregs/rulesregs/sdwa/radon/index.cfm).

Areas with elevated radon concentrations are controlled in large part by geology, generally associated with crystalline-rock aquifers located in the northeast, the central and southern Appalachians, and central Colorado [2]. The results of this study are important because they show that a large number of people may be unknowingly affected by exposure to radon. About 43 million people (or 15% of the nation's population) use drinking water from private wells, which are not regulated by the SDWA. Private well owners are responsible for testing the quality of their well water and for any water treatment that may be necessary.

A national assessment of radon was also conducted on public-supply wells located in 41 states and in parts of 30 regionally extensive, principal aquifers. Radon concentrations greater than the higher proposed MCL, were present in less than 1% of more than 900 sampled public-supply wells (generally located in deeper parts of the aquifers than domestic wells), and greater than the lower proposed MCL in 55% of the sampled wells. Public wells yielding water with radon concentrations greater than 300 pCi/L, were geographically distributed across the United States. Radon generally was lowest in public wells in the central states and in a few smaller sampled areas, such as eastern North Carolina. Similar to domestic wells, elevated radon was most common in crystalline-rock aquifers located in the Northeast [23]. Elevated concentrations of naturally occurring trace elements in groundwater are also widespread across the United States [49]. About 20% of untreated water samples from selected public, private, and monitoring wells across the nation contain concentrations of at least one trace element at levels of potential health concern. The trace elements in groundwater that most frequently exceeded USEPA human-health benchmarks were arsenic, uranium, and manganese. Patterns of trace element occurrence relate to factors associated with the geologic sources of trace elements and to features that affect their mobility, such as climate, land use, and geochemistry.

The landscape can also be a natural geologic source of contamination to surface water. The USGS reports, for example, that naturally occurring geologic sources of phosphorus contribute to elevated levels of streambed-sediment phosphorus levels in many watersheds in Florida, Kentucky, and Tennessee [50]. This USGS study characterizes the potential contributions of phosphorus to streams from naturally occurring geologic materials, based on the spatial distribution of phosphorus levels in streambed sediment from 5560 sampling sites in small, relatively undisturbed basins throughout the southeastern United States.

An important finding is that more than 75% of the phosphate ore mined in the United States comes from the southeastern United States [50]. Understanding the spatial variation in potential watershed contributions of total phosphorus from geologic materials, can assist water resource managers in developing nutrient criteria that account for natural variability in phosphorus contributions from weathering and erosion of surficial geologic materials.

2.3.2 Water Quality in Urban Areas

Water quality varies among different land uses, including agricultural, urban, and pristine land-use settings. USGS assessments show that water quality varies among the different settings; insecticides, for example, are more frequently detected at higher concentrations in urban streams than in agricultural streams. Water conditions also vary among settings with different land-use practices; phosphorus, sediment, and selected pesticides, for example, are at higher concentrations in streams draining agricultural fields with furrow irrigation than in agricultural fields with sprinkler irrigation [34,51–54].

A study by Cuffney et al. [55] examined the effects of urbanization on algae, aquatic insects, fish, habitat, and aquatic chemistry in urban streams in nine metropolitan areas across the country: Boston (Massachusetts), Raleigh (North Carolina), Atlanta (Georgia), Birmingham (Alabama), Milwaukee-Green Bay (Wisconsin), Denver (Colorado), Dallas-Fort Worth (Texas), Salt Lake City (Utah), and Portland (Oregon). Study findings showed that even at low levels of development (levels often considered protective for stream communities) the number of native fish and aquatic insects, especially those that are pollution sensitive, showed declines in urban and suburban streams. Specifically, the study reported that in watershed areas with impervious cover of 10%, many types of pollution-sensitive aquatic insects declined by as much as one-third, compared to streams in undeveloped forested watersheds. As such, even minimal or early stages of development can negatively affect aquatic life in urban streams [55]. The declines are in part related to rapid rises and falls of streamflow and changes in temperature during storms and high runoff. Stormwater from urban development can also contain fertilizers and insecticides that have been used along roads and on lawns, parks, and golf courses, which can be harmful to aquatic organisms.

Comparisons among the nine areas show that stream response to urban development varies across the country [56,57]. Differences occur mostly because stream quality and aquatic health reflect a complex combination of land and chemical use, land and stormwater management, population density and watershed development, and natural features, such as soils, hydrology, and climate. For example, aquatic communities in urban streams in Denver, Dallas-Fort Worth, and Milwaukee did not decline in response to urbanization because the aquatic communities were already degraded by previous agricultural land-use activities. In contrast, aquatic communities declined in response to urbanization in metropolitan areas, where forested land was converted to urban land—areas such as Boston and Atlanta.

These USGS studies represent an integrated approach to understanding the physical, chemical, and biological water-quality effects associated with urbanization. This integration is critical for prioritizing strategies for stream protection and restoration and in evaluating the effectiveness of those strategies over time. Stream protection and management is a top priority of state and local officials, and the findings are valuable in recognition of the unintended consequences that development can have on aquatic resources. The information is useful to predict and manage the future impacts of urban development on streams and reinforces the importance of having a green infrastructure to control stormwater runoff and protect aquatic life.

2.3.3 Water Quality in Agricultural Settings

The effects of the use of agricultural chemicals and other practices associated with agriculture on the quality of streams and groundwater are well known; however, less is known about how those effects may vary across different geographic regions of the nation. USGS studies on the transport and fate of agricultural chemicals in agricultural settings across the country, using comparable and consistent methodology and study designs, highlight how environmental processes and agricultural practices interact to affect the movement and transformation of agricultural chemicals in the environment [35,58–61]. The studies address major hydrologic compartments, including surface water, groundwater, the unsaturated zone, the streambed, and the atmosphere, as well as the pathways that interconnect these compartments. The study areas represent major agricultural settings, such as irrigated diverse cropping in the West and corn and soybean row cropping in the Midwest, and therefore, findings are relevant throughout much of the nation.

Findings from these studies show how environmental processes and agricultural practices act together to determine the transport and fate of agricultural chemicals in the environment. For example, agricultural chemicals were transported more quickly in areas of hydrologic and landscape modifications, such as those associated with irrigation, tile drains, and drainage ditches. In addition, rates of movement of agricultural chemicals over the land surface to groundwater, depended on the characteristics of the chemical (such as solubility, sorption characteristics, and susceptibility to biochemical transformation), the timing of chemical application relative to irrigation and precipitation, and the volume of water inputs relative to evapotranspiration rates. Instream processes such as photosynthesis and respiration can change nitrate loads in surface water, while nitrate attenuation in groundwater depends on flow rates, the value of water moving through reactive zones, and the presence of organic substrates needed to support biological processes.

2.3.4 Water Quality as Related to Land- and Water-Management Practices

Water quality and the ecologic productivity of streams are highly interconnected with hydrology and the amount and timing of water flowing in a stream. A recent national study by the USGS assessed relations between water quality, aquatic health, and streamflow alterations, [62] identified over 1000 unimpaired streams, to use as reference points to create streamflow models. The models were applied to estimate expected flows for 2888 additional streams where the USGS had streamgages from 1980 to 2007. The estimated values for the 2888 streams were compared to actual, measured flows to determine the degree to which streams have been altered.

Findings show the pervasiveness of streamflow alteration resulting from land and water management across the United States, the significant impact of altered streamflow on aquatic organisms, and the importance of considering this factor for sustaining and restoring the health of the Nation's streams and ecosystems [62]. Specifically, the amount of water flowing in streams and rivers has been significantly altered in nearly 90% of waters that were assessed in the study. Flow alterations are a primary contributor to degraded river ecosystems and loss of native species.

Flows are altered by a variety of land- and water-management activities, including reservoirs, diversions, subsurface tile drains, groundwater withdrawals, wastewater inputs, and impervious surfaces, such as parking lots, sidewalks, and roads. Altered river flows lead to the loss of native fish and invertebrate species whose survival and reproduction are tightly linked to specific flow conditions. These consequences can also affect water quality, recreational opportunities, and the maintenance of sport-fish populations. For example, in streams with severely diminished flow, native trout, a popular sport fish that requires fast-flowing streams with gravel bottoms, is replaced by less desirable non-native species, such as carp. Overall, the work by Carlyle et al. [62] indicated that streams with diminished flow contained aquatic communities that preferred slow-moving currents more characteristic of lake or pond habitats.

Understanding the ecological effects of these flow alterations helps water managers develop effective strategies to ensure that water remains sufficiently clean and abundant to support fisheries and recreation opportunities, while simultaneously supporting economic development. Annual and seasonal cycles of water flows (particularly the low- and high flows) shape ecological processes in rivers and streams. An adequate minimum flow is important to maintain suitable water conditions and habitat for fish and other aquatic life. High flows are important because they replenish floodplains and flush out accumulated fine sediment in channels that can degrade habitat.

Carlyle et al. [62] showed that the severity and type of streamflow alteration varies among regions as a result of varying natural landscape features, land practices, degree of development, and water demand. Differences are especially large between arid and wet climates. In wet climates, watershed management is often focused on flood control, which can result in lower maximum flows and higher minimum flows. Extremely low flows are the greatest concern in arid climates, in large part due to groundwater withdrawals and high water use for irrigation.

Salinity-control projects have been implemented since the mid-1970s by the Bureau of Reclamation, U.S. Department of Agriculture, and the Bureau of Land Management to control the salinity of water delivered to Mexico, per the 1974 Colorado River Basin Salinity Control Act. A landmark USGS study by [63] showed the positive effect that this type of informed-management strategy can have in improving water quality. The study, based on USGS salinity monitoring in streams and groundwater in the Southwest during more than 30 years, used the SPARROW model to relate the salinity to natural and human factors. The study describes salinity levels in streams and groundwater in parts of Arizona, California, Colorado, Nevada, New Mexico, Utah, and Wyoming, and concludes that although salinity varies widely throughout the region levels have generally decreased in many streams [63]. Specifically, findings showed that dissolved solids decreased from 1989 through 2003 at all sites downstream from salinity-control projects, and that the decreases were greater than decreases upstream from projects. For example, estimated annual loads of dissolved solids in the Gunnison River in the Upper Colorado River Basin decreased by about 162,000 tons per year downstream from the lower Gunnison salinity-control unit, in contrast to a decrease of only 2880 tons per year upstream from the unit. This net decrease is about 15% of the annual load in the lower Gunnison River.

This example shows how changes in land and water use, reservoir management, transbasin exports, and the implementation of salinity-control projects, including using low water-use irrigation systems and the redirection of saline water away from streams, has improved water quality in the Colorado River Basin by lowering salinity.

The USGS study also documents the variability of salinity throughout the region, from 22 to 13,800 mg/L in streams. Finally, the study shows that both natural factors and human activities affect salinity. Through new geostatistical modeling techniques, the USGS was able to show that land- and water-use activities, primarily associated with pasture and cultivated land, contribute more than half (56%) of the salinity to streams, whereas natural geologic materials provide the remaining 44% [63].

2.3.5 Water Quality and Seasonal Variation

Water quality issues, regardless of land use, management practices, or natural geographic variability, as described above, are complicated by seasonal variations. These seasonal variations are associated with climate and human factors, such as irrigation. For example, at many sites studied by the USGS in the eastern United States, total nitrogen concentrations were highest in the spring when streamflow is highest and when fertilizer is applied, while total phosphorus concentrations were highest in the summer and autumn when streamflow is lowest and less water is available to dilute effluents from point sources. At other sites, particularly in the upper Midwest, both nitrogen and phosphorus concentrations were greatest during high streamflow in the spring. In the western states, seasonal patterns were less distinct due to the highly variable topography and climate and the widespread use of dams, reservoirs, and canals [19].

USGS pesticide assessments also reveal effects of seasonality, generally showing low concentrations of pesticides in streams for most of the year—lower than most standards and guidelines established to protect aquatic life and human health [34]. However, in the case of the insecticide diazinon in a California stream, the assessments also showed pulses of elevated concentrations that exceeded standards and guidelines during times of the year associated with rainfall and applications of chemicals (Figure 2-13). Domagalski et al. [64] examined concentrations of diazinon over a 2-year period from 1996 to 1998 relative to a guideline for the protection of aquatic life (shown on Figure 2-13; guideline set by the International Joint Commission). No local water districts draw water directly from this creek, but the levels were frequently high enough to be toxic to water insects that are essential food for fish. These types of data demonstrate the need for targeted sampling over seasons to determine when peak contaminant concentrations may affect drinking-water supplies and critical life stages of aquatic organisms.

2.3.6 Water Quality Over the Long Term

A critically important reason for any governmentto sustain a scientific mission, such as the USGS, is its ability to collect and interpret water-quality data across many decades using equivalent techniques, so that the data can be compared. A classic example was

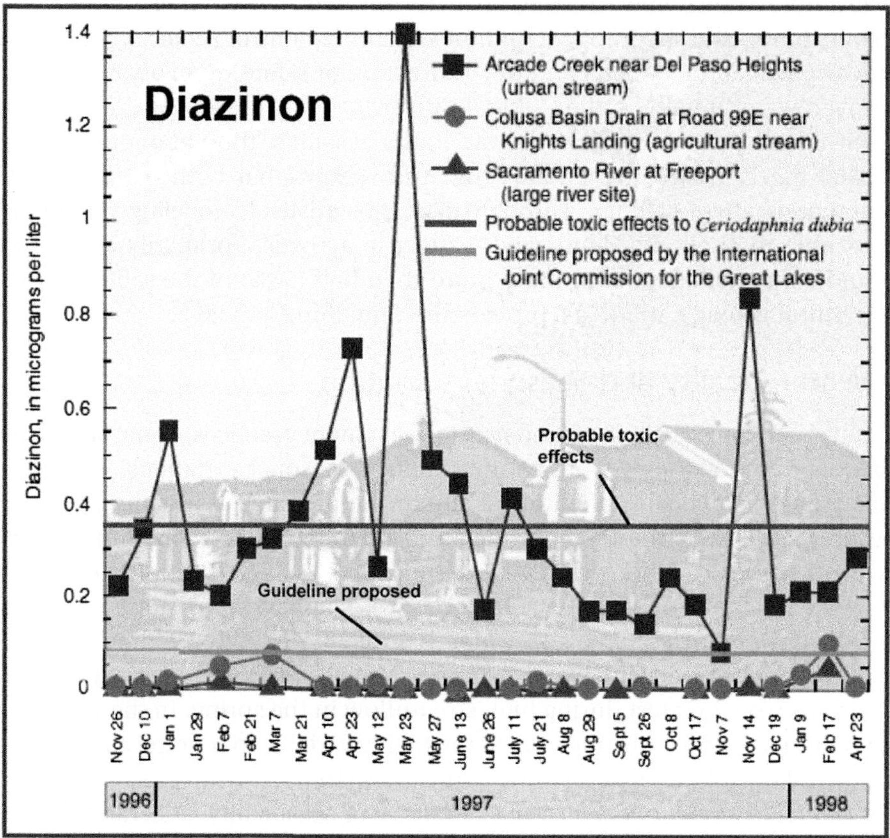

FIGURE 2-13 Pulses of elevated concentrations show the importance of tracking seasonal variation. *From Ref. [64].* (For color version of this figure, the reader is referred to the online version of this book.)

published by Goolsby and Battaglin [65]. They described nitrogen in the Mississippi River, principally nitrate and organic nitrogen (dissolved and particulate). Nitrate is the most soluble and mobile form of nitrogen. Using USGS historical records dating back as early as 1903, the authors showed that the average concentrations of nitrate in the Mississippi River and some of its tributaries increased several fold after the early 1900s, in parts of Iowa, Illinois, Indiana, Minnesota, and Ohio. They noted that concentrations increased by a factor of about 2.6 between 1905 and 1907 and 1980 and 1996, and that most of the increase in the lower Mississippi River main stream occurred between the late 1960s and the early 1980s. During that period, the average annual nitrate concentration in water flowing to the Gulf of Mexico more than doubled.

As this example indicates, without comparable data collected over time, assessments cannot distinguish long-term trends from short-term fluctuations—nor can natural fluctuations be distinguished from the effects of human activities. USGS assessments show that water quality continually changes. The changes can be relatively quick—within days, weeks, or months, such as in streams in the Midwest where types of herbicides used on corn and soybeans have changed; or changes can be relatively slow, such as in groundwater beneath the Delmarva Peninsula where nitrate concentrations are beginning to decrease after 10 years of improved management of nitrogen fertilizers [34,66].

The need for long-term data is well demonstrated in the High Plains aquifer, also known as the Ogallala aquifer, which is the Nation's most heavily used groundwater resource. Most of the extracted water is used for irrigation, but nearly 2 million people also depend on the aquifer as a source of drinking water. The eight states that use water from the High Plains aquifer are Colorado, Kansas, Nebraska, New Mexico, Oklahoma, South Dakota, Texas, and Wyoming. Nebraska hosts the largest areal extent of the water source.

USGS findings show that heavy use of water for irrigation and public supply, as well as leakage down inactive irrigation wells, are resulting in long-term gradual increases in concentrations of contaminants (such as nitrate and dissolved solids) from the water table to deeper parts of the High Plains aquifer where drinking-water wells are screened [67].

USGS scientists analyzed water for more than 180 chemical compounds and physical properties in about 300 private domestic wells, 70 public-supply wells, 50 irrigation wells, and 160 shallow, monitoring wells sampled between 1999 and 2004. The study, by Gurdak et al. [67] also assessed the transport of water and contaminants from land surface to the water table and deeper zones used for supply, to predict changes in concentrations over time.

Water produced by the High Plains aquifer is generally acceptable for human consumption, irrigation, and livestock watering. However, the increase in contaminant concentrations over time has important implications for the long-term sustainability of the High Plains aquifer as a source of drinking water. Once contaminated, the aquifer is unlikely to be remediated quickly because of slow rates of contaminant degradation and slow groundwater travel times in the aquifer; deep water in some parts of the aquifer is about 10,000 years old [67].

2.3.7 The Value of Water Quality Modeling

Models are essential if we are to better understand the status and trends of chemical and other constituents in the environment and to forecast future conditions. Models, integrated with monitoring, help us to apply our understanding of water quality, the hydrologic system, and the landscape to broader areas, including entire stream reaches and aquifers, large river basins, ecoregions, states, and the nation [30]. A map showing the distribution of average atrazine concentrations at USGS stream sites and the intensity of its use on crops, illustrates this point (Figure 2-14). The highest concentrations were found in the Midwest Corn Belt where use is most intense (data from Gilliam et al. [34]).

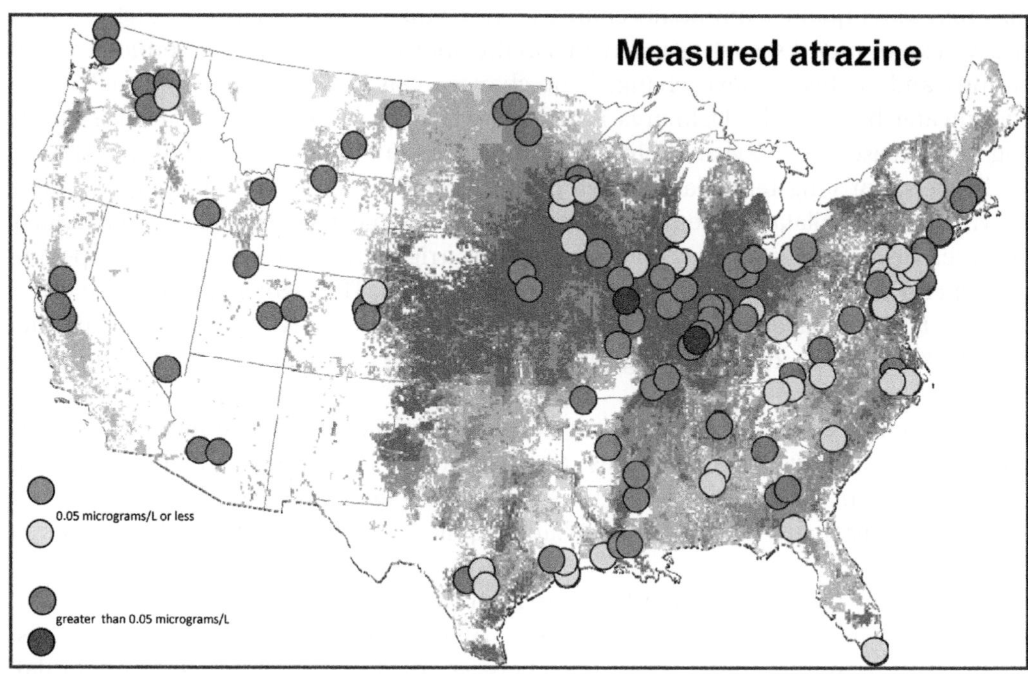

FIGURE 2-14 Atrazine measured in streams, 1992–2001. Blue and yellow circles are streams with low concentrations—0.05 μg/L or less; orange and red circles have higher concentrations—0.05 to greater than μg/L. Green shading indicates the intensity of atrazine use on crops. *From Ref. [34].* (For interpretation of the references to color in this figure legend, the reader is referred to the online version of this book.)

To extend monitoring data to a more complete national assessment, USGS scientists developed a statistical model that uses the measured data collected in streams (e.g., Figure 2-14) with information on pesticide use and land use, climate and soil characteristics, and other watershed characteristics to predict concentrations for streams that have not been sampled. This statistical model allows scientists to map predicted average atrazine concentrations for more than 60,000 streams nationwide (Figure 2-15). The development of this type of predictive method allows scientists and resource managers to extend information from relatively few sites with direct measurements to the rest of the nation, and it is used by regulatory agencies such as USEPA and state equivalents to anticipate elevated concentrations of atrazine in streams.

Models have also been developed to better understand sources and transport of contaminants. One such example, noted earlier, is SPARROW, a spatially explicit, data-driven model that relates major pollutant sources to instream measurements [30]. At the national scale, SPARROW includes agricultural land uses and nutrient inputs (including those from the atmosphere), treated sewage, and crop and livestock production. It also includes watershed characteristics that control transportation, such as stream size, soils, and slope.

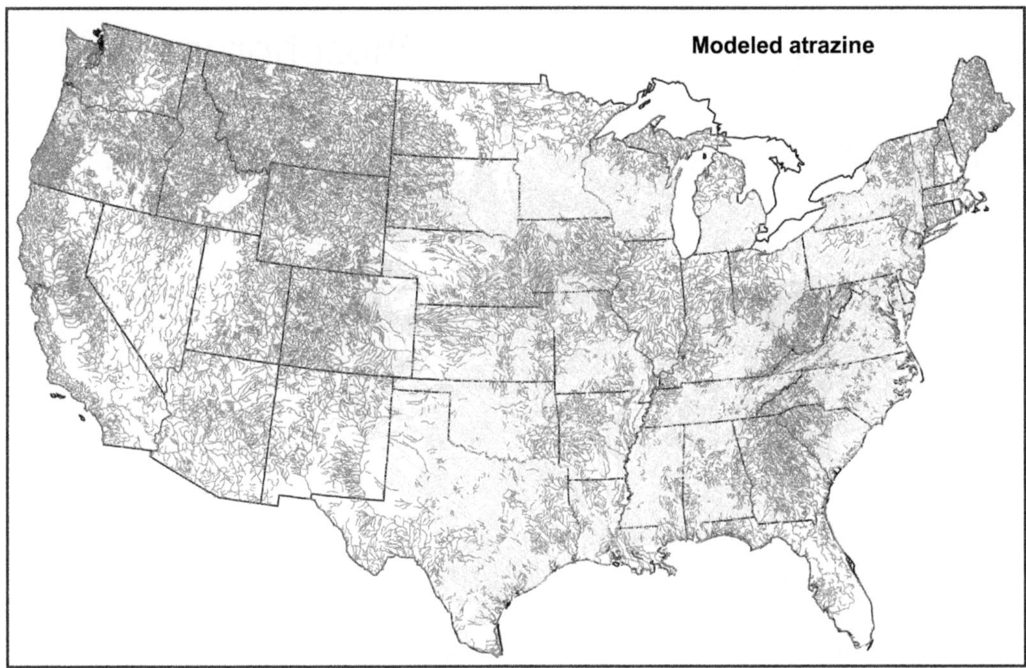

FIGURE 2-15 Prediction of atrazine in streams. Blue and yellow streams (most of the U.S. outside of the larger channels in the Mississippi River watershed) indicate low concentrations; red (mainly the Mississippi River and principal tributaries) indicates high concentrations. (For interpretation of the references to color in this figure legend, the reader is referred to the online version of this book.)

SPARROW was used by USGS scientists to show the major source areas and delivery of phosphorus from the Mississippi River Basin (an area including 31 states) to the Gulf of Mexico (Figure 2-16). The results indicated that a considerable amount of phosphorus delivered to the Gulf originates in distant watersheds, such as in the Ohio and Tennessee Rivers [30]. Similar findings are evident for nitrogen delivery. As discussed above (Figure 2-3), USEPA, the U.S. Department of Agriculture (USDA), states, and other members of the Gulf Hypoxia Task Force use this information to identify priority watersheds and agricultural management practices in the watersheds delivering the most nutrients to the Gulf [31].

The USGS has developed a regionally focused set of SPARROW water-quality models to assist with the interpretation of water-resources data and provide predictions of water quality in unmonitored streams. These regional SPARROW models incorporate geospatial data on the geology, soils, land use, fertilizer, manure, wastewater-treatment facilities, temperature, precipitation, and other watershed characteristics derived from USGS, USEPA, USDA, and the National Oceanic and Atmospheric Administration. These data are then linked to measurements of streamflow from USGS streamgages and water-quality monitoring data from approximately 2,700 sites operated by 73 monitoring agencies [68].

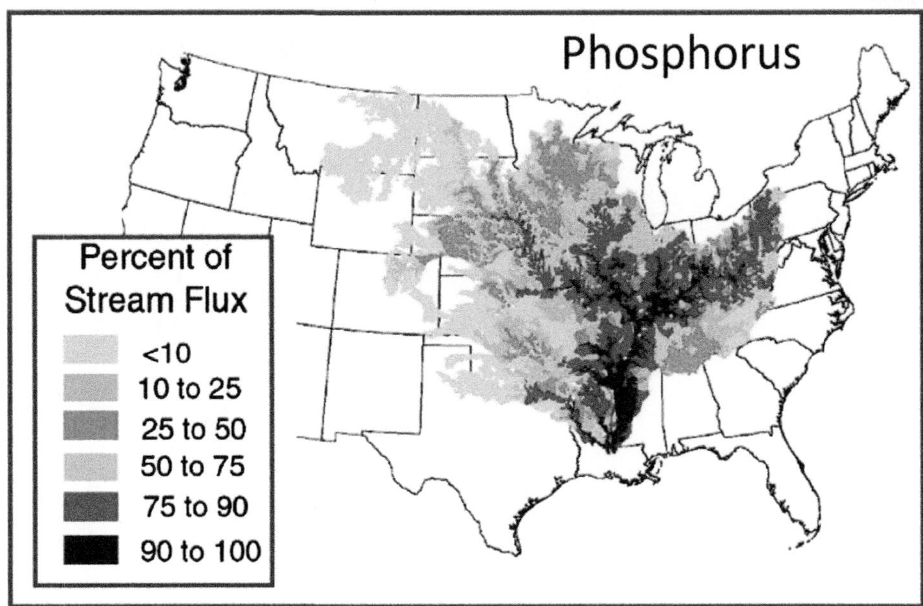

FIGURE 2-16 Estimated delivery of phosphorus from the Mississippi River Basin to the Gulf of Mexico. *From Ref. [30]* (For color version of this figure, the reader is referred to the online version of this book.)

Six USGS regional models were developed in the Northeast, Southeast, Upper Midwest (Great Lakes), Missouri River, Lower Midwest (Gulf of Mexico), and Pacific Northwest, using the national SPARROW modeling framework. Results detailing nutrient conditions in each region and description of a support system for decision making are published in a set of papers [68,69]. These authors indicate that each region and locality has a unique set of nutrient sources and characteristics that determine how those nutrients are transported to streams. For example, based on the six regional model results, wastewater effluent and urban runoff are significant sources of nutrients in the Northeast and Mid-Atlantic, while agricultural sources like farm fertilizers and animal manure contribute heavily to nutrient concentrations in the Midwest and the central regions of the nation. Atmospheric deposition is the largest contributor of nitrogen in many streams in the eastern United States, and naturally occurring geologic sources are a major source of phosphorus in many areas. Additionally, the six models show that the amount of nutrients transported varies greatly among the regions; for example, nutrients can be removed in reservoirs or used by plants before they reach downstream waters. Temperature and precipitation variations across the country also affect the rates of nutrient movement and nutrient loss on the land and in streams and reservoirs.

Using a web-based support system for decision making built on these models, users can evaluate combinations of source-reduction scenarios that target one or multiple sources of nutrients and see the change in the amount of nutrients transported to downstream waters—a capability that has not been widely available in the past. For example, the web-based support system for decision making indicates that reducing wastewater discharges throughout the Neuse River Basin in North Carolina by 25% will reduce the amount of nitrogen transported to the Pamlico Sound from the Neuse River Basin by 3%, whereas a 25% reduction in agricultural sources, such as fertilizer and manure, will reduce the amount of nitrogen by 12%.

Similar models are applied to assess groundwater quality problems and to predict the vulnerability of groundwater, across the nation, to nitrate [66]. The concentration of nitrate contamination in shallow groundwater and in drinking-water wells across the country can be predicted on the basis of (1) nitrate concentrations measured by the USGS at nearly 2000 wells, and (2) national data sets on sources of nitrogen. The model prediction demonstrates moderate to severe nitrate contamination (greater than the drinking water standard of 10 mg/L, as indicated in red on Figure 2-17) at locations in the High Plains, Mid-Atlantic region, and California. The high concentrations are generally related to high nitrogen inputs, high water input, well-drained soils, fractured rocks or those with high effective porosity, and lack of attenuation processes. The estimated number of people served by wells with nitrate concentrations severely contaminated by nitrate greater than 10 mg/L is about 467,000 in the United States. The model predicts that exposure risk is reduced by seeking deeper water supplies.

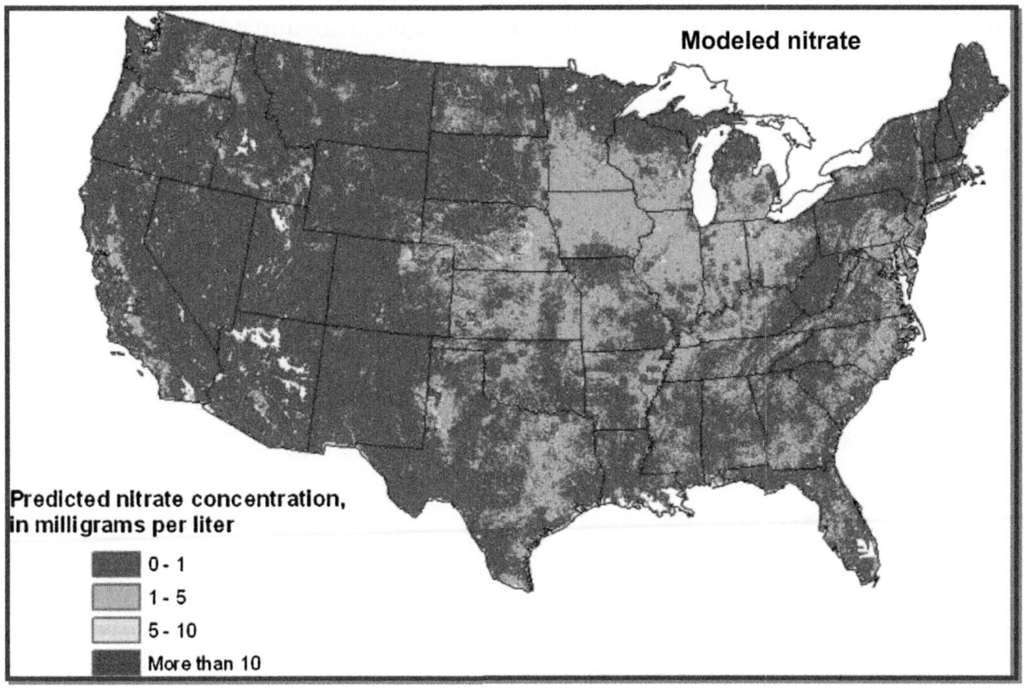

FIGURE 2-17 Vulnerability of shallow groundwater and drinking-water wells to nitrate contamination. *From Ref. [66].* (For color version of this figure, the reader is referred to the online version of this book.)

2.3.8 Water Quality and Climate Change

Climate change is a major challenge for global water-resources management [70]. Furthermore, climate change is increasingly recognized as a factor not only in water availability, but also in the quality of water [6,62]. Although numerous impacts are anticipated for the future, climate change is already affecting water resources in a variety of ways [3–5,11]. According to Dettinger [71]; "earlier spring snowmelt has been well documented in the western US. Warmer winters and springs in the Western United States have contributed to widespread hydrologic changes, including trends toward more precipitation as rain rather than as snow, less snowpack overall, earlier greening of vegetation, and earlier snowmelt" (Figure 2-18). These types of annual and decadal changes in precipitation can lead to long-term changes in surface runoff or instream dilution, contributing to long-term changes in nutrient concentrations in streams [71]. A few additional examples of water quality impacts already observed and documented, as well as others that are predicted, are described below.

Increased frequency of heavy rainfall is an expected outcome of a warmer atmosphere [9,73]. Storm rainfall is a water-treatment challenge because it increases sediment and pathogen loads, urban stormwater runoff, and combined sewer overflows [74]. Droughts are also expected to be more common and intense and to last longer [73]. A drought followed by runoff associated with heavy precipitation results in higher concentrations of contaminants and increased nutrient loads [75]. These higher levels of nutrients, in turn, increase the potential for algal blooms and the associated taste- and odor- and algal-toxin

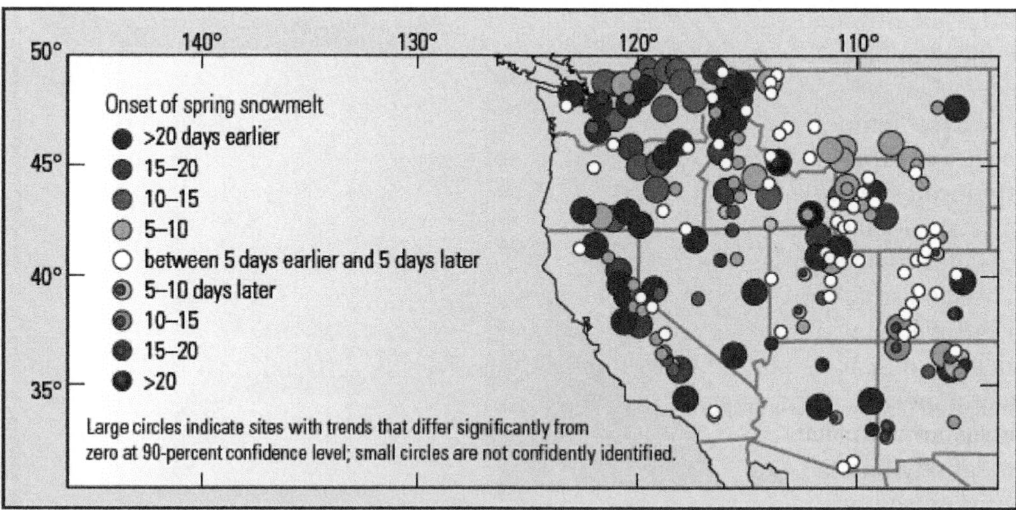

FIGURE 2-18 Trends for the onset of spring snowmelt are shown for the western United States. This is the date on which snowmelt runoff adds large water volume to rivers—the onset occurs 1 week to almost 3 weeks earlier now than in the middle of the twentieth century [71,72]. Changes in the timing of streamflow have implications for agricultural water management and hence nutrient transport. (For color version of this figure, the reader is referred to the online version of this book.)

problems [75,76]. As such, the quality of source water is expected to be affected by more frequent algal blooms, changes in types and abundance of watershed vegetation, and increased water temperature, particularly in riparian areas [74].

In their review of potential impacts of climate change on groundwater resources Earman and Dettinger [3], state that "changes in the Earth's climate have the potential to affect both the quality and quantity of available groundwater, primarily through impacts on recharge, evapotranspiration and (indirectly) on pumpage and abstraction."

A well-documented aspect of the impact of climate change on water quality pertains to saline intrusion in coastal aquifers. Globally, sea levels have risen by about 22 centimeters in the twentieth century. It is expected that sea level will continue to rise in response to global warming, as ocean waters warm and expand and major ice sheets melt into the seas. In coastal areas, these higher sea levels are likely to increase the potential for the intrusion of ocean water into freshwater aquifers, increasing groundwater salinity [3].

One impact of changing patterns of precipitation on small streams is that they are sustained through substantial parts of the year by groundwater flow [3]. These streams are relatively shallow compared to the depth of the aquifers that supply the groundwater inflow, and consequently receive inflows mainly from the uppermost parts of the contributing aquifers. As groundwater recharge decreases, water tables drop and contribution to streamflow declines [3].

The effects of reduced groundwater inflow on water quality and ecosystems have also been described. One effect is the reduced dilution of contaminants and warmer stream temperature. Groundwater is typically cooler than streamflow [3] because surface water has traveled overland to (and through) stream channels. Warmer stream temperatures are likely to have significant impacts on species viability, as described in two examples below.

Surface temperatures of rivers and oceans have all increased significantly [4,9]. Warmer streamflow is a stressor on many endemic species and facilitates colonization by introduced species. Additionally, spring warming of temperate lakes disrupts the timing between zooplankton and the phytoplankton food supply [4].

Model projections by Cloern [4] suggest that "climate-driven changes to the San Francisco Estuary-Watershed could lead to a diminishing water supply, continued shifts toward wetter winters and drier summers, sea level rising to higher levels than were projected only a few years ago, salt water intrusion, reduced habitat quality for native aquatic species." Warmer water temperature has had a negative effect on fish endemic to the Sacramento Delta, such as the Delta Smelt [4]. These fish are adapted to cool, turbid, low-salinity habitats and as delta waters warm and become more saline, their already-compromised habitat may be further degraded.

Twentieth century warming has already affected the Salmonid habitat, with unfavorable changes to thermal and hydrologic properties of aquatic systems, according to a study of risks to coldwater fish as a result of climate change [77]. In their work on Native Trout and Grayling within 11 western states, the authors note that climate model output

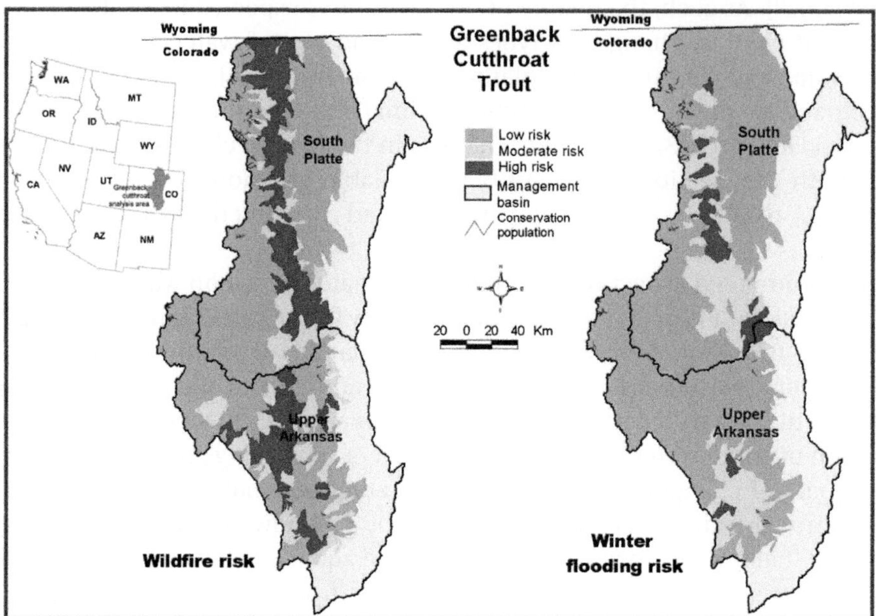

FIGURE 2-19 Risk of summer temperature increases and drought for greenback cutthroat trout in their current range in the drainages of the South Platte Basin and the Upper Arkansas Basin in Colorado. *From Ref. [77].* (For color version of this figure, the reader is referred to the online version of this book.)

indicates that these trends will continue and even accelerate until at least the middle of the twenty-first century [77]. Drought is the most pervasive threat, with 40% or more of the historic range of seven taxa at high risk. Greenback cutthroat trout are an example, in Colorado, where population pressure and land-use change have eliminated the fish from much of its former habitat, and most of the remaining populations are found in higher-elevation streams and lakes [77]. Drought is a relatively high-risk factor for this fish; additionally, increasing summer temperature and wildfire are also higher risks within its historic habitat in the South Platte Basin as compared to where they are found now (Figure 2-19) [77].

2.4 Conclusions

This brief set of mainly USGS-derived examples of water-quality status and trends in the United States indicates that to sustain water supplies of high quality to meet human and ecological needs, a continual advance in our understanding of an increasingly complex set of underlying controlling factors is required. Furthermore, constant development of new analytical tools and models to inform our approach will be essential. Lastly, the scientific community needs to synthesize and translate the results of this work so that resource managers and decision makers can effectively use the scientific information that we provide.

Water-quality issues facing the United States have changed since the implementation of the CWA and are in large part, focused on nonpoint sources of pollution and other environmental stressors. A major challenge for the monitoring of water-quality status and trends is to improve understanding of the many nonpoint-related sources and stresses that affect water quality, now and over the long term. This requires credible, objective, and interdisciplinary data on, and interpretation of, the physical, chemical, and biological conditions of water bodies and the natural landscape; and also of natural factors such as climate, climate change, and human activities, that may contribute to those conditions.

The water quality issues highlighted here are complex, and explanations for degraded water quality include interactions among the natural environment, climate change, and human land and water use that vary spatially and temporally over a variety of scales. These complexities present challenges to scientists and policy makers who are trying to understand and manage water-quality issues such as: changing patterns in flow conditions; increasing water temperature and salinity; and increases in contamination from a variety of naturally occurring and man-made chemicals. Complex mixtures of organic compounds, such as pesticides, pharmaceuticals, and hormones, can, even at very low concentrations, adversely affect the health or reproductive success of aquatic organisms; and nutrient enrichment can lead to algal blooms, low oxygen, and impaired ecosystems.

Given these complexities, achieving sustainable high-quality water across the nation to meet human and ecosystem needs requires sustained and comprehensive efforts in several key areas:

- Understanding the relations between water quality and the natural landscape, hydrologic processes, climatic variability, and human activities;
- Evaluating water quality in the context of physical, chemical, and biological systems;
- Monitoring over short-term and long-time scales;
- Integrating modeling with monitoring to help apply our understanding of water-quality conditions and hydrologic systems to unmonitored, yet comparable areas.

Such commitments and investments will continue to provide the critical and improved scientific basis for decision makers to effectively manage and protect our waters across the United States and in specific geographic areas, now and in the future. Science will help to sort out and prioritize the multitude of decisions involving competing demands for safe drinking water, irrigation, sustainable ecosystems, energy development, and recreation.

Acknowledgments

The authors thank the many USGS scientists and hydrologic technicians whose work contributed to the body of literature summarized in this paper. The manuscript was improved by peer reviews and recommendations received from Diane McKnight, University of Colorado, and Satinder Ahuja, Ahuja Consulting, Calabash, North Carolina.

References

[1] Davis WS. Biological assessment and criteria: building on the past. In: Davis WS, Simon TP, editors. Biological assessment and criteria: tools for water resource planning and decision making. Boca Raton: Lewis Publishers; 1995. p. 15–30.

[2] DeSimone LA, Hamilton PA, Gilliom RJ. Quality of water from domestic wells in principal aquifers of the United States, 1991–2004—overview of major findings, U.S. Geological Survey Circular 1332 2009.

[3] Earman S, Dettinger MD. J Water Clim Change 2011;2:213–29.

[4] Cloern JE, Knowles N, Brown LR, Cayan D, Dettinger MD, Morgan TL, et al. PLoS One 2011;6:e24465.

[5] Knowles N, Dettinger MD, Cayan DR. J Clim 2006;19:4545–59.

[6] Larsen MC. Water Resources Impact. American Water Resources Association 2012;14(5):3–7.

[7] Hirsch RM, Ryberg K. Hydrol Sci J, http://www.tandfonline.com/doi/abs/10.1080/02626667.2011.621895; 2011.

[8] Mote P, Brekke L, Duffy PB, Maurer E. EOS, Transactions. American Geophysical Union; 2011; 92:257–258.

[9] Hansen J, Ruedy R, Sato M, Lo K. Rev Geophys 2010;48:RG4004.

[10] Milly PCD, Dunne KA, Vecchia AV. Nature 2005;438:347–50.

[11] Milly PCD, Betancourt J, Falkenmark M, Hirsch RM, Kundzewicz ZW, Lettenmaier DP, et al. Science 2008;319:573–4.

[12] National Research Council. Colorado River Basin water management—evaluating and adjusting to hydroclimatic variability. Washington, D.C: The National Academies Press; 2007.

[13] Averett RC, McKnight DM, editors. Chemical quality of water and the hydrologic cycle. Chelsea, Michigan: Lewis Publ. Inc.; 1987. p. 387.

[14] U.S. Geological Survey. Biological indicators, U.S. Geological Survey Techniques of Water-Resources Investigations, Book 9, Chap. A7, http://pubs.water.usgs.gov/twri9A/; 2011 [accessed 19.02.12].

[15] Yoder CO. Policy issues and management applications for biological criteria. In: Davis WS, Simon TP, editors. Biological assessment and criteria: tools for water resource planning and decision making. Boca Raton: Lewis Publishers; 1995. p. 327–44.

[16] Hirsch RM, Hamilton PA, Miller TL. J Environ Mon 2006;8:512–8.

[17] Myers MD, Ayers MA, Baron JS, Beauchemin PR, Gallagher KT, Goldhaber MB, et al. Science 2007;318:200–1.

[18] National Research Council. Toward a sustainable and secure water future—a leadership role for the U.S. Geological Survey. Washington, D.C: The National Academies Press; 2009.

[19] Dubrovsky NM, Burow KR, Clark GM, Gronberg JM, Hamilton PA, Hitt KJ, et al. The quality of our Nation's waters—nutrients in the Nation's streams and groundwater, 1992–2004. U.S. Geological Survey Circular 1350, 2010; p. 174.

[20] Dubrovsky NM, Hamilton PA. Nutrients in the Nation's streams and groundwater—national findings and implications, U.S. Geological Survey Fact Sheet 2010–3078; 2010, p. 6.

[21] Ward MH, deKok PL, Brender J, Gulis G, Nolan B, VanDerslice J. Environ Health Perspect 2005;113:1607–14.

[22] Townsend AR, Howarth RW, Bazzaz FA, Booth MS, Cleveland CC, Collinge SK, et al. Front in Ecol Environ 2003;1:240–6.

[23] Toccalino PL, Hopple JA. The quality of our Nation's waters—quality of water from public-supply wells in the United States, 1993–2007—overview of major findings. U.S. Geological Survey Circular 1346, 2010; p. 58.

[24] Puckett LJ, Tesoriero AJ, Dubrovsky NM. Environ Sci Technol 2011;45:839–44.

[25] Jagucki ML, Landon MK, Clark BR, Eberts SM. Assessing the vulnerability of public-supply wells to contamination—High Plains aquifer near York, Nebraska. U.S. Geological Survey Fact Sheet 2008–3025, 2008; p. 6.

[26] Mississippi River/Gulf of Mexico Watershed Nutrient Task Force. A science strategy to support management decisions related to hypoxia in the northern Gulf of Mexico and excess nutrients in the Mississippi River Basin: prepared by the Monitoring, Modeling, and Research Workgroup of the Mississippi River/Gulf of Mexico Watershed Nutrient Task Force. U.S. Geological Survey Circular 1270, 2004; p. 58.

[27] Committee on Environment and Natural Resources. Scientific assessment of hypoxia in U.S. coastal waters, Interagency Working Group on Harmful Algal Blooms, Hypoxia, and Human Health of the Joint Subcommittee on Ocean Science and Technology, Washington, D.C. 2010.

[28] Aulenbach BT, Buxton HT, Battaglin WA, Coupe RH. Streamflow and nutrient fluxes of the Mississippi-Atchafalaya River Basin and sub-basins for the period of record through 2005. U.S. Geological Survey Open-File Report 2007–1080, 2007; http://toxics.usgs.gov/pubs/of-2007-1080/.

[29] Rabalais NN, Diaz RJ, Levin LA, Turner RE, Gilbert D, Zhang J. Biogeosciences 2010;7:585–619.

[30] Alexander RB, Smith RA, Schwarz GE, Boyer EW, Nolan JV, Brakebill JW. Environ Sci Technol 2008;42(3).

[31] Robertson DM, Schwarz GE, Saad DA, Alexander RB. J Am Water Res Assoc 2009;45:534–49.

[32] Sprague LA, Hirsch RM, Aulenbach BT. Environ Sci Technol 2011;45(7):7209–16.

[33] Hirsch RM, Moyer DL, Archfield SA. J Am Water Res Assoc 2010;46(5):857–80.

[34] Gilliom RJ, Barbash JE, Crawford CG, Hamilton PA, Martin JD, Nakagaki N, et al. Pesticides in the Nation's streams and ground water, 1992–2001. U.S. Geological Survey Circular 1291, 2006; p. 180.

[35] Bayless ER, Capel PD, Barbash JE, Webb RMT, Hancock TLC, Lampe DC. J. Environ Qual 2008;37:1064–72.

[36] Sullivan DJ, Vecchia AV, Lorenz DL, Gilliom RJ, Martin JD. Trends in pesticide concentrations in corn-belt streams, 1996–2006. U.S. Geological Survey Scientific Investigations Report 2009–5132, 2009; p. 75, http://pubs.usgs.gov/sir/2009/5132/.

[37] Kolpin DW, Furlong ET, Meyer MT, Thurman EM, Zaugg SD, Barber LB, et al. Environ Sci Technol 2002;36:1202–11.

[38] Barber LB, Brown GK, Nettesheim TG, Murphy EW, Bartell SE, Schoenfuss HL. Sci Total Environ 2011;409:4720–8.

[39] Van Metre PC, Callender E, Fuller CC. Environ Sci Technol 1997;31:2339–44.

[40] Van Metre PC, Mahler BJ, Furlong ET. Environ Sci Technol 2000;34:4064–70.

[41] Mahler BJ, Van Metre PC, Callender Edward. Environ Toxicol Chem 2006;25(7):1698–709.

[42] Callender E, Van Metre PC. Environ Sci Technol 1997;31:424A–8A.

[43] Van Metre PC, Mahler BJ. Environ Sci Technol 2005;39:5567–74.

[44] Mahler BJ, Van Metre PC. Coal-tar-based pavement sealcoat, polycyclic aromatic hydrocarbons (PAHs), and environmental health. U.S. Geological Survey Fact Sheet 2011–3010, 2011; p. 6.

[45] Van Metre PC, Mahler BJ. Sci Total Environ 2010;409:334–44.

[46] Scudder BC, Chasar LC, Wentz DA, Bauch NJ, Brigham ME, Moran PW, et al. Mercury in fish, bed sediment, and water from streams across the United States, 1998–2005, U.S. Geological Survey Scientific Investigations Report 2009–5109, 2009; p. 74, http://pubs.usgs.gov/sir/2009/5109/.

[47] Robinson KW, Flanagan SM, Ayotte JD, Campo KW, Chalmers A, Coles JF, et al. Water quality in the New England coastal basins, Maine, New Hampshire, Massachusetts, and Rhode Island, 1999–2001. U.S. Geological Survey Circular 1226; 2004, p. 38.

[48] Chalmers AT, Argue DM, Gay DA, Brigham ME, Schmitt CJ, Lorenz DL. Environ Mon Assess 2011;175:175–91.

[49] Ayotte JD, Gronberg JM, Apodaca LE. Trace elements and radon in groundwater across the United States, 1992–2003. U.S. Geological Survey Scientific Investigations Report 2011–5059, 2011; 115.

[50] Terziotti S, Hoos AB, Harned DA, Garcia AM. Mapping watershed potential to contribute phosphorus from geologic materials to receiving streams, southeastern United States. U.S. Geological Survey Scientific Investigations Map 3102, 1 sheet, 2010; http://pubs.usgs.gov/sim/3102/.

[51] Capel PD, McCarthy KA, Barbash JE. J Environ Qual 2008;37:983–93.

[52] Domagalski JL, Ator S, Coupe R, McCarthy K, Lampe D, Sandstrom M, et al. J Environ Qual 2008;37:1158–69.

[53] Green CT, Fisher LH, Bekins BA. J Environ Qual 2008;37:1073–85.

[54] Webb RMT, Wieczorek ME, Nolan BT, Hancock TC, Sandstrom MW, Barbash JE, et al. J Environ Qual 2008;37:1145–57.

[55] Cuffney TF, Brightbill RA, May JT, Waite IR. Ecol Appl 2010;20:1384–401.

[56] Kashuba R, Cha Y, Alameddine I, Lee B, Cuffney TF. Multilevel hierarchical modeling of benthic macroinvertebrate responses to urbanization in nine metropolitan regions across the conterminous United States. U.S. Geological Survey Scientific Investigations Report 2009–5243, 2010; p. 88. http://pubs.usgs.gov/sir/2009/5243/.

[57] Qian SS, Cuffney TF, Alameddine I, McMahon G, Reckhow KH. Ecology 2010;91(2):355–61.

[58] Hancock TC, Sandstrom MW, Vogel JR, Webb RMT, Bayless ER, Barbash JE. J Environ Qual 2008;37:1086–100.

[59] Puckett LJ, Zamora C, Essaid H, Wilson JT, Johnson HM, Brayton MJ, et al. J Environ Qual 2008;37:1034–50.

[60] Steele GV, Johnson HM, Sandstrom MW, Capel PD, Barbash JE. J Environ Qual 2008;37:1116–32.

[61] U.S. Geological Survey. A whole-system approach to understanding agricultural chemicals in the environment. U.S. Geological Survey Fact Sheet 2009–3042, 2009. p. 6, http://pubs.usgs.gov/fs/2009/3042/.

[62] Carlisle DM, Wolock DM, Meador MR. Front Ecol Environ 2011;9:264–70.

[63] Anning DW, Bauch NJ, Gerner SJ, Flynn ME, Hamlin SN, Moore SJ, et al. Dissolved solids in basin-fill aquifers and streams in the southwestern United States. U.S. Geological Survey Scientific Investigations Report 2006–5315, version 1.1, 2010.

[64] Domagalski JL, Knifong DL, Dileanis PD, Brown LR, May JT, Connor V, et al. Water quality in the Sacramento River basin, California, 1994–1998, U.S. Geological Survey Circular 1215, 2000, p. 36.

[65] Goolsby DA, Battaglin WA. Nitrogen in the Mississippi Basin—estimating sources and predicting flux to the Gulf of Mexico. U.S. Geological Survey Fact Sheet 135–00: 2000; p. 6.

[66] Nolan BT, Hitt KJ. Environ Sci Technol 2006;40:7834–40.

[67] Gurdak JJ, McMahon PB, Dennehy KF, Qi SL. Water quality in the high plains aquifer, Colorado, Kansas, Nebraska, New Mexico, Oklahoma, South Dakota, Texas, and Wyoming, 1999–2004. U.S. Geological Survey Circular 1337 2009; p. 63.

[68] Preston SD, Alexander RB, Wolock DM. J Am Water Res Assoc 2011;47:887–90.

[69] Booth NL, Everman EJ, Kuo I-Lin, Sprague L, Murphy L. J Am Water Res Assoc 2011;47:1136–50.

[70] UNESCO. The impact of global change on water resources—the response of UNESCO's International Hydrological Programme, SC/HYD/2011/PI/H/1, 2011; http://unesdoc.unesco.org/images/0019/001922/192216e.pdf.

[71] Dettinger MD. Effects of precipitation and climate on streamflow and temporal variability of nutrient concentrations. In: Dubrovsky NM, Burow KR, Clark GM, Gronberg JM, Hamilton PA, Hitt KJ, et al, editors. 1992–2004, U.S. Geological Survey Circular The quality of our Nation's waters—Nutrients in the Nation's streams and groundwater. vol. 1350. 2010. p. 38–39.

[72] Intergovernmental Panel on Climate Change. Climate change 2007—impacts, adaptation and vulnerability. Contribution of Working Group II to the Fourth Assessment Report of the IPCC. Cambridge: Cambridge University Press; 2007.

[73] Rayburn C. Drinking water research. Climate Change 2008;(Special Issue):2–4.

[74] Albert J. Drinking water research. Climate Change 2008;(Special Issue):8–10.

[75] Graham JL, Loftin KA, Meyer MT, Ziegler AC. Environ Sci Technol 2010;44:7361–8.

[76] Haak AL, Williams JE, Isaak D, Todd A, Muhlfeld CC, Kershner JL, et al. The potential influence of changing climate on the persistence of salmonids of the inland west. U.S. Geological Survey Open-File Report 2010–1236, 2010; p. 74. http://pubs.usgs.gov/of/2010/1236/.

[77] Dettinger MD. Changes in streamflow timing in the western United States in recent decades. U.S. Geological Survey Fact Sheet 2005–3018, 2005; p. 4. http://pubs.usgs.gov/fs/2005/3018/.

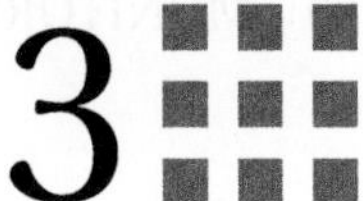

Rivers in Africa Are in Jeopardy

Shem O. Wandiga

DEPARTMENT OF CHEMISTRY, UNIVERSITY OF NAIROBI, NAIROBI, KENYA

CHAPTER OUTLINE

3.1 Introduction

The challenges presented with regard to water in Africa come in three forms: Water is either too much, too little, or too dirty. The rainfall pattern has changed because of climatic variability and change. The drought and flooding cycle has been reduced from a frequency of 7 years to a frequency of 2.5–3 years [1]. The water is unacceptably dirty because of many factors that may include natural processes (such as seasonal trends, underlying geology and hydrology, and weather and climate), as well as human activities, including domestic, agricultural and industrial use, and environmental engineering. Some of these are analyzed below. Africa has 84 rivers, of which 17 are major ones, each with catchments over 100,000 km^2. There are more than 160 lakes larger than 27 km^2. In addition, Africa has vast wetlands and widespread groundwater. It receives plentiful rainfall, with annual precipitation equal to that of Europe and North America in different regions. Water withdrawal is mainly for three purposes—for agricultural, domestic, and industrial use at 3.8% of total annual renewals. There are large disparities of water

Monitoring Water Quality, http://dx.doi.org/10.1016/B978-0-444-59395-5.00003-0

availability amongst regions, with 50% of surface water in the Congo basin and 75% of water found in eight major basins.

The causes for Africa's water pollution may be found in high population growth standing in 2009 at 1×10^9 [2] and growing at a rate of about 2.4%; overstocking and overgrazing; poor agricultural practices; desertification; atmospheric deposition; industrialization; and lack of sustainable sanitary facilities and alternatives. The high load of soil sediments deposited [3] into the river systems was last estimated at 54 tons per hectare per year of crop soil. Human and animal waste deposited in water arises from the fact that most rural and peri-urban dwellings have no sewage-treatment plants. A further cause is lack of efficient municipal and industrial waste collection, treatment, and disposal. Other contributing causes are weak legal instruments coupled with weak enforcement of the existing laws. The lack of quality data on water to assist policy formulation is another major contributor to weak laws. The high load deposits of large quantities of nitrogen and phosphorus plant nutrient arising possibly from biomass burning, sewage, industrial effluents, and dust contribute to eutrophication of rivers and other water-receiving bodies.

Water- and sanitation-related diseases collectively account for 80% of sicknesses in developing countries. The World Health Organization [3] estimates that globally about 2.2 million children die from diseases related to inadequate water supply, poor sanitation and poor hygiene annually—more than from AIDS, malaria, and TB combined. Diarrheal diseases accounted for 16% mortality of children under five in Kenya in the year 2002–2003 [4]. This is more than malaria (14%) and HIV/AIDS (15%). Diarrheal diseases killed 24,000 Kenyans in 2002 and constituted 7% of all deaths among people of all ages. The waterborne diseases caused by pathogenic bacteria, viruses and parasites (e.g. protozoa and helminths) are the most common and widespread health risk associated with drinking water. The public health burden is determined by the severity of the illnesses associated with pathogens, their infectivity, and the population exposed.

The coliform counts in most river waters are often very high. Figure 3-1 gives the coliform loads in the Nairobi River [5]. The readings during the wet season were above the scale of the instrument used. Rainwater is the environmental cleansing agent during the wet season. Hence all human and animal wastes in the bush are carried by storm water into the river. This is the cause of the high coliform load during the rainy season. During the dry season, the storm drains are used to carry sewage water from residential houses. This is contrasted with the water spring source at Ondiri, Swamp in Kenya, which has low coliform load in both seasons.

3.2 The Pollution of River Systems

Several human activities lead to the deterioration of river systems. Some of the most obvious activities are deforestation and biomass burning of the trees in watershed areas; agricultural activities such as food and horticultural crops growing next to the riverbed; dumping of agricultural residues into the water; soil erosion; and inflows of sanitary, animal, and industrial wastes into the river system. These activities lead to overloading of

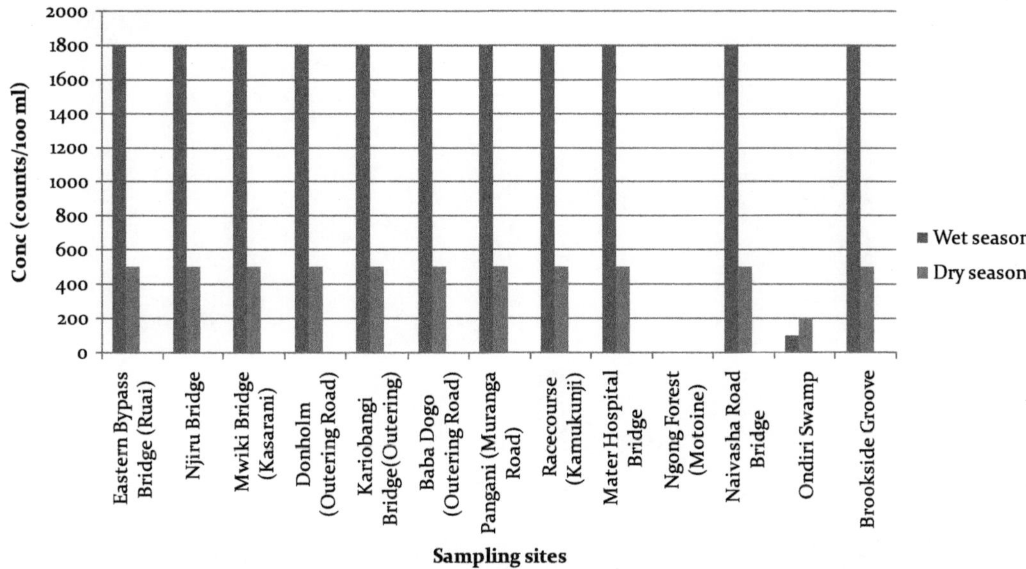

FIGURE 3-1 Coliform load in the Nairobi River (All readings during the wet season were above the scale of the instrument used.). (For color version of this figure, the reader is referred to the online version of this book.)

the nutrients phosphorus and nitrogen, with subsequent depletion of water oxygen resulting in eutrophication of the water body. Most river systems have seen their ecological systems changed by the increasing encroachment of settlements into their riparian course ways. The increased settlements have brought with them high withdrawals of water for agricultural practices, household use, and industrial activities; and the return of untreated wastewater into the rivers. Figure 3-2 shows some of these activities.

Figure 3-3 further illustrates some of the bad practices along riverbeds. The photos show a weir used to raise the water level to pumpable height. Above and below the weir there are human settlements. In addition, a road runoff directed into the river, directs storm waters along the road into the river. The combined discharges of raw sewage, storm water, and riparian agricultural wastes, accelerate the deterioration of the water systems. Most rivers pass through urban centers with artisan activities. For example, effluents from the informal sector, garages, markets, residential areas, and hotels, as well as from industries, enter the river systems untreated.

Burning of motor vehicle tires and plastics is a common activity at riverside garages. Furthermore, solid wastes from markets, hotels, and residences are frequently dumped along the riverbed or into the river (Figure 3-4). In large bodies of water, sighting of dead animals and humans floating in the river is common. Tillage from mining and prospecting activities, such as gold prospecting, add mercury and cyanide contents into river waters.

The deterioration of quality in large and small water systems has evolved since independence of the developing countries, because of neglect and inaction by the local government and citizens whose livelihoods depend on the water. Yet these rivers provide life-support for the livelihoods of innumerable riparian communities. The destruction of

FIGURE 3-2 Human pressures along a riverbed. (For color version of this figure, the reader is referred to the online version of this book.)

the water catchments through deforestation and poor agricultural practices has resulted in increased soil erosion, as well as release of agricultural chemicals into the rivers. Riparian communities dispose of their wastes into the rivers as if the waterways were their cesspools, with little regard for the downstream user. As a result, health has deteriorated in the downstream riparian communities. Anthropogenic abuses have combined forces with nature in the destruction of the rivers through global climate change and variability. Shortage of fresh water that flows through the rivers may also be attributed to changes in the rainfall pattern and intensity. Continued climate change and variability may worsen the water situation in the future. It is for these considerations that river basin restoration and water-management systems need to be established.

3.3 Climate Variability and Change

The Intergovernmental Panel on Climate Change (IPCC) 4th Assessment Report [6] states that "climate change and variability have the potential to impose additional pressures on water availability, water accessibility, and water demand in Africa. Even in the absence of climate change, present population trends and patterns of water use indicate that more African countries will exceed the limits of their 'economically usable and land-based

FIGURE 3-3 Weir for water- and sewer-exhaustion next to a riverbed. (For color version of this figure, the reader is referred to the online version of this book.)

water resources before 2025.' In some assessments, the population at risk of increased water stress in Africa, for the full range of SRES scenarios, is projected to be 75–250 million and 350–600 million people by the 2020s and 2050s, respectively." Climate variability and change affect food availability and food accessibility by its variation and through extreme events, such as droughts, floods, and heat waves. There is indeed in the Sahel region a strong correlation between the years of poor rainfall and years of food crises. The years 1973, 1985, 1996, 1998, 2001, 2005, and 2010 that have been years of drought and also years of food crises or famine in some areas or the whole of the Sahel region [7–9], illustrate this correlation. The climate-related risks for food security could increase in the future because of climate change that could alter rainfall patterns, increase the frequency and the intensity of climate shocks, and induce transformations of ecosystems [10]. According to the latest report of the IPCC, some impacts of climate change on food security in Africa could be large by 2020; in some countries, yields from rain-fed agriculture could be reduced by up to 50%; and by 2080, an increase of 5–8% in arid and semiarid land in Africa is projected, under a range of climate scenarios.

For Africa, 2010 was the warmest year on record. Temperatures in 2010 averaged 1.29 °C above the long-term average for Africa. From 1980 to 1998, global maize and wheat yields decreased by 3.8% and 5.5% respectively, because of increasing temperatures [10].

FIGURE 3-4 Waste dumping along a riverbed. (For color version of this figure, the reader is referred to the online version of this book.)

3.4 Agricultural and Industrial Chemical Pollution

Africa's agriculture is the least developed because of low levels of investment. Agriculture accounts for only 4% of most African government budgets compared to 14% for Asia. Agricultural yield declined by 30% between 1967 and 2007 on the African continent. On the other hand, the yield in South Asia increased two-fold, while in East Asia it was three times higher. Africa relies on rain-fed agriculture as only 6.5% of its farmland is irrigated, compared to 40% for Asia. The continent's cultivable land stands at 60% compared to 31% for Latin America. It uses one tenth of the fertilizer per hectare, compared to European levels. Given these levels of agricultural activities, one would expect that African rivers would be much cleaner than they are. However, as Africa needs to increase its agricultural productivity by at least 70% in order to feed its burgeoning population, one worries about the possible demise of life in all the river systems in the continent, if the present agricultural practices are maintained.

Various pesticides are in wide public use in Kenya. Rivers receive the pesticides from storm waters that drain into the rivers. Figures 3-5 and 3-6 give the concentrations of pesticides in the Nairobi River during the dry seasons and the wet seasons. Heptachlor and γ-hexachlorohexane were the widely used pesticides along the catchments of this river.

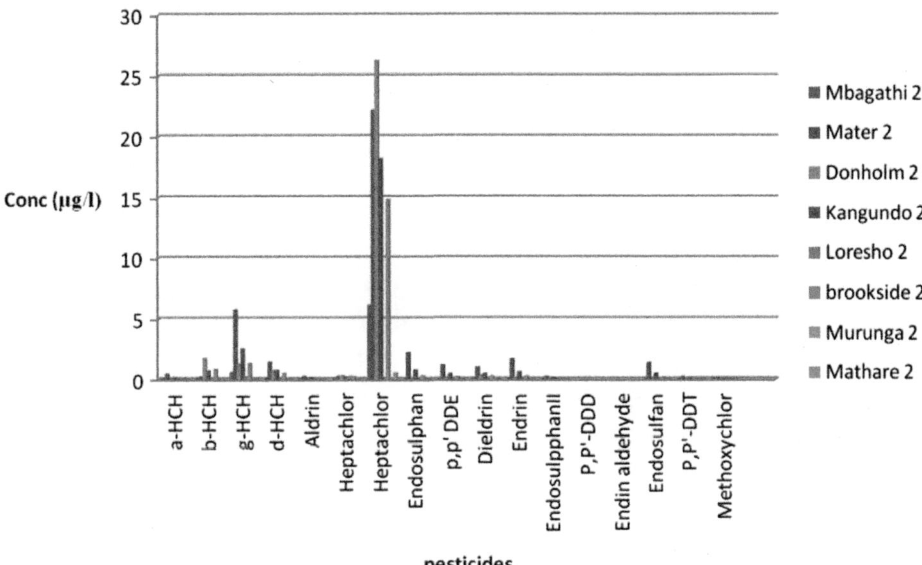

FIGURE 3-5 Concentrations of pesticides in Ngong River during the dry season. (For color version of this figure, the reader is referred to the online version of this book.)

Rivers also contain harmful trace metals like cadmium, mercury, lead, and chromium, to mention a few. The sources of these metals are attributed to industrial artisanal usage, agriculture, vehicles using leaded gasoline and the general mass circulation system. Figure 3-7 gives the concentration of lead in the Nairobi and Ngong rivers during the dry

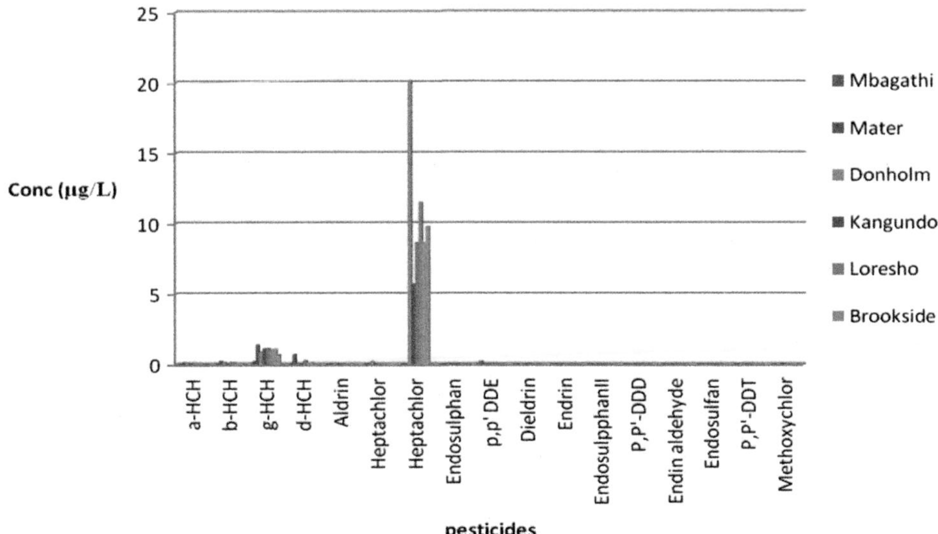

FIGURE 3-6 Concentration of pesticides in Ngong River during the wet season. (For color version of this figure, the reader is referred to the online version of this book.)

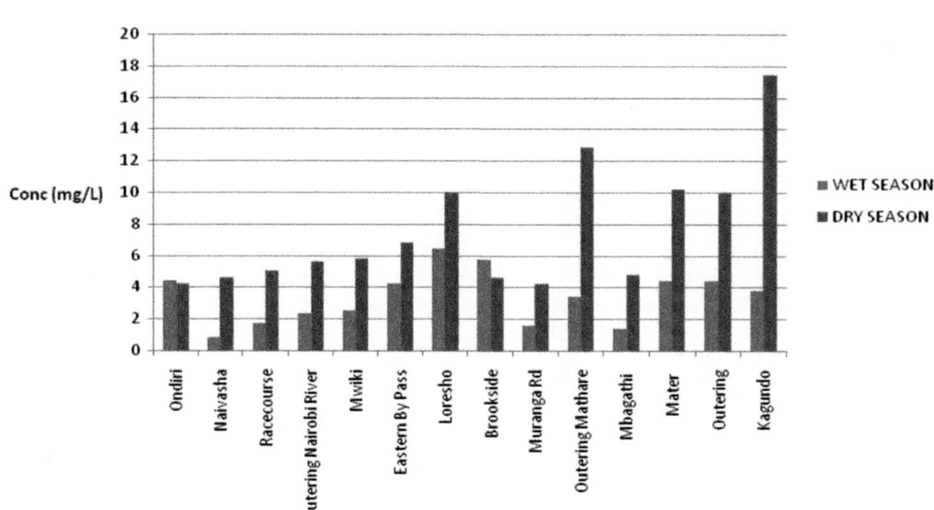

FIGURE 3-7 The concentration of lead in the Ngong and Nairobi rivers water during the wet and dry seasons. (For color version of this figure, the reader is referred to the online version of this book.)

and wet seasons. The concentration of this metal during the wet and dry seasons exceeded the WHO limit of 10 μg/L at all sites. These sites are located at the source and along the stream of the rivers, implying high traffic and industrial activity. It is also possible that the storm water cleanses the deposits from the drainages and soils into the river.

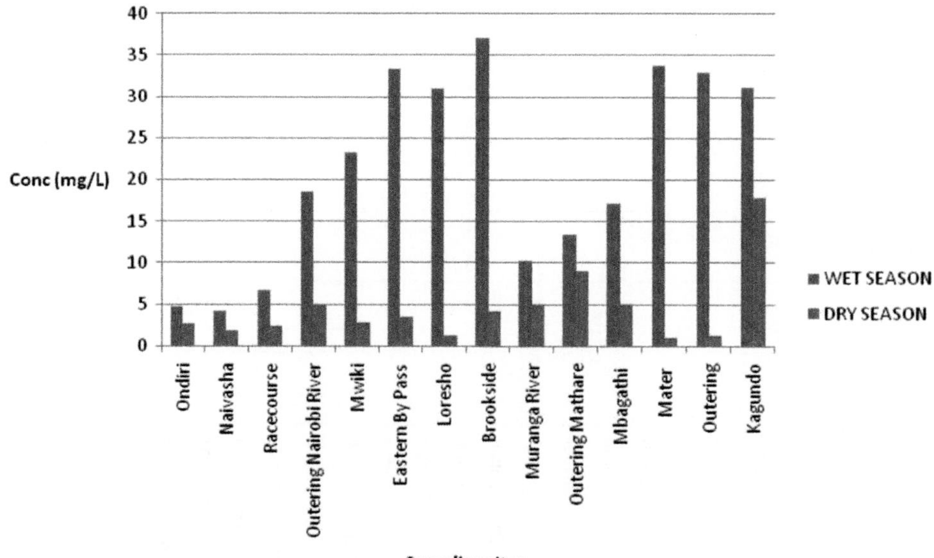

FIGURE 3-8 The concentration of chromium in Nairobi River water samples during wet and dry seasons. (For color version of this figure, the reader is referred to the online version of this book.)

Figure 3-8 presents the concentrations of chromium along the Nairobi River. The wet season concentrations are much higher than the dry season concentrations. The concentrations generally increase as one moves from the source of the river to the residential and industrial areas. The dry-season concentration at the Kagundo site reflects the impact of the sewage-treatment plant outlet. The metal is used mainly for chrome plating, leather processing, and for making metal alloys.

In addition to the sanitary wastes deposited into the rivers, the rivers carry high deposits of various chemicals and sediments. Some of the chemicals found in the water have developmental consequences to the consumers of such waters. Figure 3-9 gives the levels of polychlorinated biphenyls (PCBs) some of which are known endocrine disruptors.

The PCB concentration levels are highest in the industrial area of Nairobi city through which the Ngong and Nairobi rivers flow. The upper parts of these rivers and residential areas, show low levels of these compounds. However the concentration picks up as one approaches the busy roads and after passing the city sewage-treatment plant. The pattern of the dry-season river content is different from the wet-season pattern. Figure 3-10 gives the concentration of PCBs during the dry season. Selected congeners of PCBs are found to be more prevalent at sites with either high traffic or industrial activities. PCBs 28, 52, 118, and 105 are more prevalent in measurable quantities at these sites. The source of these chemicals is speculated to be old electric transformers that are still in use, burning of plastics, high traffic levels, and general mass circulation of wind that may import these chemicals from other countries. The presence of the chemicals, though at low concentrations at the World Meteorological Organization (WMO) station at Mt. Kenya confirms that wind is capable of transporting these chemicals over long distances [11].

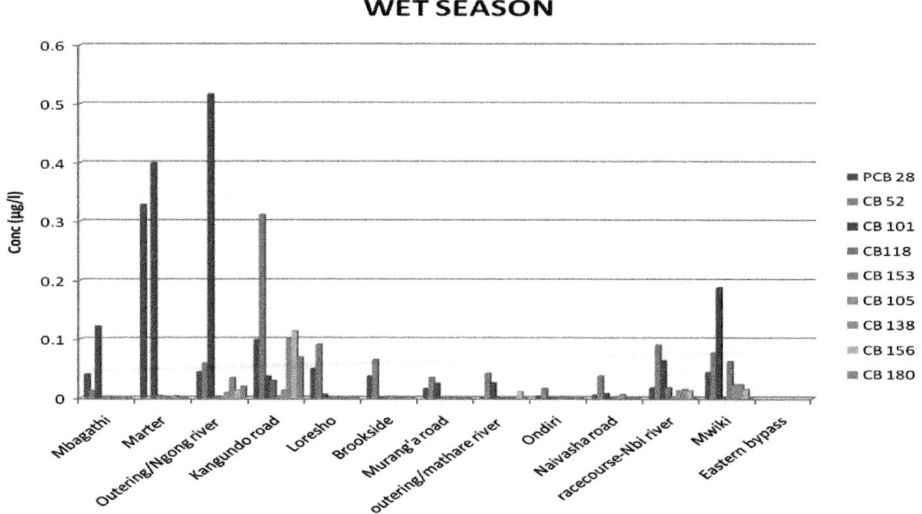

FIGURE 3-9 Concentration of PCBs in Ngong and Nairobi rivers during wet season. (For color version of this figure, the reader is referred to the online version of this book.)

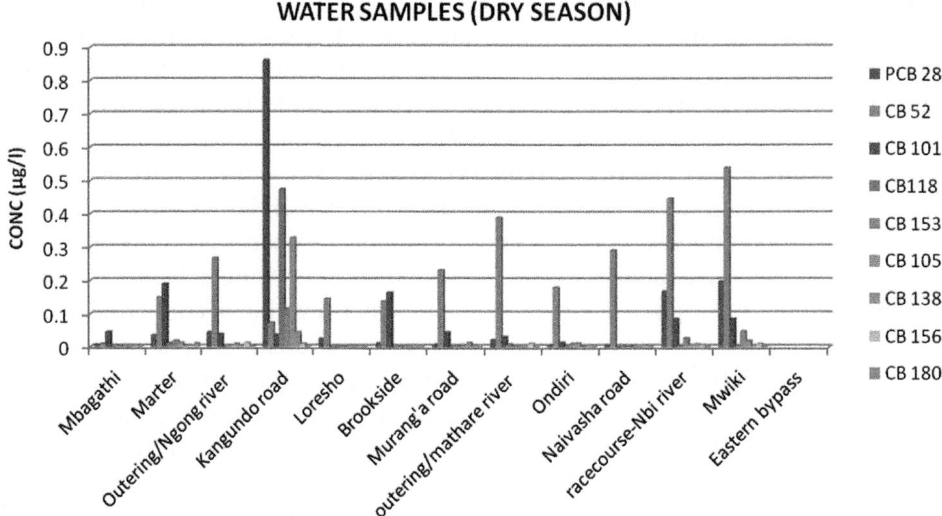

FIGURE 3-10 Concentrations of PCBs during the dry season in Ngong and Nairobi rivers. (For color version of this figure, the reader is referred to the online version of this book.)

3.5 Sediments and Other Physicochemical Parameters

Most rivers carry high loads of sediments because of soil erosion, poor farming practices that allow plowing up to the riverbanks, and lack of construction of terraces to slow down the water runoff. As a result of this, most rivers have water whose color is muddy brown. Measurement of total dissolved solids and electrical conductivity indicates that the nature of dissolved substances in the water is within the USEPA standards of 500 mg/L and 900 µS/cm respectively, except for six sites of high industrial artisan activity. Figure 3-11 presents the Nairobi River water conductivity at various sampling points.

On the other hand, measurements of the biochemical oxygen demand (BOD) and dissolved oxygen indicate that the rivers have low water quality with respect to these two parameters. Figures 3-12 and 3-13 give the results of the measurements. The BOD levels are much higher in areas of industrial activity, and the situation deteriorates during the dry season when there is less water for dilution. The levels of dissolved oxygen are worse during the dry season and imply that during this period, higher aquatic species may not survive in the river water. A dissolved oxygen level of 4–8 mg/L is normally accepted as good for supporting life-form ecosystems.

Agricultural fertilizers are not excluded from the water system. Some of these come as a result of rainwater runoff from crop fields. Some leach through the groundwater into the river system. The Nairobi River passes through some agricultural fields and receives agricultural chemicals like phosphates, nitrates, sulphates and pesticides from storm water and field runoffs.

Figure 3-14 gives the level of phosphates detected in the water during the wet and dry seasons. The phosphate levels are low at the upper portion of the river, but increase after

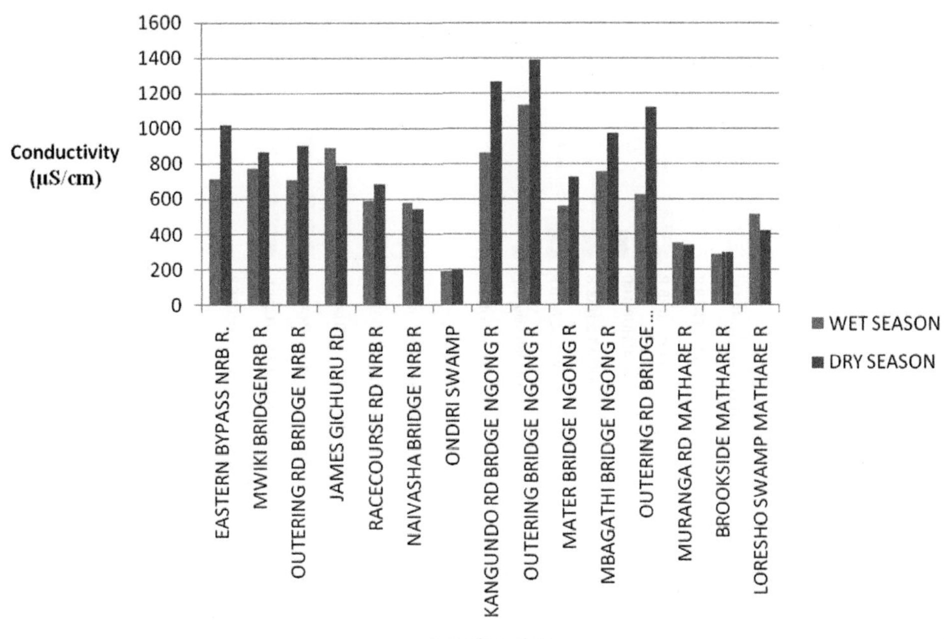

FIGURE 3-11 Nairobi River basin water conductivities during the wet and dry seasons. (For color version of this figure, the reader is referred to the online version of this book.)

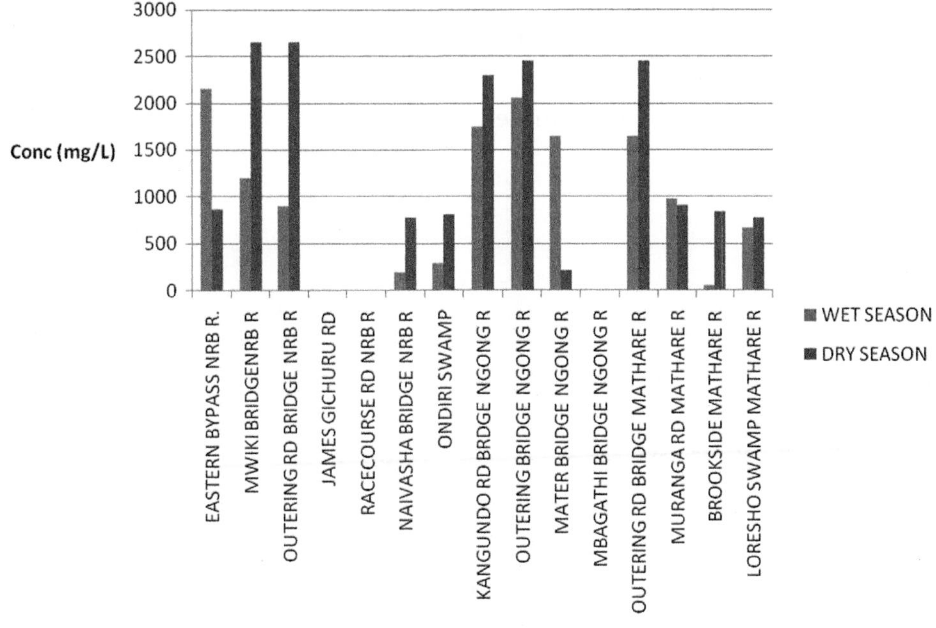

FIGURE 3-12 The BOD of the river waters during the wet and dry seasons. (For color version of this figure, the reader is referred to the online version of this book.)

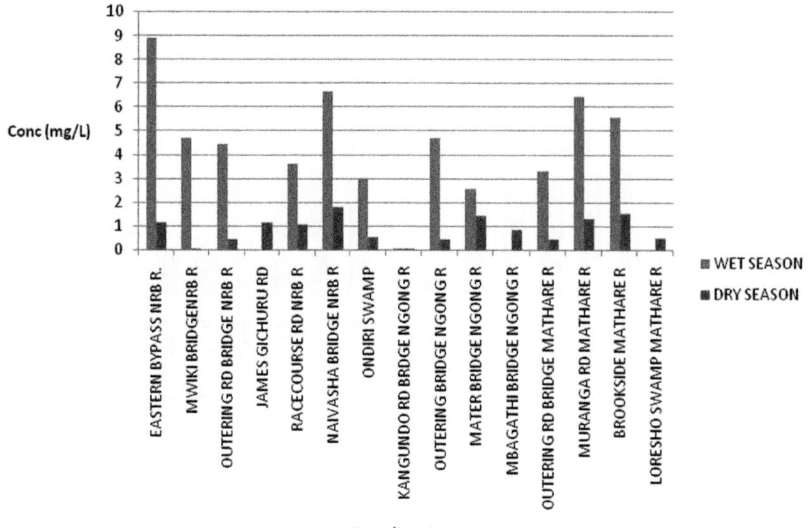

FIGURE 3-13 Dissolved oxygen in the river waters during the wet and dry seasons. (For color version of this figure, the reader is referred to the online version of this book.)

passing the residential area. The high levels observed at the lower reaches of the river indicate that the industrial area also adds significant amounts of fertilizers into the water. The increased levels could be attributed to the use of phosphate detergents in the washing of clothes and also in urban farming. The marked difference between the seasons is the dilution effect during the rainy season.

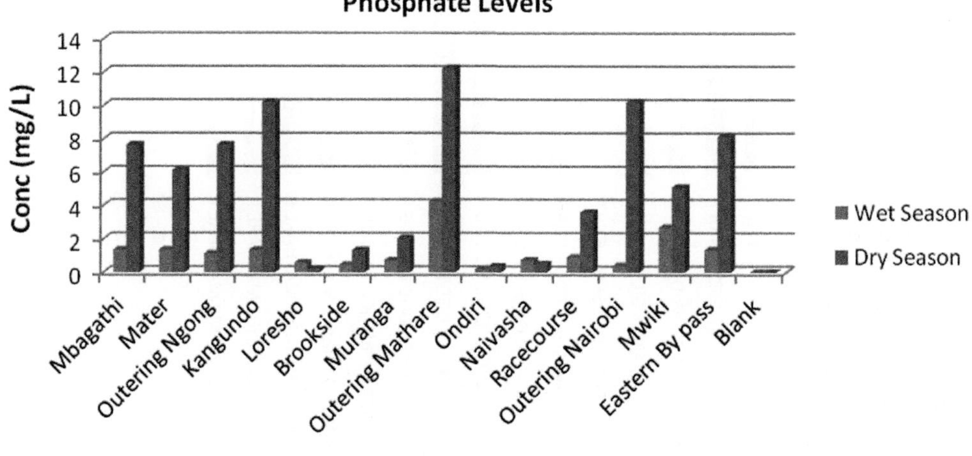

FIGURE 3-14 The concentration of phosphates in Nairobi River basin waters. (For color version of this figure, the reader is referred to the online version of this book.)

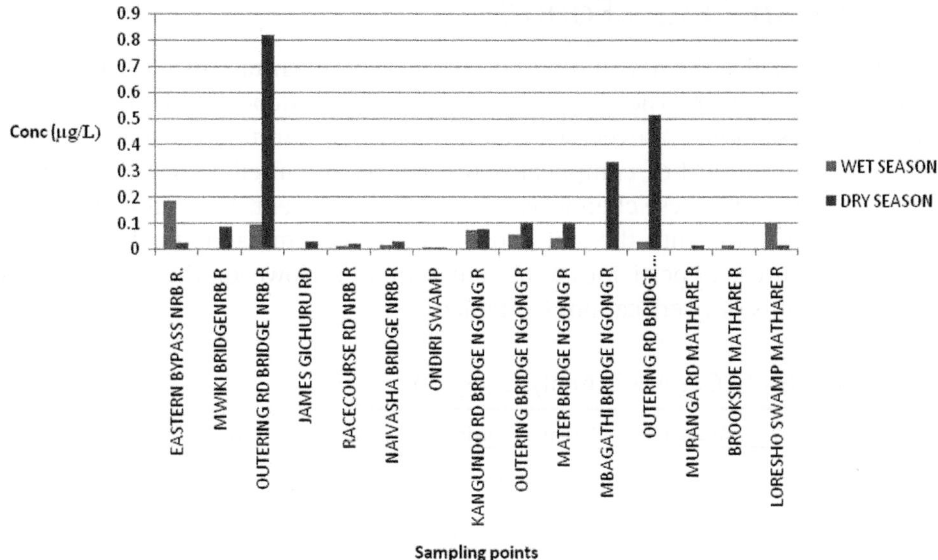

FIGURE 3-15 The concentration of oil and grease in Nairobi River water. (For color version of this figure, the reader is referred to the online version of this book.)

The artisan mechanics that repair motor vehicles have no proper disposal system for oil and grease. They either dispose of the oils directly into the river or into storm-water drainage pipes. Figure 3-15 gives the concentration of oil and grease in the Nairobi River during the experimental period. The dry-season concentration is generally much higher than concentrations in the wet season. The sites with the highest level correspond to the density of artisans working in these areas.

3.6 Atmospheric Deposition

Pesticides and polycyclic aromatic hydrocarbons (PAHs) have also been detected in atmospheric depositions at various points monitored in Kenya [10]. These deposits add to the load of chemicals in the rivers. The deposition of persistent organic pollutants (POPs) was measured at four sites in Kenya. These were located at the WMO Mt. Kenya site, which served as a reference point, the Kabete campus of the University of Nairobi, which represents an agricultural environment with less industrial and vehicle activity, the Dandora, City of Nairobi waste dump site, and at Kitengela, an old pesticide storage site for the Desert Locust Organization. All deposition measurements were done using passive air samplers with polyurethane foam, mounted on containers that allow air to flow through, on stands that are about 2 m above the ground. Each filter was replaced after three months of exposure and analyzed in the laboratory. Similar deposits have recently been reported [11] in several African and Asia Pacific countries. The long-range transport of chemicals has been established and reported [11].

3.7 POPs and PCBs Found

Tables 3-1 and 3-2 give the maximum and minimum concentrations of POPs and PCBs found at sites in Nairobi. The detection of atmospheric deposits of POPs and PCBs confirms the grasshopper movement of the pesticides through air-mass circulation around the globe. It further complicates the efforts mounted under the Stockholm Convention to eliminate these chemicals from the environment. The cumulative effect of wet and dry deposition onto the soil and the water systems adds to the difficulty of identifying the sources of pesticides in the environment. However, their contribution to the worsening status of rivers cannot be denied.

Table 3-1 Results of POPs in Air Analysis (ng/μL)

	Average	Median	Geomean	Min	Max	SD	%RSD	n/45
Orins								
Aldrin	38	35	36	22	78	13	34	21
Dieldrin	39	37	36	14.20	92	16	40	25
Endrin	38	40	35	7	61	13	35	20
Chlordanes								
trans-Chlordane	36	34	35	20	68	11	31	16
cis-Chlordane	38	38	36	13	68	12	32	17
trans-Nanochlor	45	41	42	27	81	18	41	10
Oxychlordane	48	45	44	29	70	20	43	4
Heptachlor	60	42	47	25.5	155	54	90	5
cis-Hepta-chlorepoxide	38	34	36	20	71	12	32	20
trans-Hepta-chlorepoxide	36	32	34	22	90	16	45	15
DDTs								
P,p'-DDT	40	38	36	8	67	15	38	21
O,p'-DDT	37	34	34	13	71	16	44	10
P,p'-DDT	53	37	40	14	420	76	143	26
P,p'-DDE	38	36	37	21	68	11	30	14
O,p'-DDE	45	40	40	9	85	19	43	22
P,p'-DDD	41	37	40	29	75	14	33	14
O,p'DDD	40	38	36	8	67	15	38	21

Table 3-2 Results of PCBs in Air Analysis (ng/μL)

PCBs	Average	Median	Min	Max	SD	%RSD	n/44
PCB #28	1.20	1.23	0.61	1.87	0.29	24	17
PCB #52	1.20	1.21	0.52	2.00	0.32	27	18
PCB #101	1.17	1.15	0.59	2.15	0.34	29	18
PCB #118	0.42	0.04	0.001	1.12	0.57	134	5
PCB #138	1.26	1.10	0.10	3.82	0.81	64	18
PCB #153	1.22	1.22	0.29	2.19	0.39	32	19
PCB #180	1.22	1.06	0.07	3.49	0.68	56	19

3.8 Polycyclic Aromatic Hydrocarbons (PAHs)

PAHs can find their way into soil and water from the air. Sixteen PAHs were detected [11] in air samples from three sites in Nairobi, using passive air-samplers. Concentrations of the PAHs in air are reported in nanograms per filter (ng/filter). This provides information on time-integrated values based on a weighted 28-day scale, which is approximately equivalent to 1000 m^3 of air obtained during one day of active sampling. PAHs were detected in all air samples collected in the Nairobi industrial area, a municipal dump site in Dandora, and the urban background in Kabete. The total 16 PAHs (ng/filter) were lowest in Kabete (1512.0) during the wet month of March and highest in Dandora (8467.4) in July during the dry season. A summary of the maximum and minimum levels of PAHs detected over the 6-month period at the three sites are given in Table 3-3 below.

The levels of PAHs were highest in the industrial area until the beginning of the long rainy season in March. There was a general increase in the levels of PAHs between March and May in Dandora and the industrial area. In Kabete, PAH levels dropped from 1908.8 ng/filter in February to 1512 ng/filter in March, and then rose to 2082 ng/filter in the next month. This was followed by a decline at all sites in May. However, during the cool dry season, levels of PAHs in the air in the industrial area and Kabete remained relatively constant within the 6000 and 2000 ng/filter, respectively. In Dandora, there was a more marked increase from 8402 ng/filter in June to 9676 ng/filter (Figure 3-16).

The fairly constant level of PAHs in Kabete and more marked variations in Dandora and the industrial area suggest point source inputs of PAHs in Nairobi air. The Kabete

Table 3-3 Maximum and Minimum Levels of PAHs Detected in Nairobi Air (ng/filter)

		Sampling Sites					
		Kabete		Dandora		Industrial Area	
	PAHs	Min	Max	Min	Max	Min	Max
1	Naphthalene	84.0	310.7	143.0	334.0	161.0	322.0
2	Acenaphthylene	9.3	84.0	46.4	123.0	50.6	86.0
3	Acenaphthene	17.0	181.2	56.0	88.5	42.5	69.1
4	Fluorene	153.0	245.8	579.3	774.0	395.6	528.5
5	Phenanthrene	667.0	977.9	2881.3	3867.0	2138.8	2896.6
6	Anthracene	28.0	46.7	230.8	359.0	195.0	297.5
7	Fluoranthene	265.0	426.2	1142.2	1591.0	1180.0	1727.3
8	Pyrene	204.0	340.1	940.4	1429.0	1118.0	1695.4
9	Benzo(a)anthracene	10.0	19.7	74.3	100.0	83.1	115.9
10	Chrysene	23.0	46.7	92.7	137.0	119.0	164.1
11	Benzo(b)fluoranthene	6.0	12.4	24.1	38.0	25.4	47.3
12	Benzo(k)fluoranthene	4.0	8.3	10.6	23.0	16.0	21.7
13	Benzo(a)pyrene	1.7	5.2	4.4	21.0	12.3	19.9
14	Benzo(ghi)perylene	1.8	3.0	7.7	18.0	11.4	18.1
15	Indeno(1,2,3-cd)pyrene	bd	bd	0.9	2.0	1.8	2.0
16	Dibenz(ah)anthracene	1.8	6.8	8.7	24.0	13.1	20.8

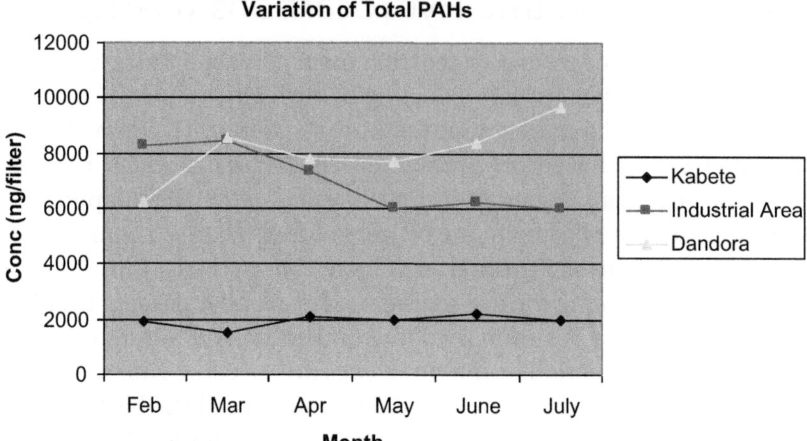

FIGURE 3-16 Graphs showing variations in concentrations of total 16 PAHs in air. (For color version of this figure, the reader is referred to the online version of this book.)

site has minimum human settlement and there are no industrial activities. The Dandora dump site is subjected to inputs from municipal dumping and burning of waste. The possible inputs of PAHs in the industrial area are from industries and motor vehicles.

At all sites, phenanthrene was the most abundant polycyclic aromatic hydrocarbon in air samples. Levels between 667.0 and 3896.0 ng/filter were detected in Kabete and Dandora in March, during the rainy season. Concentration ranges of phenanthrene were widespread across the seasons (Kabete 667.0–997.9 ng/filter; industrial area 2138.8–3278.0 ng/filter; Dandora 2881.3–3867.0 ng/filter). Phenanthrene is a source marker for incineration and for motor vehicle emissions. There is an indication that the burning of municipal waste in Dandora and the emission from motor vehicles in the industrial area are the most probable sources of phenanthrene at these sites.

The concentrations of the other three-ringed PAHs: naphthalene, acenaphthylene, acenaphthene, fluorine, and anthracene were much lower at all sites. The lowest average concentrations were those of acenaphthylene (17.1 ng/filter in Kabete; 70.6 ng/filter in the industrial area; 72.2 ng/filter in Dandora) and acenaphthene (21.6 ng/filter in Kabete; 62.2 ng/filter in the industrial area; and 71.1 ng/filter in Dandora). Concentrations of the simplest PAH, two-ringed naphthalene, were 188.7, 246.2 and 201.3 ng/filter in Kabete, the industrial area and Dandora, respectively.

The levels of six-ringed PAHs benzo(ghi)perylene, indeno(1,2,3-cd)pyrene and dibenzo(a,h)anthracene averaged between below the limit of quantitation and 24.0 ng/filter. Passive sampling using polyurethane foam captures the low-molecular-weight PAHs more effectively than the heavier ones. There may be an underestimation of levels of heavier PAHs numbers (10–16) that calls for further investigation. The average concentrations of PAHs in the air at the three sites are provided in Figure 3-17.

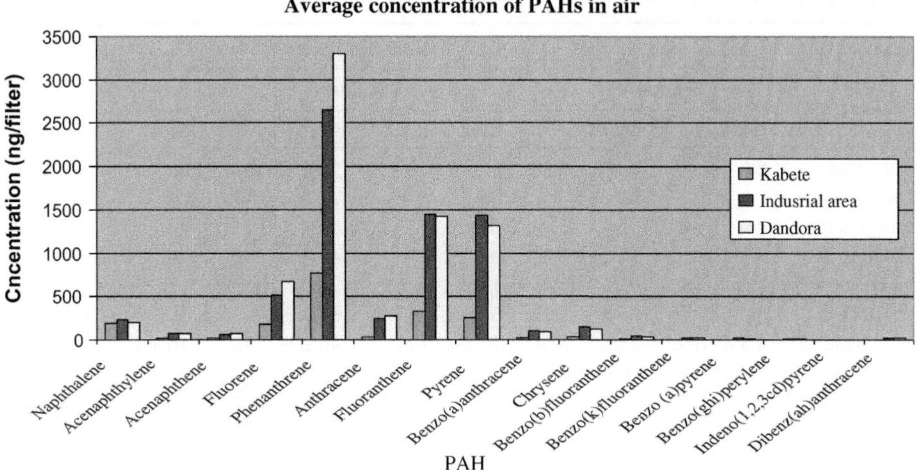

FIGURE 3-17 Average concentrations of PAHs in Nairobi air. (For color version of this figure, the reader is referred to the online version of this book.)

3.8.1 Levels of PAHs Detected in Soils

All the 16 targeted PAHs in the survey were detected in soils in Nairobi. However, only eight of these were quantified because of shortage of standards in the local laboratory. The Research Centre for Environmental Chemistry and Ecotoxicology (RECETOX) laboratory analyzed a mixture of soil for the first three months only, since soil analysis was complementary to air. The integrated concentrations of PAHs are highlighted at the end of this section. Levels reported are for dry soils obtained by treating soils with baked-out sodium sulfate.

Levels of PAHs in soils were generally lowest in Kabete soils compared with Dandora and the industrial area. Acenaphthene was the least abundant compound at all three sites. The compound was not detected in soils at the three sites, except in March when the concentration was 0.02 ± 0.01 ng/g in Dandora; and in June at the industrial area site (1.5 ± 0.31 ng/g). Phenanthrene was remarkably abundant at the Dandora dump site with concentrations ranging between 14.9 ± 2.76 ng/g in February and 526.7 ± 8.62 ng/g in July. This was unlike other PAHs whose concentrations were within the same range of magnitude. The possibility of higher loadings because of localized contamination requires further investigation. Ranges of PAHs at the three sites are given in Table 3-4.

Marked variations in levels of the eight PAHs were noted at the Dandora and the industrial area. Levels of the eight PAHs quantified in soils were fairly constant in Kabete, where the lowest concentrations were also detected (17.3 ± 1.19 to 62.2 ± 0.97 ng/g). Throughout the study period, levels of the total eight PAHs were lowest in Kabete and highest in Dandora (Table 3-5).

There was a general increase in the concentrations of PAHs at the Dandora and the industrial area sites between April and July. At the Dandora dump site, the lowest levels of

Table 3-4 Minimum and Maximum Levels of PAHs Detected in Soils (ng/g)

		Sampling Sites					
		Kabete		Dandora		Industrial Area	
		Min	Max	Min	Max	Min	Max
1	Naphthalene	bd	10.2 ± 0.59	bd	12.0 ± 4.50	bd	9.4 ± 1.52
2	Acenaphthylene	bd	bd	bd	2.0 ± 0.10	bd	39.4 ± 1.03
3	Acenaphthene	bd	bd	bd	0.03 ± 0.01	bd	1.5 ± 0.31
4	Fluorene	bd	bd	bd	146.4 ± 7.52	bd	bd
5	Phenanthrene	bd	0.92 ± 0.03	14.9 ± 2.76	526.7 ± 18.62	12.8 ± 0.44	19.8 ± 1.05
6	Anthracene	bd	0.92 ± 0.03	68.1 ± 15.77	162.7 ± 18.70	16.1 ± 3.44	41.6 ± 1.15
7	Fluoranthene	4.3 ± 1.76	33.1 ± 2.81	18.6 ± 0.11	40.2 ± 2.90	16.1 ± 3.44	41.6 ± 1.15
8	Pyrene	7.2 ± 0.49	16.6 ± 3.45	7.8 ± 0.17	81.9 ± 9.58	19.1 ± 1.33	25.4 ± 1.44

bd = below detection.
The number of filters used for analysis is 3.

Table 3-5 Levels of Total Eight PAHs Detected in Soils (ng/g ± SD)

Sampling Sites	February	March	April	May	June	July
Kabete	40.3 ± 0.84	32.1 ± 0.46	38.4 ± 1.52	51.9 ± 1.28	62.2 ± 0.97	117.3 ± 1.28
Dandora	204.2 ± 1.27	281.4 ± 3.56	175.8 ± 2.46	141.4 ± 1.03	211.9 ± 2.01	726.6 ± 4.50
Industrial area	86.2 ± 1.30	56.3 ± 1.29	116.3 ± 4.6	109.7 ± 4.3	99.9 ± 0.74	138.7 ± 1.93

The number of filters used for analysis is 3.

(175.8 ± 2.46 ng/g) were recorded in the wet month of April. This was followed by an increase in levels of PAHs in the next two months to the highest concentration of 726.5 ± 4.50 ng/g in July, a dry cool month. In the industrial area soil, the lowest levels of PAHs (63.4 ± 1.29 ng/g) were detected in March and the highest levels in July, (138.7 ± 1.93 ng/g) as illustrated in Figure 3-18.

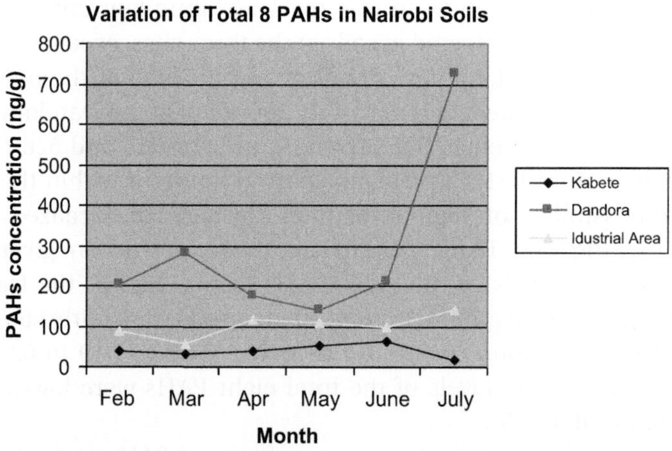

FIGURE 3-18 Graphs showing variation in concentrations of total eight PAHs in soil. (For color version of this figure, the reader is referred to the online version of this book.)

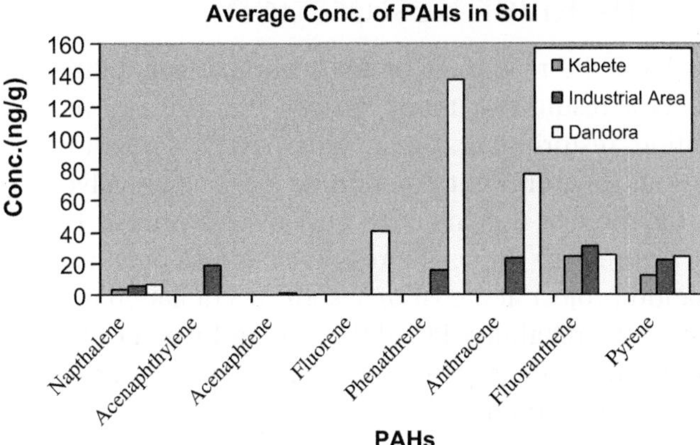

FIGURE 3-19 Variation of average concentrations of PAHs in Nairobi soils. (For color version of this figure, the reader is referred to the online version of this book.)

On the average, phenanthrene was the most abundant of the eight PAHs in soils at the Dandora site, where 137.4 ± 5.09 ng/g was detected. At the industrial area site, fluoranthene levels were the highest, averaging 30.4 ± 2.42 ng/g. Levels of PAHs were generally lowest in Kabete, but there were appreciable quantities of fluoranthene and pyrene at this site, whose average concentrations (ng/g) at the site were 24.7 ± 1.86 and 11.7 ± 2.00, respectively.

Notable also are the more marked levels of the heavier four- to five-ringed PAHs than the lighter two- to three-ringed PAHs in the soils. Concentrations of acenaphthene were particularly low at all sites, with none detected in Kabete and Dandora soil samples and 1.5 ± 0.05 ng/g in the industrial area (Figure 3-19).

3.8.2 Correlation of Levels of PAHs in Air and Soil

The main limitation of comparing actual concentrations is the difference in the matrices, which also results in a difference in the units of measurement. For comparability of data, only PAHs 1–8 that were quantified in both matrices were considered. Correlation was done using Pearson's correlation coefficient from MS Excel 2007.

The Pearson's correlation coefficients of levels of the total of eight PAHs in soils and air were 0.7976, 0.5527, and 0.1796 at Kabete, Dandora, and the industrial area respectively. The relatively strong correlation at the Kabete site indicates balanced exchanges of PAHs between the air and the soil. Lower correlation coefficients at the Dandora and the industrial area sites suggest sudden changes in the local input of PAHs in soils or air.

These studies of atmospheric deposition convince us to believe that the levels of chemicals detected in water are fortified by atmospheric sources. Cleaning up the rivers would require stringent enforcement of anti-air-pollution measures, which are unavailable in most African countries.

3.9 The Cost of River Restoration

Very few countries have the resources or the technical capabilities to undertake river restoration works. As a result, not many African countries have undertaken projects aimed at rehabilitating, restoring, and managing river systems. Kenya, in collaboration with the United Nations Environment Programme, has been involved in the Nairobi River Basin Programme for the last 6 years. The cost of this project has been estimated at 200,000 US dollars. Figure 3-20 shows the results that have been achieved by the project.

A similar restoration project at a cost of 4×10^9 US dollars has been proposed by the Indian Government for the rehabilitation of the Ganges River. For financial reasons alone, it is preferable to prevent the degradation of the river system rather than wait, and then restore it after it has been decimated.

3.10 Conclusions

Water quality in Africa is generally impacted by natural processes—seasonal trends, underlying geology and hydrology, atmospheric deposition, and human activities. It is estimated that 75% of Africa's drinking water comes from groundwater, and it is often used with little or no purification. If such water is contaminated by microbiological pollutants, diseases like dysentery, cholera, and typhoid will easily spread. Sources of water with chemicals that are endocrine disruptors and/or carcinogenic, or teratogenic, have disease consequences for the long-term users of such waters. It is estimated that, globally, about 884 million people have no access to potable drinking water and about 2.5 million people globally have no sustainable sanitation. Access to clean drinking water and basic sanitation, including toilets, wastewater treatment, and recycling, affects a country's developmental progress in terms of human health, education, and gender equality. The provision of sustainable drinking water and sanitation are inadequate across many parts of Africa, and, where available, water supply and sanitation services are differentiated according to urban, rural, or informal settlements.

Food production comprises a large proportion of Africa's economy, bringing in, on average, 20% of the gross domestic product (GDP) for each country. In sub-Saharan Africa, it accounts for 67% of jobs. Africa has the highest rate of rural poverty in the world and the poorest people remain in agriculture, with women taking the leading role in agricultural activities. Improving and sustaining agriculture in Africa will therefore be fundamental to economic development and the alleviation of poverty.

It follows therefore that Africa needs a strong integrated system for water-quality monitoring. A strong focus on developing and improving technologies to conserve and reuse water for agriculture is required, including optimizing water use, treating contaminated water, recycling water, desalinating water, and harvesting water for irrigation. Rapid analytical techniques fit for these purposes are required. The analytical instrument which has robust, portable, automated instrumentation; has contamination-free environment (includes reagents, containers, sampling apparatus, transportation);

FIGURE 3-20 The Nairobi River, before and after restoration. (For color version of this figure, the reader is referred to the online version of this book.)

has sensitive and selective detection with removal of matrix interferences, e.g. sea salts; has system stability (reagents, standards, pumps, detector); has onboard filtration and prevention of bio-fouling; and has remote calibration, validation, and maintenance. African scientists call for collaboration and assistance in building such systems.

Acknowledgments

I am grateful to the Nairobi River Basin Programme team, and to UNEP Regional Office for Africa and the University of Nairobi for financial support.

References

[1] Wandiga Shem O, Opondo Maggie, Olago Daniel, Githeko Andrew, Githui Faith, Marshall Michael, et al. Vulnerability to epidemic malaria in the highlands of Lake Victoria basin: the role of climate change/variability, hydrology, health and socio-economic factors. Clim Change J(No.3–4):473–97, http://dx.doi.org/10.1007/s10584-009-9670-7, 2010;99. ISSN 0165-0009(print) 1573-1480 (Online).

[2] State of world population, 2009. United Nations Population Fund; 2009.

[3] Safer water, better health, WHO, 2008: http://www.whqlibdoc.who.int/publications/2008/9789241596435_eng.pdf, page 12; The Global Fund to Fight AIDS, Tuberculosis and Malaria, "HIV/AIDS Background." http://www.theglobalfund.org/en/hivaids/background/.

[4] Wandiga SO, Mavuti MK, Oduor FDO, Kariuki DK, Mirikau CW, et al. UNEP, reconnaisence survey report: Nairobi River Basin Programme, 2009.

[5] IPCC (Intergovernmental Panel on Climate Change). Climate change 2007: Synthesis report. Contribution of working groups I, II and III to the fourth assessment report of the Intergovernmental Panel on Climate Change, IPCC, Geneva, Switzerland. 2007b.

[6] Nicholson SE. Climatic and environmental change in Africa during the last two centuries. Clim Res 2001;17:123–44.

[7] CILSS (Comité Permanent inter états de lutte contre la sécheresse au Sahel). Vingt ans De prévention des crises alimentaires au Sahel, Bilan et perspectives. 2004. p. 88.

[8] Janin P. La lutte contre l'insécurité alimentaire au Sahel: permanence des questionnements, évolution des approches, Institut de Recherche pour le Développement, UMR 201 « Développement et sociétés », IEDES (Université de Paris I), hal-00475265, version 1–21 April, 2010. p. 11.

[9] IPCC (Intergovernmental Panel on Climate Change). Climate change 2007: the physical science basis. Contribution of Working Group I to the fourth assessment report of the Intergovernmental Panel on Climate Change. In: Solomon S, Qin D, Manning M, Chen Z, Marquis M, Averyt KB, et al., editors. Cambridge: Cambridge University Press; 2007a. p. 996.

[10] Situma DS. Assessment of polycyclic aromatic hydrocarbons (PAHs) in air and soil from Nairobi. MSc Thesis, University of Nairobi, Department of Chemistry; 2010.

[11] UNEP report. Supporting the implementation of the Global Monitoring Plan of POPs in eastern and southern African countries. 2010.

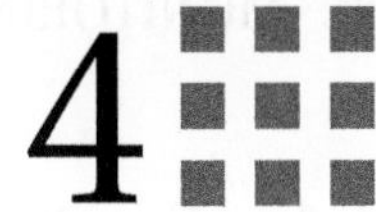

Septic Systems in the Coastal Environment: Multiple Water Quality Problems in Many Areas

Michael A. Mallin

UNIVERSITY OF NORTH CAROLINA WILMINGTON, CENTER FOR MARINE SCIENCE, WILMINGTON, NC, USA

CHAPTER OUTLINE

Monitoring Water Quality, http://dx.doi.org/10.1016/B978-0-444-59395-5.00004-2

4.1 Septic Systems and Their Prevalence in the United States

The septic system is a relatively simple means of treating human sewage, also known as wastewater. These systems are the major subset of small-scale wastewater-treatment processes that fall under the acronym OWTSs, or on-site wastewater-treatment systems. They play an important role in wastewater treatment because they provide a means of pollution abatement in rural areas, outlying suburban areas, and in areas that are distant from centralized sewage-treatment systems. A centralized sewage-treatment system requires a large capital expenditure (millions of dollars) and a consistent tax base to support its construction and maintenance [1]. In the areas mentioned above, one or both of the required financial support requirements may not be available. Septic systems can be small or large in that they have been used to treat human sewage from individual homes, multifamily structures, businesses and even hotels in both urbanized and rural areas. According to the United States Environmental Protection Agency (USEPA) [2], nearly one-fourth (23%) of homes in the US use septic systems, and they are particularly abundant in the Carolinas, Georgia, Alabama, Kentucky, West Virginia, Maine, Vermont and New Hampshire. Vermont has the highest septic-system usage (55%), and California has the lowest (10%). Reliance on septic tanks for wastewater treatment is often significant in waterfront areas in other coastal and inland areas [3]. This discussion will focus on septic systems for individual homes; for detailed information on multifamily or institutional systems, see Ref. [2].

How does a septic system function? At its simplest (Figure 4-1), a septic system consists of a septic tank and an overflow pipe connecting to a set of pipes that discharge liquid waste into a drain field [2,4]. Wastewater from the commode, sink, shower, and washing machine enters the septic tank, where solids are settled, grease is retained, and some anaerobic microbial digestion of sewage occurs. The remaining liquid waste (supernatant) passes through a filtered valve into a set of perforated pipes that discharge the waste in a diffuse manner into a drain field, which is also known as a distribution trench, leach field, soil absorption field, or subsurface wastewater infiltration system (Figure 4-1). The drain field is generally located 0.3–0.6 m (1–2 ft) below the ground surface and lined with gravel. In cold climates it may be from 0.9–1.2 m (3–4 ft) below the surface, to prevent freezing. The gravel and pipes are covered with a porous cloth which, in turn, is covered with soil and planted with grass. The drain field sits above a zone of aerated soil, called the vadose zone. The vadose zone is perched at some distance above the upper groundwater table, which is the eventual destination of the treated wastewater.

4.2 Septic System Pollutants and Treatment Processes

Wastewater contains pathogenic and other microbes (viruses, fecal bacteria, protozoans); organic material that exerts a biochemical oxygen demand (BOD) during decomposition; plant nutrients—total nitrogen (TN) and total phosphorus (TP) are of particular

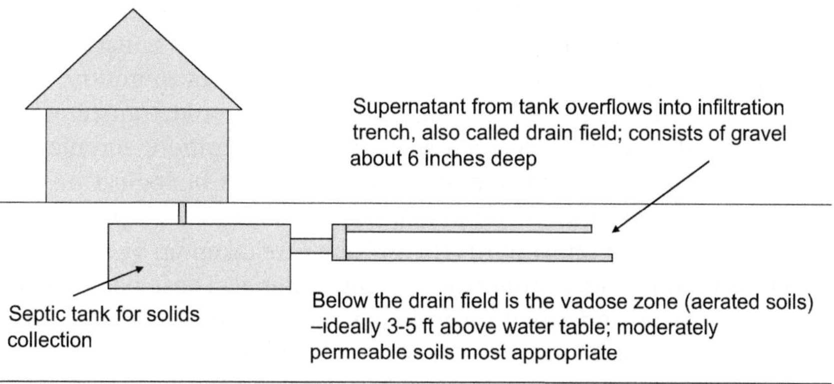
Supernatant from tank overflows into infiltration trench, also called drain field; consists of gravel about 6 inches deep
Septic tank for solids collection
Below the drain field is the vadose zone (aerated soils) –ideally 3-5 ft above water table; moderately permeable soils most appropriate
Upper or surface water table receives treated leachate, which will flow in the direction of the groundwater movement
Fecal microbes, BOD and some nutrient species are treated in drain field by a combination of filtration, microbial consumption, microbial transformation, protozoan consumption; in underlying soil layer, adsorption of phosphate and ammonium can occur as well. Nitrate is not significantly removed in the septic-system process.

FIGURE 4-1 Stylized depiction of a septic system (not to scale). (For color version of this figure, the reader is referred to the online version of this book.)

importance to receiving aquatic ecosystems; surfactants from soaps and detergents; heavy metals; chemical cleaning agents; and other chemical environmental contaminants, such as personal care products including prescription medications, over-the-counter drugs and vitamins, makeup products, and anything else that people flush down sinks and toilets (Table 4-1). After wastewater enters the septic tank, the solids settle and the tank requires to be regularly pumped out to prevent problems. The remaining liquid

Table 4-1 Approximate Concentrations of Typical Pollutants in Raw Domestic Wastewater

Parameter	Concentration (mg/L)
Total solids	500–880
Total suspended solids	155–330
Biochemical oxygen demand (BOD5)	155–286
Ammonium-N	4–23
Nitrate-N	<1
Total nitrogen	26–75
Soluble phosphorus-P	7
Total phosphorus	7–12
Surfactants	9–18
Fecal coliform bacteria (as CFU/100 mL)	10^6–10^8

From Refs. [2,47]; City of Wilmington, NC, records.

waste passes through a screened valve and is transported to the drain field. As the liquid waste is discharged over the gravel in the drain field, many of the constituent pollutants undergo removal or reduction. Some dewatering occurs by evaporation, especially in warm climates or seasons, and there is also some uptake and transpiration through overlying grasses. Pollutant removal occurs as a result of filtration, aerobic decomposition, chemical transformation and microbial interactions. A biological mat (biomat or biofilm) forms over the gravel in the drain field of a regularly used septic system. Within the biomat, nitrogen (N) and phosphorus (P) nutrients are taken up by soil microbes, and thus converted from inorganic nutrients into living material. These bacteria, in turn, are consumed by protozoans and small macrofauna such as nematodes. Microbial transformations of elements also occur, such as nitrification, which is the oxidation of ammonium to nitrite and nitrate under aerobic conditions facilitated by the bacteria *Nitrosomonas* and *Nitrobacter*. Another process is called denitrification, which is the microbial conversion of nitrate to elemental nitrogen gas, N_2. This process is considered beneficial because it involves the microbial conversion of the pollutant, nitrate, into elemental nitrogen gas, N_2. Little denitrification occurs, however: this process is most efficient under anaerobic conditions, which are the least efficient conditions for removal or reduction of the remainder of the pollutants in wastewater.

Labile carbon, which creates the majority of the BOD in wastewater, is used as a substrate by bacteria. As the liquid percolates from the drain field down through the vadose zone, other wastewater components such as orthophosphate, ammonium, and some metals are adsorbed by the soil particles. Some of the fecal bacteria are consumed by protozoans and other grazers; others are filtered within the biomat, and still others are adsorbed by soil particles beneath the drain field. Death to enteric microbes also occurs over time because they are removed from their optimal (intestinal tract) growth conditions into a hostile environment characterized by different conditions of temperature, DO concentrations, pH, etc. Treatment processes and efficacies for constituents such as nutrients, fecal microbes, and BOD are well documented, but effectiveness of treatment of residues from many personal care products is poorly documented for septic systems.

Aerated soils provide more effective treatment of fecal microbes than anaerobic soils. Regarding treatment efficacy, the greater the contact between wastewater and the soil particles in this aerated zone, the greater is the degree of treatment [4]. Under ideal circumstances septic systems can achieve nearly complete removal of fecal bacteria and BOD [2]. Unfortunately, under many circumstances septic systems and other OWTS have caused significant water-pollution problems involving microbial contamination of drinking water and bathing waters. Septic-tank leakage is the most frequently reported cause of groundwater (well water) contamination [5]. Such problems are largely caused by placing septic systems in locations inappropriate for adequate treatment.

What are the proper hydrological and geographical requirements for optimal septic-system usage? First, the soil should have moderate percolation capabilities. If the soils are very rapidly drained, like many of the coastal sands, there will be little attenuation and adsorption of pollutants and many will pass through to reach groundwater. Alternatively, if the soil is too resistant to percolation (such as many fine-grained clays), surface ponding may result, wherein poorly treated wastewater sitting above the system is unable to drain through. Surface ponding obviously presents a health hazard to humans, pets, livestock, and wildlife. Thus, in permitting a site for septic-system usage, state agencies rely on soil tests of permeability to avoid surface ponding. A second key factor is the groundwater table. The USEPA [2] recommends that the vadose zone (aerated soil layer) should be at least 0.6 m (2 ft) and, ideally, up to 1.5 m (5 ft) in depth, but individual states often permit less depth to the water table. High water tables commonly lead to poorly treated sewage and pollution of nearby water bodies. An important consideration is that the depth of the upper groundwater table typically changes seasonally. In summer, the active terrestrial vegetation uses large quantities of groundwater and also transpires large volumes to the atmosphere. Evaporation is greatly enhanced in summer relative to winter; for example, in the Coastal Plain of North Carolina, a warm temperate state, evaporation increases by a factor of 6.5 between December and July [6]. Thus, in winter the seasonal groundwater table is normally closest to the surface relative to warmer seasons. Periods of rain cause the groundwater table to rise closer to (or even above) the surface, whereas drought lowers the groundwater table.

A third important factor influencing optimal septic-system usage is proximity to surface water bodies. In coastal areas characterized by relatively porous soils, plumes from septic-system drain fields move downslope until they encounter a body of water. Setback requirements vary widely among states and can be arbitrary [2]. For example, an analysis of off-site nutrient movement from septic systems on St. George Island, Florida, found that the legal setback limit of 23 m was inadequate to protect nearby surface waters from receiving elevated nutrients, and a 50 m setback was recommended for full protection [7]. Local ordinances should include science-based setback requirements from drinking wells, streams and lakes, and property lines to prevent off-site pollution.

In a given area, features such as the soil texture, the land slope, and the depth to the water table, control the area's carrying capacity for septic systems. Too many septic systems in the area will overwhelm the carrying capacity and render treatment ineffective [2,8,9]. Where there is excessive crowding, individual septic plumes may intermingle to pollute large areas of groundwater. Duda and Cromartie [9] working in six tidal watersheds in coastal North Carolina, found significant positive correlations between fecal and total coliform levels and increased densities of septic tank drain fields in the watersheds. They also reported that in watersheds where septic-system density exceeded 0.62 drain fields per ha (0.25 drain fields per acre), shellfish beds were closed due to high fecal bacteria counts; and in watersheds where drain field densities were less than 0.37 per ha (0.15 per acre), water quality was acceptable and shellfish beds were open. The USEPA has designated areas with greater than 0.15 septic tanks per ha (0.06 septic tanks per acre) as potentially problematic [5].

4.3 Septic Systems and Fecal Microbial Pollution

According to the USEPA (2002) [2], average water use in the US ranges from 150 to 265 L/day (40–70 gallons per day). Thus, areas with high septic-system density can contribute considerable pollutant loading to groundwaters and surface waters, if the septic effluent is not effectively treated by the soil. In the US, various documented disease outbreaks have been traced to drinking well contamination by fecal bacteria or viruses from septic-system drain fields [5]. In addition to the direct contamination of drinking waters, microbial pollution from septic-system drain fields was implicated as contributing to about one-third (32%) of the shellfish bed closures in a 1995 survey of US state shellfish managers (National Oceanic and Atmospheric Administration's State of the Coast Report) [10]. Among coastal areas of the US, this type of shellfish-bed pollution was noted as particularly problematic in coastal areas of the Gulf of Mexico and the West Coast [10].

The location of septic systems in improper soils is a major fecal microbial pollution issue in surface waters and shallow groundwaters. Improper septic-system location occurs where drain fields are situated on porous soils such as sand, combined with a high water table. Research has demonstrated that at least 0.6 m of aerated soil is needed for adequate treatment of fecal microbes [4,11,12]. Accordingly, the USEPA (2002) [2] recommends that 0.6–1.5 m (2–5 ft) of aerated soil is needed to achieve nearly complete treatment of the wastewater before it enters the groundwater table. It is important to note that where soils are sandy, porous, and waterlogged, microbial pollutants such as fecal bacteria and viruses can flow through the soils laterally via the shallow groundwater to enter surface waters. For example, Scandura and Sobsey [3], found that septic-sourced viruses can travel rapidly in waterlogged soils—approximately 35 m in two days.

Another major microbial pollution problem occurs when septic systems are placed in soils that are too impermeable to permit proper percolation. This is primarily an inland phenomenon, where clay soils are common. As mentioned, in such circumstances, the polluted liquid from the drain field will seep to the surface, causing ponding. The high fecal microbial concentrations in such ponded wastewater make it an immediate health hazard to humans and animals that contact it. While ponded on the soil surface, the wastewater is also subject to rainfall, leading to storm water runoff that contains very high fecal bacteria concentrations [13].

4.4 Examples of Coastal Microbial Pollution from Septic Systems

4.4.1 Brunswick County, North Carolina

This area is characterized by sandy coastal soils that are poorly suited for wastewater treatment [8]. Nevertheless, it contains excessive densities of septic systems (20 per ha, or

8 per acre) and sampling within estuarine watersheds yielded the highest fecal coliform concentrations in association with septic tanks. Cahoon et al. [8] also reported high levels of N and P in septic-system-rich areas, with pollutant movement facilitated by ditching and draining activity.

4.4.2 Charlotte Harbor, Florida

Charlotte Harbor, the second largest estuary in Florida, is located on the west coast of that state. Approximately 37,000 septic systems are located within Charlotte County [14]. A study was undertaken to investigate the relationship between tidal movement and microbial water quality in a residential canal. The greatest abundances of fecal coliforms, *Enterococci* bacteria, and fecal coliphages were found at low- to outgoing tides, because the action of outgoing tidal water on the saturated soils drew septic-system-polluted groundwater into the canal on the falling tide. A further study of seven watersheds in the Charlotte Harbor system [15], found that fecal microbes were concentrated in areas of low salinity and high septic-system density. The highest pollution levels in this study were reported for a 292-ha (672-acre) watershed that contained 252 septic systems at a density of 0.94 septic tanks/ha (0.38/acre).

4.4.3 The Florida Keys

The Florida Keys presents a different problem with underlying soils. The Keys contain over 30,000 septic systems and injection wells into which raw sewage is disposed [16]. However, the soils are karst (limestone) and, thus, very porous [17]. Field experiments showed that fecal viruses injected into injection wells flow out through the porous soils into coastal waters within 10–53 h depending on location in the Keys, with movement as fast as 19 m/h in the Key Largo area [18]. Outgoing tides exacerbate this offshore movement of fecal waste [18]. Furthermore, microbial contaminants associated with human sewage have been isolated in mucus layers of coral heads in waters of the Keys [16], indicating that inadequate treatment by the abundant septic systems in that region can pose a health threat to humans (e.g., recreational divers) and beneficial aquatic life, even in marine waters away from the immediate coastline.

4.5 Septic Systems and Nutrient Pollution

Coastal water bodies, including tidal creeks, lagoons, riverine estuaries and the coastal ocean can be highly sensitive to nutrient pollution (i.e. eutrophication). This sensitivity to nutrient inputs is evidenced by formation of toxic and otherwise noxious phytoplankton blooms [19–21], overgrowths of periphyton and/or macroalgae that can smother seagrass beds and generate hypoxia (see Refs. [22,23] and references therein), and direct toxicity to sensitive seagrass species by nitrate or ammonium inputs (see Ref. [22]; and references therein). Production of noxious high-biomass algal blooms in many coastal waters has resulted in increased BOD which contributes to hypoxia [24].

Creation of more bacterial biomass provides more food for harmful algal species that can use mixotrophy for their nutrition (see Ref. [25]; and references therein). Mixotrophy is the concomitant use by algae of photosynthesis and heterotropy to obtain nutrition. Eutrophic conditions typically are characterized by abundant dissolved organic substances from high organic loads and excreted or lysed materials, together with abundant particulate living matter such as bacteria, small phytoplankton, and protozoans, whose growth has been stimulated by the nutrient loading. These organisms are available for phagotrophy. Thus, all of these prey items and dissolved substances are available for consumption by mixotrophic harmful algal species (see Ref. [25]; and references therein).

In many coastal systems, N is the principal nutrient controlling phytoplankton growth (e.g., Refs. [26,27]), although P can limit or colimit algal growth in some situations [28]. Septic-system leachate can have high concentrations of ammonium in the tank (Table 4-1); in the drain field and beyond; inorganic nitrogen can be in the form of ammonium or nitrate, depending upon redox conditions.

The concentrations of nitrate known to cause enhanced algal growth can be quite low in these naturally oligotrophic coastal waters. Results of nutrient-addition bioassay experiments have shown that additions of 50–100 μg N/L (3.5–7.0 μM) of nitrate can cause significant chlorophyll *a* increases over controls in systems as diverse as the coastal ocean [29], the lower Neuse Estuary [30], and North Carolina tidal creeks [31]. Mesocosm experiments have demonstrated that nitrate concentrations as low as 100 μg N/L (7 μM) can cause direct toxicity to the seagrass *Zostera marina* (see Ref. [22]; and references therein). In coral reef systems, such as those near the shorelines of Caribbean islands, concentrations of inorganic nitrogen of 14 μg N/L (1 μM) and concentrations of phosphorus of inorganic P of 3.1 μg P/L (3.2 μg/L, or 0.1 μM) can promote eutrophication and coral reef degradation [32,33]. Thus, in coastal areas where septic systems are the dominant form of sewage treatment, excessive nutrients entering creeks, lagoons, and bays can cause serious environmental damage.

With sufficient aerated soil beneath the drain field, N in the wastewater is nitrified to nitrate, which is highly soluble and moves readily through soils. Plumes of nitrate sourced from septic systems have been documented to move more than 100 m from their source [2]. Under reducing conditions where nitrification is suppressed, elevated ammonia concentrations occur in the septic plumes [34]. While not as mobile in the soil as nitrate, under sandy, porous soil and waterlogged conditions, groundwater ammonia plumes can adversely affect nearby surface waters and increase eutrophication.

Nitrate pollution can be a human health problem where septic systems are placed near drinking-water wells. High concentrations of nitrate accumulate in groundwater, and can exceed national standards (USEPA, Canada) of 10 mg nitrate N/L to protect against methemoglobinemia (blue-baby syndrome; [35]). Studies have shown nitrate concentrations well in excess of this standard in groundwater plumes draining septic-system drain fields [4,12,34,36]; and drinking wells have been contaminated by nitrate from septic systems [37].

Phosphate in the wastewater plume binds readily to soils and is much less mobile than nitrate [4], such that considerable phosphate sorption occurs in the vadose zone [34]. Even so, under sandy soil conditions or conditions where long usage has led to saturation of phosphate sorption capacity in soils, phosphate concentrations exceeding 1.0 mg P/L in septic-system plumes have been documented as far as 70 m from point of origin [34]. For septic systems serving homes or cottages along lake shores, the phosphate from septic effluent leachate can contribute to algal blooms and the eutrophication of fresh waters. Karst regions (e.g., in Florida) and coarse-textured soils low in aluminum, calcium and iron present the biggest risk of phosphate movement and water contamination [2]. Thus, in Florida it has been estimated that nearly three-fourths (74%) of the soils have severe limitations to conventional septic-system usage [2].

4.6 Examples of Coastal Nutrient Pollution from Septic Systems

4.6.1 Florida Keys

The principal means of wastewater disposal in the Florida Keys is OWTSs—primarily 30,000 conventional septic tanks [16,17]. The soils are porous limestone in many areas and, as noted, subject to rapid transmission of septic-sourced nutrients. The vadose zone over the groundwater table is, on average, only 40 cm [17]. Numerous constructed residential canals, built to facilitate boaters' access to the ocean, provide inadequate space between septic drain fields and surface waters. Lapointe et al. [17] determined that groundwaters adjacent to septic drain fields were enriched 400-fold in dissolved inorganic N, and by 70-fold for soluble P. The wet season increases hydraulic head, and this, in combination with tidal movement, forces nutrients into the canals. The canal waters and nearby seagrass meadows outside of the canals, receive elevated N and P loading from septic leachate and show eutrophication symptoms of algal blooms and hypoxia relative to other, nearby ecosystems [28].

4.6.2 Chesapeake Bay

Seagrass beds are critical coastal habitat for production of commercially and ecologically important finfish and shellfish, and protect the sediments from erosion by wind, wave and currents (see Ref. [22] and references therein). In Chesapeake Bay a statistical analysis of factors influencing seagrass growth and survival in 101 subestuaries showed that a sharp decline in seagrass coverage occurred where watershed septic-system density exceeded 39/km^2 (0.39/ha or 0.16/acre) [38]. Another set of septic-system-impact studies was conducted at three Virginia Chesapeake Bay tributary watersheds [39]. This study found significant attenuation of TP and fecal microbes in the drain field, but the dissolved inorganic N concentrations reaching the estuarine shoreline were 50- to 100-fold higher than in nearby open waters. The groundwater loads of dissolved N to shoreline waters were estimated to be equivalent to groundwater loads from row crop agriculture.

4.6.3 Maryland

A statistical study of land use factors potentially impacting nitrate concentrations in community well-water systems was undertaken in three Maryland provinces: the Eastern Shore, Southern Maryland, and the Piedmont [40]. A highly significant relationship was reported between community-well-water nitrate concentrations and the number of county septic systems. The models indicated that a 1% increase in the number of septic systems in a county was associated with a 1.1% increase in well-water nitrate concentration.

4.7 North Carolina's Outer Banks—Idyllic, but Not for Septic Systems

4.7.1 Background

This section describes in more detail a study conducted by the author.

The Outer Banks of North Carolina consist of a string of long, narrow islands stretching from the Virginia border south along the coast of the Atlantic Ocean for over 260 km. In historic times the Banks figured heavily in shipping commerce, but presently they are best known as premier vacation destinations for tourists, with abundant motels, rental homes, and second or third homes for inlanders with means. Most of the land is presently owned by the US National Park Service and is within the bounds of either Cape Hatteras National Seashore or Cape Lookout National Seashore. However, several municipalities that are densely populated (on a seasonal basis) abut Federal Park property. These population centers include the villages of Nags Head on South Bodie Island to the North, and the villages of Corolla, Rodanthe, Duck, Waves, Frisco, Hatteras and Ocracoke farther South. These villages are linked by Highway 12, which in some areas that are owned by the National Park Service is the only North-South road. Almost all of the human sewage produced on these islands is treated by septic systems.

As mentioned above, septic systems are most common in areas that are rural or otherwise decentralized, and the Outer Banks villages fall into this category. However, the hydrological and geological characteristics of these islands are not favorable to septic-system usage. The predominant soils in this area are Duckston fine sand and Corolla fine sand and have very rapid permeability above the water table, averaging from 0.3 to 0.5 m (12- to more than 20 inches) per hour [41,42]. Coupled with rapid soil permeability is a high local water table, which in much of South Bodie Island ranges from less than 0.3–1.2 m (less than 1–4 ft) from the soil surface [42]. In some areas of the Outer Banks wetlands are abundant, and some are subject to tidal water movement. These characteristics lead to poorly treated waste and enhance the movement of pollutants to nearby waterways.

4.7.2 Sampling and Analysis

In 2007 a water sampling program was completed on surface waters within and near the northern portion of Cape Hatteras National Seashore (CAHA), where it borders the village of Nags Head on South Bodie Island. Nine stations were sampled on six dates between April and November, of which seven stations are described here. One is a control site; a pond located well away from human habitations designated CP. Another station (called URD) was near the terminus of a large drainage ditch (holding water continuously) that runs along the main road through urban Nags Head for several kilometers; the ditch subsequently empties into the Atlantic Ocean at a swimming beach located at CAHA ramp. The other sites were five large ditches, some of which were modified tidal creeks that flow westward from Nags Head into CAHA, then flow under Highway 12 and drain into Roanoke Sound. These ditches are strongly tidally influenced. They were designated as sites D1–D5 southward along Highway 12. The upper ends of these ditches are within a few meters of the last row of houses along the park border in Nags Head.

The sites were sampled for physical parameters including water temperature, salinity, turbidity, pH, dissolved oxygen (DO, sampled on-site using a YSI 6920 Multi-parameter Water Quality Sonde probe linked to a YSI 650 MDS display unit). DO concentrations were low on several dates at the ditch sites, especially D4 and D5. The two latter sites maintained mean and median DO levels below the North Carolina brackish water standard of 5.0 mg/L to protect aquatic life. Overall, DO concentrations in the drainage ditches were below the state standard on 33–83% of the sampling dates, depending on the site.

At each location water samples were collected from either shore or bridges for analysis of nutrients including TN (TN = TKN + nitrate), ammonium, nitrate + nitrite (hereafter referred to as nitrate), TP and orthophosphate. Analytical methods (see Refs. [43–45]) were as follows: TKN (USEPA 351.1), nitrate + nitrite (SM 4500-NO3-F), ammonia (SM 4500-NH3-H), TP (SM 4500-P-E) and orthophosphate (SM 4500-P-F). Chlorophyll *a* samples (in triplicate) were collected at all sites as a measure of phytoplankton biomass, and analyzed fluorometrically (USEPA 445.0). Septic leachate is a direct source of BOD, and as excess phytoplankton, periphyton, and macroalgae die, the decaying organic matter becomes another source of BOD which can subsequently depress DO [24]. Therefore, five-day BOD (BOD5) was sampled as well, and was analyzed using method SM 5210 B. Samples for two indicators of fecal pollution, fecal coliform bacteria and *Enterococcus*, were analyzed using membrane filtration methods (SM 922 D and USEPA 1600, respectively).

The data were subjected to summary statistical analysis in Excel (mean, standard deviation, median, range, geometric mean for fecal bacteria). Datasets were tested for normality using the Shapiro–Wilk test, with log-transformation required for subsequent normalization of most parameters. Correlation analyses were used to investigate potential relationships between sampling parameters or between sampling parameters and environmental variables. These correlations included nutrient concentrations versus monthly municipal water use for the Town of Nags Head, with water-use data kindly provided by the Town of Nags Head, Department of Public Works, Water Operations. To investigate impacts of storm water runoff, fecal bacteria counts and nutrient concentrations versus local rainfall were tested in three ways (counts versus rainfall the day of sampling, counts versus rainfall the day of sampling plus total rain fallen within the previous 24 h, and counts versus rainfall the day of sampling plus cumulative rain fallen within the previous 48 h). Rainfall data for the nearest location, Manteo Airport on Roanoke Island (10 km from the sampling area), were obtained from the North Carolina State Climate Office in Raleigh, NC. Correlation analysis was used to investigate potential factors causing hypoxia, including BOD5 versus potential drivers including chlorophyll *a*, total organic carbon, and ammonium. A significance level of $\alpha = 0.05$ was used for all correlations except those involving fecal bacteria counts; due to the inherent high variability with bacterial counts (especially when caused by storm water), the significance level was set at $\alpha = 0.10$.

4.7.3 Findings

Average nitrate-N concentrations at most sampling sites were low to moderate (23–71 μg/L) and lowest at the control site CP (12 μg N/L). However, average ammonium concentrations in our sites were substantially greater, (175–247 μg N/L) with average ammonium at three ditches (D1, D5 and URD) exceeding 200 μg N/L. Average ammonium concentrations in those three sites were comparable to the average ammonium concentrations found in tidal Calico Creek, which accepts sewage effluent from a wastewater-treatment plant in Carteret Count, NC [46]. Thus, ammonium was the dominant inorganic N form, and was present in substantial concentrations in the sampled areas. Since, as discussed, an aerated vadose layer is required beneath septic-system drain fields for nitrification of sewage N to nitrate to occur, the high water table in the area [42] may have prevented appreciable nitrification of septic system waste. Ammonium is a by-product of sewage (see Ref. [47]; and also Table 4-1) and it was hypothesized that the primary source of ammonium loading to these waters was septic leachate (decaying organic matter is also a source of ammonium). Additionally, periodic low DO concentrations in the ditches may favor the predominance of ammonium (a reduced N compound) rather than the oxidized nitrate form of N. Thus, the elevated ammonium signal suggests that septic-system leachate impacts area groundwater and ditches. It predominates as the principal inorganic N form sourced from barrier island septic systems in sandy soils with a high water table [12]. Seasonally, ammonium concentrations peaked in mid-summer and showed a strong correlation ($r = 0.834$, $p = 0.039$) with monthly water use in Nags Head (Figure 4-2). Ammonium concentrations in summer, when human presence is highest and septic-tank usage is greatest, rose to 3- to 5-fold higher than levels in spring and fall.

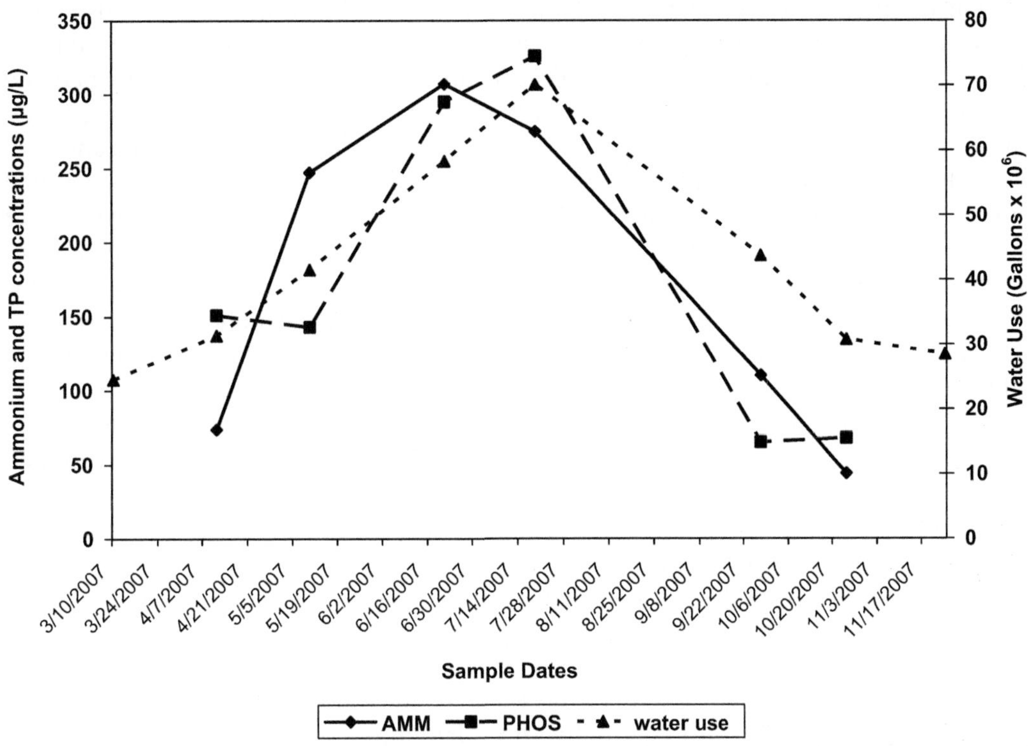

FIGURE 4-2 Ammonium-N and total phosphorus (TP) concentrations over time in South Bodie Island stations in and near Cape Hatteras National Seashore, compared with monthly municipal water use for the town of Nags Head. Nutrient data are averages of nine stations.

Orthophosphate concentrations were generally high, and highest in the first two ditches crossing Highway 12 (D1, 250 µg P/L; D2, 176 µg P/L) and the urban ditch in Nags Head (178 µg P/L). For all sites combined, the average and median TP concentrations were 249 µg P/L and 247 µg P/L, respectively. In comparison with stream and lake data, TP concentrations at all sites except the control pond (TP – 57 µg P/L) would be considered to be representative of eutrophic waters [48,49]. Seasonally, the highest concentrations were in June and July, and TP concentrations were strongly correlated with human water use in the Town of Nags Head ($r = 0.865$, $p = 0.026$; Figure 4-2). These data would also support the premise that septic leachate, which contains elevated phosphorus, is contributing significantly to water bodies near the populated areas.

Ratios of N to P in water bodies can provide insight into sources of nutrients; for example, fertilizer has generally high N/P ratios, so that if elevated N/P ratios are found in a lake or stream, the nutrients may originate from lawn, ornamentals or crop fertilization. Historically, N/P ratios much less than 16 (the Redfield ratio) were generally considered indicative of situations where N limited algal growth, while P was considered to be limiting when ratios were considerably higher than 16. The control site in this study, control pond CP, had a median inorganic N/P ratio of 17.7, very close to the Redfield Ratio, while the TN/TP ratio was 11.3, indicating little sewage influence. This site had one of the lowest TN concentrations, and by far the lowest P concentration of any site in the dataset. However, at all of the other sites, the excessive nutrient concentrations and very low N/P ratios suggest stoichiometric imbalance (see [50,51]). Median inorganic N/P molar ratio (i.e. ammonium, nitrate, and orthophosphate data) for the five Highway 12 ditches considered collectively, was 5.4, whereas the median inorganic N/P molar ratio for the runoff ditch URD was 3.3. Median TN/TP ratios ranged from 1.5 to 5.5, except for the control site. Sewage is rich in P relative to N, and has inorganic molar N/P ratios of 1–7 (see Table 4-1); thus, the low N/P ratios found at most sites indicate strong sewage influence.

Phytoplankton blooms were a frequent occurrence particularly in some of the ditches passing under Highway 12 (Table 4-2). The North Carolina State standard for acceptable algal biomass as chlorophyll *a* is less than or equal to 40 μg/L [52]. Stations D3 and D5 exceeded the State standard on three of six sampling dates. All of the sites sustained at least one algal bloom exceeding the State standard during the study, while the control pond and the urban runoff ditch maintained the lowest median chlorophyll *a* concentrations (Table 4-2). Average BOD5 concentrations ranged from 3.5- to 6.9 mg/L, with high concentrations in the ditches along Highway 12, and the lowest concentration in the beach drainage ditch (Table 4-2). There was a significant relationship between BOD5 and chlorophyll *a* ($p < 0.001$), with chlorophyll *a* explaining 43% of the variability in BOD5.

Table 4-2 Chlorophyll *a*, BOD5, Dissolved Oxygen, Fecal Coliform Bacteria and *Enterococcus* Bacteria Concentrations at Several Sampling Sites on South Bodie Island, North Carolina. Data Presented as mean ± standard deviation/median, Fecal Bacteria Presented as Geometric mean/median. Units are μg/L for Chlorophyll *a*, mg/L for BOD5 and DO, and Colony Forming Units (CFU)/100 mL for Fecal Bacteria

Site	Chlorophyll *a*	Five-day BOD	DO	Fecal Coliforms	*Enterococcus*
D1	31 ± 15	4.1 ± 1.5	5.6 ± 3.2	194	596
	33	4.0	5.4	312	632
D2	28 ± 13	4.8 ± 3.3	7.1 ± 4.5	213	443
	30	4.0	6.4	198	463
D3	38 ± 18	4.4 ± 1.8	6.0 ± 3.7	216	348
	41	5.0	5.5	225	297
D4	36 ± 32	4.8 ± 2.1	4.3 ± 3.0	305	272
	23	5.0	3.8	213	317
D5	83 ± 99	6.5 ± 1.6	3.7 ± 1.9	95	367
	49	6.0	3.1	163	358
URD	25 ± 16	3.5 ± 2.3	7.9 ± 3.0	204	6267
	20	2.5	7.6	188	4560
CP	27 ± 19	4.3 ± 1.9	9.0 ± 2.4	29	80
	21	4.5	9.3	83	60

The analysis suggests that stimulation of algal blooms in the ditches by labile N from septic leachate, created a BOD load that contributed to the low DO levels found in the ditch/tidal creek sites.

High fecal coliform bacteria counts characterized most of the sampling locations. Of the 54 samples collected, 26 exceeded the state freshwater standard for human contact (200 CFU/100 mL) [45] for an exceedence rate of 48%. The geometric mean fecal coliform bacteria count for the five Highway 12 ditches considered collectively was 192 CFU/100 mL (Table 4-2). As mentioned, the ditch site URD was sampled just upstream from where the outfall enters the swimming beach at Ramp 1 of CAHA. The control pond maintained the lowest geometric mean count of 29 CFU/100 mL (Table 4-2).

Enterococcus samples indicated even greater fecal bacterial pollution than the fecal coliform counts. Geometric mean *Enterococcus* counts for the five ditch sites on Highway 12 considered collectively were 391 CFU/100 mL. The geometric mean for the drainage ditch URD was 6267 CFU/100 mL, while control pond CP had a geometric mean of 80 CFU/100 mL. The North Carolina (and USEPA) beach water *Enterococcus* standard of 104 CFU/100 mL was exceeded on 44 of 54 occasions for an 82% exceedence rate. Some sites (D1, D2, URD) exceeded the beach water standard on all sampling dates.

Rainfall had a strong influence on fecal bacteria abundance and movement. There was a strong positive correlation between rainfall the day of sampling and *Enterococcus* counts at URD, the roadside ditch within the developed area ($r = 0.869$, $p = 0.020$), although there was no significant correlation ($p > 0.05$) between rainfall lagged 24 or 48 h. The rapid response of fecal bacteria in the beach drainage ditch URD to same-day rainfall indicates that it received urban storm water runoff that likely was contaminated with fecal microbes from dogs, cats, and urban wildlife. Based on the high ammonium concentrations, this beach drainage ditch may also have received septic system leachate from dwellings located east of the ridgeline in Nags Head, but urban storm water runoff appeared to be the major source of fecal microbial pollution.

The westward flowing ditches also maintained high fecal bacteria counts, well in excess of the fecal bacterial densities in the control pond. Correlation analysis between rainfall and the mean *Enterococcus* bacterial counts in the five westward flowing ditches showed that, in contrast to the urban ditch, there was no significant correlation between *Enterococcus* and rainfall on the day of sampling. However, *Enterococcus* counts were strongly correlated with lagged rainfall (for 24 h, $r = 0.938$, $p = 0.006$; for 48 h, $r = 0.938$, $p = 0.005$). Some of these ditches began within a few meters of the last row of houses in the developed area (Figure 4-3), yet there was no same-day effect of rain. It was hypothesized that surface storm water runoff had little impact on these ditch/creek sites. Given the strong delayed (24–48 h) rain signal, the rainfall apparently saturated the surface and applied vertical head to the shallow groundwater table, thus forcing lateral movement of groundwater (and associated septic leachate) through the porous soils into the drainage ditches flowing westward through the park. Since most of these drainage ditches were influenced by the tides, the outgoing tide tended to draw fecal bacteria from septic-influenced saturated

FIGURE 4-3 Close proximity of housing in the town of Nags Head, North Carolina, to drainage ditches/tidal creeks within Cape Hatteras National Seashore. (For color version of this figure, the reader is referred to the online version of this book.)

soils into waterways toward and possibly into the sound. The presence of drainage ditches exacerbates the movement of fecal microbes from polluted groundwater near the septic systems [8]. The present-day extensive septic system usage within Nags Head, shallow groundwater, porous soils, and lack of centralized sewage treatment promote pollution of nearby public waters with human-derived fecal microbes.

There is evidence that fecal bacteria find an environment conducive to survival and even growth in river and creek sediments, due to elevated nutrients and protection from UV radiation. For instance, following an animal waste spill [21] and a human sewage spill [53] fecal bacteria concentration were found to persist for months in sediments, as opposed to days in the water column. It is likely that the N- and P-enriched groundwater in septic plumes also provide a similar refuge environment for fecal bacteria in water-logged soils in coastal areas.

Microbial contamination of the waterways near Nags Head is also increased by local human water consumption. Abundance of *Enterococcus* in the sampling sites was positively correlated with town of Nags Head water use ($r = 0.336$; $p = 0.013$). This finding supports the premise that with more people in this resort area, more water is used, more wastewater is subsequently discharged into septic systems, and more *Enterococci* appear in local waterways.

In summary, the ditches and altered tidal creeks on South Bodie Island are polluted with high concentrations of fecal bacteria, ammonium and phosphate.

These contaminated waters support frequent algal blooms, and have elevated BOD and hypoxia. The high ammonium and phosphate concentrations and very low N/P ratios suggest a sewage source, notably septic effluent leachate. The delayed (24–48 h) fecal coliform response to rainfall in the ditches suggests lateral forcing of septic effluent leachate and polluted groundwater into the ditches. The strong correlation between ammonium, phosphate, and *Enterococcus* counts at the sampling sites with municipal water use also supports the premise that septic system pollution is impacting human health (fecal microbial) and the ecosystem (algal blooms, hypoxia) along Cape Hatteras National Seashore. Standard septic systems are clearly inappropriate for the porous soils and high water tables in this environment. North Carolina Environmental Health regulations requires septic-system drain fields to be at least 0.3 m (12 inches) above the water table, and at least 0.15 m (6 inches) of soil above the infiltration trench (15A NCAC 18A.1955). In much of South Bodie Island, septic drain fields are in contact with the water table, in violation of these regulations. As mentioned, the USEPA (2002) [2] recommends 0.6–1.5 m (2–5 ft) of unsaturated, aerobic soil above the water table—which is not available in much of the Outer Banks—for effective removal of fecal bacteria and BOD in septic effluent. There are also exceedingly large numbers of septic systems per unit area in this community, considering that dwelling unit density along the beach areas in Nags Head averages about 7.5/ha (3.0/acre) [54].

Several other communities adjoin CAHA property to the south of Nags Head, and septic systems are commonly used within the communities and within the park. It is of interest to note that Cape Lookout National Seashore, just to the south of CAHA, managed to greatly reduce sewage disposal problems (either by chance or design) by purchasing all of the homes of former permanent residents, and leaving just a few seasonal cabins for tourist use.

The study described above provides an example of surface storm water runoff and septic-system pollution from a coastal community entering a large public nature park that contains extensive estuarine and marine resources. Barrier islands and many low-lying coastal mainland areas share the physical characteristics of porous sandy soils, high water tables, and drainage into adjoining estuaries and coastal waters, including areas where shellfishing is a common occupation or recreational pursuit. In the US, development of barrier and other coastal islands continues to occur, particularly in the Southeast, including sewage treatment by septic systems [55,56]. For example, regulations on septic-system usage devised by a stakeholder group and approved by the South Carolina Department of Health and Environmental Control were subsequently rejected by a legislative review committee [56], leaving formerly pristine marsh island areas subject to future pollution. Similarly, residential building on Georgia marsh hammocks is ongoing [55], and septic-system use is likely to pollute the adjacent waters. Clearly, in low-lying coastal areas it is imperative to use alternatives to standard septic systems to treat human waste in order to protect both ecosystem quality and human health.

4.8 Minimizing Pollution from Septic Systems

Under suitable conditions, septic systems can serve as efficient and safe means of disposal of human waste, but proper placement of septic systems is key. Movement of fecal microbes off-site will occur if the soils are too impervious and ponding occurs, or if the soils are too porous (sandy or karst) and the water table is too high for an appropriately aerated vadose zone. Where there is a high water table, a "mounded" system in which the drain field is essentially located above the ground surface can provide an aerated vadose zone to achieve proper pollutant removal, and this approach has been effective in coastal soils [57]. Corbett et al. [7] recommended that such a system should provide 1 m of vadose layer above the surface.

Even under the best circumstances, pollution can move off-site if there are excessive densities of septic systems in a given area. Based on the various research efforts cited above, having more than 0.61 systems/ha (0.25/acre) has been shown to be problematic in terms of fecal contamination, while densities as low as 0.39 system/ha (0.16/acre) have shown nutrient (N)-based impacts on coastal ecosystems. As seen within this chapter, some coastal areas have far higher densities. Movement of leachate-polluted groundwater toward sensitive coastal water bodies is further enhanced when an area is ditched and drained for development, providing rapid conduits for nutrients and other wastewater constituents. Finally, risks to the environment and human health are increased when septic system use is prevalent near drinking water wells, surface water bodies that are nutrient-sensitive and coastal waters where shellfishing occurs. The placement of septic-system drain fields less than 50 m from surface waters increases nutrient pollution risks [7].

4.9 Summary and Conclusions

Septic systems are very common means of wastewater treatment in mainland coastal areas and on barrier islands. Unfortunately scientific studies from a variety of locations along the east and southeast coasts of the US have demonstrated incomplete treatment of sewage and subsequent pollution of nearby waterways. The principal pollutants documented as coming from septic systems include fecal microbes and nutrients, especially ammonium and phosphorus. Thus the pollutants are both concerns to human health (i.e. infection from human contact or contamination of shellfish) and ecosystem-level (algal bloom formation, BOD, and hypoxia). The principal reasons for off-site pollution in coastal areas are hydrological and geological. The water table is often <1 m from the surface, saturating septic drain fields. The soils are sandy and porous, or else karst and porous, so that pollutants in saturated drain fields can move readily through groundwater until surface waters are reached. The continued development of coastal mainland and island habitats for human use will lead to further such pollution; thus alternative means of treatment such as mounded individual systems, or small-scale treatment plants must be more widely required and utilized.

Acknowledgments

For funding of the South Bodie Island study we thank the National Park Service, Natural Resource Program Center. The opinions expressed within are those of the author and are not necessarily those of the National Park Service. This report was accomplished under Cooperative Agreement H5000 02 0433, Task Agreement Number J2380 03 0238, Modification No. 2, with facilitation by C. McCreedy. For laboratory assistance, I thank Brad Rosov, Rena Haltom, Matthew McIver and Byron Toothman of UNC Wilmington, and Jenny James of the Center for Applied Aquatic Ecology at North Carolina State University.

References

[1] National Research Council (NRC). Managing wastewater in coastal urban areas. Washington, D.C: National Academy Press; 1992.

[2] United States Environmental Protection Agency (USEPA). Onsite wastewater treatment systems manual. Report #EPA/625/R-00/008. Washington, DC: Office of Water and Office of Research and Development, USEPA; 2002.

[3] Scandura JE, Sobsey MD. Viral and bacterial contamination of groundwater from on-site sewage treatment systems. Water Sci Technol 1997;35:141–6.

[4] Cogger CT. On-site septic systems: the risk of groundwater contamination. J Environ Health 1988;51:12–6.

[5] Yates MV. Septic tank density and ground-water contamination. Ground Water 1985;23:586–91.

[6] Robinson PJ. North Carolina weather and climate. Chapel Hill, North Carolina: The University of North Carolina Press; 2005.

[7] Corbett DR, Dillon K, Burnett W, Shaefer G. The spatial variability of nitrogen and phosphorus concentration in a sand aquifer influenced by onsite sewage treatment and disposal systems: a case study on St. George Island, Florida. Environ Pollut 2002;117:337–45.

[8] Cahoon LB, Hales JC, Carey ES, Loucaides S, Rowland KR, Nearhoof JE. Shellfish closures in southwest Brunswick County, North Carolina: septic tanks vs. storm-water runoff as fecal coliform sources. J Coastal Res 2006;22:319–27.

[9] Duda AM, Cromartie KD. Coastal pollution from septic tank drainfields. J Environ Eng Division, Am Soc Civil Eng 1982;108:l265–1279.

[10] National Oceanic and Atmospheric Administration (NOAA). "Classified shellfish growing waters" by C.E. Alexander. NOAA's State of the coast report. Silver Spring, Maryland: NOAA. Available at: http://oceanservice.noaa.gov/websites/retiredsites/sotc_pdf/SGW.PDFat:; 1998.

[11] Bicki TJ, Brown RB. On-site sewage disposal: the importance of the wet season water table. J Environ Health 1990;52:277–9.

[12] Cogger CG, Hajjar LM, Moe CL, Sobsey MD. Septic system performance on a coastal barrier island. J Environ Qual 1988;17:401–8.

[13] Reneau Jr RB, Elder Jr JH, Pettry DE, Weston CW. Influence of soils on bacterial contamination of a watershed from septic sources. J Environ Qual 1975;4:249–52.

[14] Lipp EK, Rode JB, Vincent R, Kurz RC, Rodriguez-Palacios C. Diel variability of microbial indicators of fecal pollution in a tidally influenced canal: Charlotte Harbor, Florida. Surface Water Improvement and Management Program, Southwest Florida Water Management District, Brooksville, Florida, 1999.

[15] Lipp EK, Kurz R, Vincent R, Rodriguez-Palacios C, Farrah SK, Rose JB. The effects of seasonal variability and weather on microbial fecal pollution and enteric pathogens in a subtropical estuary. Estuaries 2001;24:266–76.

[16] Lipp EK, Jarrell JL, Griffith DW, Lukasik J, Jacukiewicz J, Rose JB. Preliminary evidence for human fecal contamination in corals of the Florida Keys. Mar Pollut Bull 2002;44:666–70.

[17] Lapointe BE, O'Connell JD, Garrett GS. Nutrient couplings between on-site sewage disposal systems, groundwaters, and nearshore surface waters of the Florida Keys. Biogeochemistry 1990;10:289–307.

[18] Paul JH, Rose JB, Jiang SC, Zhou X, Cochran P, Kellogg C, et al. Evidence for groundwater and surface marine water contamination by waste disposal wells in the Florida Keys. Water Res 1997;31:1448–54.

[19] Anderson DM, Glibert PM, Burkholder JM. Harmful algal blooms and eutrophication: nutrient sources, composition and consequences. Estuaries 2002;25:704–26.

[20] Burkholder JM. Implications of harmful microalgal and heterotrophic dinoflagellates in management of sustainable marine fisheries. Ecol Appl 1998;8:S37–62.

[21] Burkholder JM, Mallin MA, Glasgow Jr HB, Larsen LM, McIver MR, Shank GC, et al. Impacts to a coastal river and estuary from rupture of a swine waste holding lagoon. J Environ Qual 1997;26:1451–66.

[22] Burkholder JM, Tomasko DA, Touchette BW. Seagrasses and eutrophication. J Exp Mar Biol Ecol 2007;350:46–72.

[23] Morand P, Briand X. Excessive growth of macroalgae: a symptom of environmental disturbance. Botanica Marina 1996;39:491–516.

[24] Mallin MA, Johnson VL, Ensign SH, MacPherson TA. Factors contributing to hypoxia in rivers, lakes and streams. Limnol Oceanogr 2006;51:690–701.

[25] Burkholder JM, Glibert PM, Skelton HM. Mixotrophy, a major mode of nutrition for harmful algal species in eutrophic waters. Harmful Algae 2008;8:77–93.

[26] Howarth RW, Marino R. Nitrogen as the limiting nutrient for eutrophication in coastal marine ecosystems: evolving views over three decades. Limnol Oceanogr 2006;51:364–76.

[27] Rabelais NN. Nitrogen in aquatic systems. Ambio 2002;31:102–12.

[28] Lapointe BE, Clark MW. Nutrient inputs from the watershed and coastal eutrophication in the Florida Keys. Estuaries 1992;15:465–76.

[29] Paerl HW, Rudek J, Mallin MA. Stimulation of phytoplankton productivity in coastal waters by natural rainfall inputs; nutritional and trophic implications. Mar Biol 1990;107:247–54.

[30] Rudek J, Paerl HW, Mallin MA, Bates PW. Seasonal and hydrological control of phytoplankton nutrient limitation in the Neuse River Estuary, North Carolina. Mar Ecol Prog Ser 1991;75:133–42.

[31] Mallin MA, Parsons DC, Johnson VL, McIver MR, CoVan HA. Nutrient limitation and algal blooms in urbanizing tidal creeks. J Exp Mar Biol Ecol 2004;298:211–31.

[32] Bell PRF. Eutrophication and coral reefs – some examples in the Great Barrier Reef Lagoon. Water Res 1992;26:553–68.

[33] Lapointe BE. Nutrient thresholds for bottom-up control of microalgal blooms on coral reefs in Jamaica and Southeast Florida. Limnol Oceanogr 1997;42:1119–31.

[34] Robertson WD, Schiff SL, Ptacek CJ. Review of phosphate mobility and persistence in 10 septic system plumes. Ground Water 1998;36:1000–10.

[35] Fan AM, Steinberg VE. Health implications of nitrate and nitrite in drinking water: an update on methemoglobinemia occurrence and reproductive and developmental toxicity. Reg Toxicol Pharmacol 1996;23:35–43.

[36] Postma FB, Gold AJ, Loomis GW. Nutrient and microbial movement from seasonally-used septic systems. J Environ Health 1992;55:5–10.

[37] Johnson CJ, Kross BC. Continuing importance of nitrate contamination of groundwater and wells in rural areas. Am J Ind Med 1990;18:449–56.

[38] Li X, Weller DE, Gallegos GL, Jordan TE, H- Kim C. Effects of watershed and estuarine characteristics on the abundance of submerged aquatic vegetation in Chesapeake Bay subestuaries. Estuaries and Coasts 2007;30:840–54.

[39] Reay WG. Septic tank impacts on ground water quality and nearshore sediment nutrient flux. Ground Water 2004;42:1079–89.

[40] Lichtenberg E, Shapiro LK. Agriculture and nitrate concentrations in Maryland community water system wells. J Environ Qual 1997;26:145–53.

[41] Natural Resources Conservation Service (NRCS). Natural resources conservation service. United States Department of Agriculture, http://soildatamart.nrcs.usda.gov/; 2010.

[42] Stone Environmental, Inc. Town of Nags Head Decentralized Wastewater Management Plan, Final Technical Report. Stone Project No. 04-1477. Montpelier, Vermont: Stone Environmental, Inc; 2005. p. 185.

[43] American Public Health Association (APHA), American Water Works Association, and Water Environment Federation. Standard methods for the examination of water and wastewater. 19th ed. Washington, DC: APHA; 1995.

[44] United States Environmental Protection Agency (USEPA). Methods for chemical analysis of water and wastes. Report #EPA-600/4-79-020. Washington, DC: Environmental Monitoring and Support Laboratory, USEPA; 1983.

[45] United States Environmental Protection Agency (USEPA). Methods for the determination of chemical substances in marine and estuarine environmental matrices. Report #EPA/600/R-97/072. 2nd ed. Cincinnati, Ohio: National Exposure Research Laboratory, Office of Research and Development, USEPA; 1997.

[46] Sanders JG, Kuenzler EJ. Phytoplankton population dynamics and productivity in a sewage enriched tidal creek in North Carolina. Estuaries 1979;2:87–96.

[47] Clark JW, Viessman Jr W, Hammer MJ. Water supply and pollution control. 3rd ed. New York, New York: IEP-A Dun-Donnelley Publisher; 1977.

[48] Dodds WK, Jones JR, Welch EB. Suggested classification of stream trophic state: distributions of temperate stream types by chlorophyll, total nitrogen, and phosphorus. Water Res 1998;32:1455–62.

[49] Wetzel RG. Limnology: lake and river ecosystems. 3rd ed. New York, New York: Academic Press; 2001.

[50] Burkholder JM, Glibert PM. Eutrophication and oligotrophication. In: Levin S, editor. Encyclopedia of biodiversity. vol. 2, 2nd ed. New York: Academic Press, in press.

[51] Glibert PM, Fullerton D, Burkholder JM, Cornwell JC, Kana TM. Ecological stoichiometry, biogeochemical cycling, invasive species and aquatic food webs: San Francisco Estuary and comparative systems. Rev Fish Sci 2011;19:358–417.

[52] North Carolina Department of Environment and Natural Resources (NC DENR). Classification and water quality standards applicable to surface waters and wetlands of North Carolina. Administrative Code Section: 15A NCAC 2B. 0100. Raleigh, North Carolina: Division of Water Quality, NC DENR; 1999.

[53] Mallin MA, Cahoon LB, Toothman BR, Parsons DC, McIver MR, Ortwine ML, et al. Impacts of a raw sewage spill on water and sediment quality in an urbanized estuary. Mar Pollut Bull 2007;54:81–8.

[54] Town of Nags Head. Town of Nags Head land use and water use plan. 2000. North Carolina: Town of Nags Head; 2000.

[55] Albers GL. Applications of island biogeography: plant diversity and soil characteristics of back-barrier islands near Sapelo Island, Georgia. M.S. Thesis. Athens, Georgia.: University of Georgia; 2004.

[56] NOAA. Regulating access to coastal islands in South Carolina. Coastal Services 10:2–3, Report #NOAA/CSC/20702-PUB. NOAA, Charleston, South Carolina. 2007.

[57] Conn KE, Habteselassie MY, Blackwood AD, Noble RT. Microbial water quality before and after the repair of a failing onsite wastewater treatment system adjacent to coastal waters. J Appl Microbiol 2012;112:214–24.

Thinking Outside the Box: Assessing Endocrine Disruption in Aquatic Life

Susanne M. Brander

DEPARTMENT OF BIOLOGY AND MARINE BIOLOGY, UNIVERSITY OF NORTH CAROLINA, WILMINGTON, NC, USA

CHAPTER OUTLINE

Monitoring Water Quality, http://dx.doi.org/10.1016/B978-0-444-59395-5.00005-4

5.1 Introduction

Exposures to environmental concentrations of endocrine disrupting compounds are now a known threat to both human and ecological health [1,2]. A large body of work has established that endocrine disrupting chemicals (EDCs) can agonize, antagonize or synergize the effects of endogenous hormones, resulting in epidemiological impacts in humans and in physiological and behavioral abnormalities in aquatic organisms [1,3,4,5]. EDCs originate from a variety of sources and are widespread in the aquatic environment [3,6]. Examples of EDC-mediated disruption in fishes include altered secondary sexual characteristics and male production of female egg proteins [7,8], while in invertebrates, EDCs appear to impact protein expression, immune response, and adversely affect development [4,9,10]. This chapter reviews historic and emerging topics in EDC research on aquatic organisms, with a focus on fish physiology. Also suggested are considerations for evaluating risk that are not currently factored into EDC assessments. This includes the role of less-considered pathways (such as those mediated by progestins, prostaglandins, and insulin-like growth factors), disruption of the neuroendocrine system, and the alteration of immune response. The importance of assessing impacts at multiple biological scales, the effects of complex environmental mixtures, differences in species-sensitivity, adaptation to pollution, epigenetic change, and the occurrence of non-monotonic responses to EDCs are also discussed.

5.2 EDC Sources

The universe of potential EDCs is expanding as new pesticides and pharmaceuticals constantly enter the marketplace. Since the establishment of the now outdated US Toxic Substances Control Act (TSCA) in 1976, upwards of 20,000 new chemical compounds have been developed for commercial use globally. There is little information available on the many of these chemicals, most of which are not subject to review prior to entering into the marketplace [11,12]. Legislation introduced in the European Union in 2006, “Registration, Evaluation, Authorization and Restriction of Chemicals (REACH),” and new United States Environmental Protection Agency

(USEPA) chemical screening initiatives, such as ToxCast™, and the Endocrine Disruptor Screening Program (EDSP), seek to address this problem either through increased regulation of chemical manufacturers or by the development of high throughput screening approaches [11–13]. However, the monumental tasks of prioritizing the backlog of compounds to be assessed and reducing their release into the environment remain. The largest sources of environmental EDCs are wastewater from sewage treatment plants and concentrated animal feed lots, and urban and agricultural runoff (Table 5-1).

5.2.1 Wastewater

Municipal and industrial wastewater outfalls are a well-documented source of EDCs [3,12–14]. While much concern regarding endocrine disruption has focused on the estrogenic effects of wastewater effluent [17,18,19], EDCs in such effluent may disrupt the actions of any number of hormones in exposed organisms, including estrogens, androgens, thyroid hormone, and glucocorticoids [20,21]. For example, the synthetic progestins and estrogens present in oral contraceptives detected downstream of outfalls, act as estrogens, antiandrogens and androgens in exposed organisms [15]. The EDCs triclosan and triclocarban (antimicrobials commonly used in consumer products such as soap, toothpaste, and plastics) are also frequently detected in the vicinity of outfalls [22].

Commonly prescribed drugs, such as selective serotonin reuptake inhibitors (SSRIs) and ibuprofen, which are known to modulate hormones involved in fish reproductive behavior [23–26] (see Section 6.7: Neuroendocrine Disruption), are also ubiquitous in aquatic ecosystems [27,28]. Pharmaceuticals that were not previously suspected to be EDCs, such as β-blockers, which interfere with the action of catecholamines, potentially impact reproduction and stress response in fishes and may alter metabolic rate and immune response in aquatic invertebrates [29]. Additionally, municipal wastewater contamination extends beyond endogenous hormones and pharmaceuticals or metabolites excreted by humans, to EDCs associated with consumer products, such as alkylphenols, plasticizers (bisphenol A, phthalates), flame retardants (polybrominated diphenyl ethers—PBDEs) and pesticides [16,19,30–32]. The endocrine-disrupting propensity of alkylphenols, bisphenol A (BPA), phthalates and PBDEs has been well-documented and continues to be an environmental health concern [33–35].

Wastewater runoff originating from dairy farms and concentrated animal feeding operations (CAFOs) also contains a mixture of hormones and pharmaceuticals. Beef and dairy cattle are typically administered a cocktail of hormones containing a mixture of growth hormone (GH), androgens, estrogens and progestins, some of which may be excreted as either parent or metabolized forms and transported via runoff to the aquatic environment [36–38]. Both endogenous and synthetic hormones detected in effluent from CAFOs have been shown to alter hormone levels, alter endocrine gene and protein expression, interfere with gonadal or secondary sex characteristic

Table 5-1 Commonly Detected Endocrine-Disrupting Compounds in Municipal or Feedlot Wastewater and Agricultural or Urban Runoff

Source	Compound	Description	Affected Hormone Pathway(s)	Biological Effect(s) in Fishes
Municipal wastewater effluent	Estradiol	Endogenous hormone	Estrogen	Alters endocrine gene, protein expression, intersex, sex ratio—feminization, immunotoxicity (reviewed in Refs. [37,241])
	Estrone	Endogenous hormone	Estrogen	Alters endocrine gene, protein expression, intersex, sex ratio (reviewed in Ref. [37])
	Estriol	Endogenous hormone	Estrogen	Alters endocrine gene, protein expression, intersex, sex ratio (reviewed in Ref. [37])
	Progesterone	Endogenous hormone	Gestagen	Alters endocrine gene, protein expression, reduced GSI, reduced fecundity, immunotoxicity [178,241]
	Testosterone	Endogenous hormone	Androgen	May alter steroidogenesis (reviewed in Ref. [37])
	Cortisol	Endogenous hormone	Glucocorticoid	May disrupt steroidogenesis, masculinization [172]
	Cortisone	Endogenous hormone	Glucocorticoid	Immunotoxicity [241]
	Dexamethasone	Synthetic hormone; antiinflammatory	Glucocorticoid	Alters endocrine gene, protein expression, decreased fecundity, malformation of offspring [201]
	Levonorgestrel	Synthetic hormone; birth control, hormone replacement therapy	Gestagen, antiestrogen	Reduced fecundity, masculinization [173]
	Drospirenone	Synthetic hormone; birth control, hormone replacement therapy	Gestagen, antiestrogen	Reduced fecundity, masculinization [173]
	Norethindrone	Synthetic hormone; birth control, hormone replacement therapy	Gestagen, antiestrogen	Alters steroidogenesis, reduces fecundity, morphological changes, masculinization [177]
	Ethinylestradiol	Synthetic hormone; birth control, hormone replacement therapy	Estrogen	Alters steroidogenesis, reduces fecundity, morphological changes, feminization, intersex, altered sex ratio [151], disrupts social hierarchy [316]
	Fluoxetine	Serotonin reuptake inhibitor, antidepressant	Neuroendocrine	Alters steroidogenesis [317], behavioral disruption, reduced fecundity, morphological changes, altered feeding, bioconcentration (reviewed in Refs. [215,318])
	Sertraline	Serotonin reuptake inhibitor, antidepressant	Neuroendocrine	Morphological changes, altered feeding, bioconcentration [318]
	Buproprion	Dopamine reuptake inhibitor; antidepressant	Neuroendocrine	Bioconcentration [318]

	Ibuprofen	Cyclooxygenase inhibitor; NSAID	Prostaglandin	Alters steroidogenesis, endocrine gene, protein expression [319], disrupts spawning activity [319] (Flippen et al., 2007), delays hatching [319]
	Bisphenol A	Plasticizer; adds flexibility to plastics	Estrogen	Alters steroidogenesis [33,320], endocrine gene, protein expression, alters hatching, development (reviewed in [321]), neuroendocrine disruption [322], alters sex ratio [19]
	Phthalates	Plasticizers; adds flexibility to plastics	Antiandrogen	Alters steroidogenesis [85], gene, protein expression [237], immunotoxicity [241]
	Nonylphenol	Surfactant in consumer products	Estrogen	Alters steroidogenesis [33,85], sex ratio [19], social behavior [323]
	Octylphenol	Surfactant in consumer products	Estrogen	Alters steroidogenesis [33], gene, protein expression [324], sex ratio [19]
	Parabens	Preservative in cosmetics	Estrogen	Alters gene, protein expression [342]
	Triclosan	Antimicrobial; consumer products	Estrogen,Thyroid	Alters gene, protein expression (reviewed in Ref. [22,343], reduces sperm count, increases hepatosomatic index [343], alters morphology, bioconcentration (reviewed in Ref. [22])
	Triclocarban	Antimicrobial; consumer products	Estrogen	Alters steroidogenesis [320], bioconcentration (reviewed in Ref. [22])
	PBDE	Flame retardant in consumer products	Thyroid	Alters thyroid hormone levels, thyroid gene expression [93,182], developmental disruption [325]
Feedlot/dairy farm runoff	Trenbolone acetate	Synthetic androgen; growth promoter	Androgen	Alters gene, protein expression, morphology, intersex, fecundity, sex ratio (reviewed in Ref. [37,326])
	Melengestrol acetate	Synthetic progestin; growth promoter	Gestagen	May alter oocyte maturation and spermatogenesis (reviewed in Ref. [37])
	Estradiol benzoate	Synthetic estrogen; growth promoter	Estrogen	Alters endocrine gene, protein expression, morphology, intersex, fecundity, sex ratio (reviewed in Ref. [37])
	Estradiol	Endogenous hormone	Estrogen	Alters endocrine gene, protein expression, intersex, sex ratio (reviewed in Ref. [37])
	Estrone	Endogenous hormone	Estrogen	Alters endocrine gene, protein expression, intersex, sex ratio (reviewed in Ref. [37])

(*continued on next page*)

Table 5-1 Commonly Detected Endocrine-Disrupting Compounds in Municipal or Feedlot Wastewater and Agricultural or Urban Runoff—*cont'd*

Source	Compound	Description	Affected Hormone Pathway(s)	Biological Effect(s) in Fishes
	Estriol	Endogenous hormone	Estrogen	Alters endocrine gene, protein expression, intersex, sex ratio (reviewed in Ref. [37])
	Testosterone	Endogenous hormone	Androgen	May alter steroidogenesis (reviewed in Ref. [37])
	Progesterone	Endogenous hormone	Gestagen	Alters endocrine gene, protein expression, reduced GSI, reduced fecundity, immunotoxicity [178,241]
Agricultural/ urbanrunoff	Permethrin	Pesticide; ion channel opening	Estrogen, Antiestrogen	Alters gene, protein expression [48,91]
	Bifenthrin	Pesticide; ion channel opening	Estrogen, Antiestrogen	Alters gene, protein expression [88,91]
	Cypermethrin	Pesticide; ion channel opening	Antiandrogen	Alters reproductive behavior [339]
	Prochloraz	Fungicide; fungal metabolism interference	Antiandrogen	Alters androgenic endocrine gene, protein expression, steroidogenesis, androgen receptor antagonist, intersex, masculinization [50,327]
	Atrazine	Herbicide; inhibits photosynthesis	Estrogen	Alters reproductive behavior [196], alters steroidogenesis [85]
	Simazine	Herbicide; inhibits photosynthesis	Estrogen	Alters reproductive behavior [196], alters steroidogenesis [340]
	Vinclozolin	Fungicide; fungal metabolism interference	Antiandrogen	Alters endocrine gene, protein expression, decreased fecundity [328]
	Dicofol	Pesticide; nervous system stimulation	Estrogen	Alters steroidogenesis [85]

development, reduce fecundity and alter sex ratio in fishes [35,37]. For example, results from a recent study showed evidence of masculinization of female fish exposed in the wild to such runoff [38].

5.2.2 Pesticides

Organochlorine legacy pesticides such as DDT, which persist in sediments decades after their last application [41], are well-established xenoestrogens [42,43]. While remnants of organochlorine contamination contribute to the overall problem of aquatic endocrine disruption and may be resuspended during restoration or dredging events concern is building over newer-use, endocrine-active pesticides. In particular, use of pyrethroid pesticides has increased considerably as organophosphate pesticides that are more toxic to mammals are phased out [44]. In addition to toxicity caused by the intended mechanism of pyrethroids, which is paralysis via sodium channel binding, recent results from in vitro and in vivo assays reveal that some pyrethroids can act as estrogens, antiestrogens and/or antiandrogens [45–48]. For example, a recent study showed that aqueous exposure to the pyrethroid bifenthrin causes expression of an estrogen-dependent protein in juvenile fish at concentrations as low as 1 ng/L [45]. Furthermore, pyrethroid metabolites are reported to have even greater endocrine activity than their parent structures [48,49]. Additionally, fungicides such as prochloraz, among others, can alter sex ratio, cause development of intersex gonads, and increase or decrease expression of estrogen-dependent genes, depending on exposure concentration, in juvenile zebrafish [50]. The effects of the above mentioned pesticides and of EDCs found in wastewater with respect to specific endocrine pathways will be detailed later in this review.

5.3 Cellular Mechanisms of Endocrine Disruption

EDCs modulate the function and response of hormones because they operate via the same cellular mechanisms, be it via binding to nuclear or cell membrane-bound receptors, or interacting with ion channels or other membrane-associated proteins such as multidrug transporters (Figure 5-1). Steroid hormones are typically transported through the bloodstream bound to a carrier protein, from which they can be displaced by an EDC [51]. Estrogens, androgens, progestogens, thyroid hormones, glucocorticoids and retinoic acid (vitamin D) and many EDCs are lipophilic, meaning they are able to cross the cell membrane to bind nuclear receptors, forming a receptor–hormone complex that acts as a transcription factor. This complex, along with a suite of other coactivators, binds to a specific region of the DNA known as a hormone response element [52–55].

EDCs that operate via this so-called classical nuclear signaling pathway (Figure 5-1, 1–6a) include components of oral contraceptives such as ethinylestradiol, growth promoters such

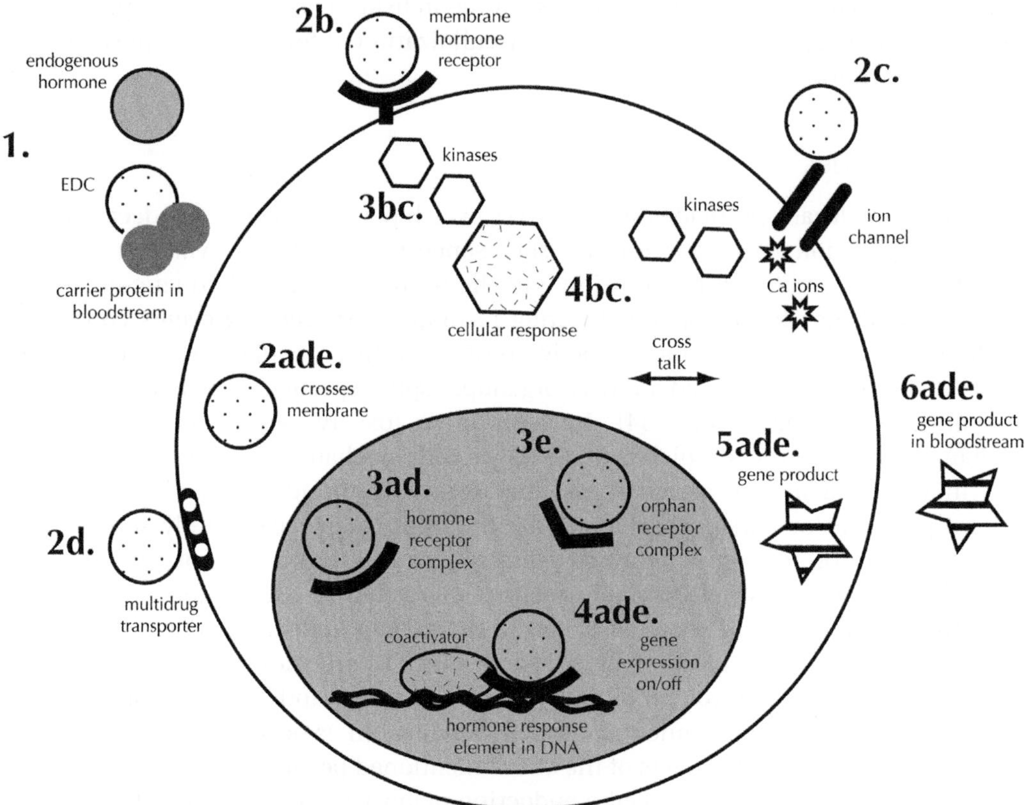

FIGURE 5-1 Summary of the cellular mechanisms by which hormones and EDCs act. Endocrine-disrupting compounds (EDCs) modulate the function and response of hormones by operating via the same cellular mechanisms as endogenous hormones. 1. EDC may displace an endogenous hormone from carrier protein. 2a–6a. An EDC crosses cell and nuclear membrane, binds to a nuclear receptor, which forms a complex that interacts with other transcription factors, and binds to a hormone-response element which results in gene expression and often the translation of a gene product (protein). 2b–4b. EDC binds to a membrane hormone receptor and causes a cellular signaling cascade mediated by protein kinases, generating a variety of downstream cellular responses. 2c–4c. An EDC interacts with an ion channel, facilitating the passage of ions (i.e., calcium) into the cell, which causes a cellular signaling cascade as described in 2b–4b. 2d–6d. An EDC bypasses multidrug transporter, entering the cell and proceeding as described in 2a–6a. 2e–4e. An EDC crosses cell and nuclear membrane, but binds to orphan receptor (PXR or CAR) instead of nuclear hormone receptor, influencing gene expression and possibly resulting in a gene product. Cross talk may occur between any of these pathways. Other mechanisms, not pictured, include alteration of transcription factors (e.g., coactivators or corepressors), altering the rate of nuclear receptor degradation, and binding to neurotransmitter receptors or transporters, resulting in alteration of signaling.

as trenbolone acetate, and the plasticizer BPA [56–58]. EDCs such as BPA appear to increase levels of transcriptional coactivators, altering response to endogenous hormones or EDCs. BPA and other EDCs may also alter of the rate of receptor degradation. Certain xenobiotics can also enhance or decrease the activity of protein kinases, leading to increased sensitivity to both endogenous hormones and EDCs via nuclear receptors [59]. In addition to binding the nuclear receptors for specific endogenous hormones, many EDCs also bind to the "orphan" receptors pregnane X receptor (PXR) and constitutive androstane receptor (CAR), which are involved in the metabolism, transport, and deactivation of endogenous hormones (Figure 5-1, 1–6e) [60].

Although there is a large degree of homology between human and fish nuclear-hormone receptors, differences in a few key amino acids situated in the ligand-binding pocket of the estrogen receptor (ER), for example, may contribute to differences in the binding affinity of endogenous hormones and EDCs between fish and human ER [61–63]. Because of a whole genome duplication event that occurred just prior to the divergence of ray-finned and lobe-finned fishes [61,64,65], there is more variation amongst teleost species in nuclear receptor sequence than there is amongst tetrapods [61], due to numerous lineage-specific gene and genome duplications followed by independent divergence [66]. This adds complexity to the task of determining which receptor(s) a particular EDC interacts with when extrapolating from mammalian cell lines to fish responses or even between different species of fishes [61,66,67,68].

In the past decade, alternative mechanisms of endogenous hormone action have been elucidated, and it is now known that steroid hormones also bind to receptors on the cell membrane surface [69,70] (Figure 5-1, 2b–4b). In addition, steroid hormones interact with ion channels to facilitate calcium-mediated cell signaling [71] (Figure 5-1, 2c–4c). Adding to this complexity is that many pharmaceuticals that disrupt endocrine function in aquatic organisms do not operate via steroid receptors, but can be influenced by the action of steroids or EDCs. For example, SSRIs such as fluoxetine bind to transporters for the neurotransmitter serotonin [24,25]. The binding of SSRIs can be affected by the simultaneous presence of estrogen or estrogenic EDCs (Nadal et al., 2005). In invertebrates, evidence is accumulating that these nonclassical pathways dominate [73]. Furthermore, cross talk between pathways activated by the same EDC or by different EDCs in an environmental mixture can occur in both vertebrates and invertebrates [74–77].

Additionally, the importance of multidrug transporters must be considered, since these membrane-bound proteins act as the cell's "bouncers," essentially preventing certain molecules from entering. Although a diversity of molecules, including hormones, are substrates for these transporters [78,79], EDCs that either have low affinity for or are not substrates for multidrug transporters may be able to accumulate in cells at higher concentrations, hence exacerbating their detrimental impact [80,81] (Figure 5-1, 2d–6d). There is also growing evidence that multidrug transporters facilitate the transport and elimination of steroid metabolites in fish, and that endocrine-disrupting compounds can interfere with this process [82]. For example, in catfish the xenoestrogen nonylphenol appears to interfere with the efflux of estradiol by such transporters, potentially leading to cellular accretion and prolonging the effects of estradiol [83]. Ultimately the binding of EDCs to endocrine targets, be they receptors, ion channels or transporters; alters hormonal gene expression, protein transcription, steroidogenesis, and elimination rates [84,85].

5.4 Impacts on Fish

5.4.1 Gene Transcription

Evaluating the expression of endocrine-sensitive genes by measuring mRNA transcript levels is an effective and common approach used to discern mechanisms of action of EDCs or EDC mixtures. It has been demonstrated in a number of studies that wild male fish captured from sites with measurable estrogenic activity express the mRNA transcripts for the female reproductive proteins vitellogenin (Vtg) and choriogenin (Chg) [84,86]. Conversely, female fish exposed to androgenic EDCs have decreased Vtg mRNA expression [87], suggesting that estrogen levels decrease concomitantly with xenoandrogen exposure. Emerging estrogenic EDCs are still being discovered; it was recently found by several research groups that the pyrethroid pesticides bifenthrin and permethrin influence the expression of estrogen-responsive genes in fish [48,88,90,91]. The influence of EDCs extends well beyond estrogen-responsive genes, however, as a recent study found that anti-androgenic and estrogenic EDCs not only modulate the expression of genes for steroidogenic enzymes, but also influence expression of genes for GH and its receptor, insulin-like growth factor and thyroid hormone [92]. Other research has demonstrated that flame retardants, ubiquitous EDCs which are known to bioaccumulate, selectively modulate the expression of thyroid transcripts in the brain of fathead minnows [93]. These genes are not only essential for reproduction but also for somatic growth [92,94,95]. With recent advances in genomic sequencing technology, microarrays have been developed for many fish species [96–100], and the ability to evaluate thousands of genes simultaneously has revealed that EDCs influence cellular functions as diverse as stress response, cell division and apoptosis [99,101,102]. Next generation sequencing technologies, which are now being employed in EDC studies, include RNA-Seq for study of the transcriptome and Chip-Seq to identify regulatory regions of DNA. RNA-Seq is a powerful technique that allows a detailed examination of gene transcription to the level of alternative splicing, providing higher resolution than a microarray [103]. RNA-Seq was recently used in concert with a microarray, to compare the transcriptome of populations of the Atlantic Killifish from polluted sites that are known to be resistant to polychlorinated biphenyls (PCBs)—an EDC—with those from reference sites [104].

In vitro transactivation assays have also been heavily utilized to determine the impact of EDCs on gene transcription. The yeast estrogen/androgen screen (YES/YAS), which consists of yeast cells stably transfected with the estrogen or androgen receptor (ER or AR) and a reporter plasmid, has been used in numerous studies to determine that compounds ranging from pesticides to plasticizers can interfere with ER and AR transcriptional activity [49,105–108]. A more sensitive alternative to the YES/YAS assays is the CALUX (Chemically Activated Luciferase Gene Expression), which consists of cells that natively express ER or AR, but are stably transfected with a reporter plasmid containing luciferase [109,110]. The CALUX assays offer an approach to measuring the overall nuclear receptor-transcriptional activity of complex environmental mixtures [19,42,99,100].

5.4.2 Protein Expression

The impact of endocrine disruption on protein expression, the products of genes, is in many ways a better predictor of tissue- and organism-level EDC effects than the mRNA transcripts. In fact, there is often lack of correlation between mRNA expression and protein amounts or activity, and transcript levels only partially explain levels of protein expression [112]. Therefore, specific proteins that are modulated by exposure to EDCs have been established for use as biomarkers. For example, the estrogen-responsive proteins Vtg and Chg have been in use as indicators of exposure to estrogenic EDCs in male fish [113], and high throughput enzyme linked immunosorbent assays (ELISAs) have been developed for both Vtg and Chg for a number of different species and sample types [57,75,76,113]. An androgen-responsive protein marker, spiggin, has also been identified, although its applicability is much reduced in comparison to Vtg and Chg, since it can only be used in one fish species. Expression of this nest glue protein that is normally produced by male three-spined sticklebacks (*Gasterosteus aculeatus*) is increased in males and also produced in females in the presence of androgenic EDCs and decreased in males in the presence of estrogenic EDCs [115,116]. Probably partly because of the considerable investment of time and expense associated with antibody development, other widely-used estrogen and androgen-responsive protein biomarkers have not been developed. However, the use of proteomics in ecotoxicology is rapidly expanding, allowing the high throughput analysis of hundreds of proteins simultaneously via several mass spectrometry-supported methods [117–119]. Although the application of this technology is in its infancy, the use of proteomic, metabolomic and transcriptomic approaches in tandem will eventually allow for a better mechanistic understanding of the links between EDC-influenced changes in protein and gene expression [120].

5.4.3 Morphological Change

Usually occurring concomitantly with changes in gene transcription and/or protein expression, alterations in secondary sexual characteristics or gonad morphology are common in cases of moderate to severe endocrine disruption correlated with reduced reproductive performance [121]. In the fathead minnow (*Pimephales promelas*), a commonly utilized bioindicator species, exposure to estrogenic EDCs causes reduction or elimination of masculine characteristics, such as breeding tubercles, fat pads, and breeding coloration in males. Conversely, exposure to androgenic EDCs causes development of these masculine traits in females [122,123].

Additionally, the appearance of histological sections of gonads, coupled with analysis of endocrine-responsive protein expression, is considered one of the most sensitive end points for determining whether endocrine disruption has occurred [124]. Exposure to both estrogenic and androgenic compounds has been shown to reduce gonadal somatic index (gonad weight relative to total body weight) and increase the incidence of germ cell necrosis, which may represent an interruption of spermatogenesis [125–127]. In cases of severe intersexuality, the presence of both male and female gonads in one organism or a complete sex reversal can occur [123,128,129]. Intersexuality in fish has been correlated with a significant reduction in fecundity [121].

5.4.4 Behavioral Change

Behavioral changes that occur as a result of exposure to EDCs can be more sensitive than and are often seen in conjunction with physiological disruption [130,131]. In fishes, a number of studies have evaluated the behavior of either estrogen- or androgen-exposed males in two pair-breeding species: fathead minnows and three-spined sticklebacks. Changes in aggression levels, ability to maintain spawning substrate, courtship frequency and nest-building behavior were observed [132–137]. Fewer studies have been conducted on group-spawning fish, but findings in zebrafish exposed to estrogenic EDCs indicate that environmentally relevant concentrations disrupt reproductive hierarchies by altering aggression and courtship behavior [138–140]. Subtle changes elicited by EDCs in male secondary sexual characteristics can also decrease the response of females to altered males [141]. The latest research findings demonstrate that non-reproductive behaviors can also be impacted by estrogenic EDCs. For example, in zebrafish (*Danio rerio*), shoaling and anxiety-related behaviors were modified in the presence of ethinylestradiol [142].

5.4.5 Population-Level Effects

Both mathematical models and empirical data confirm that EDCs can cause declines in fish populations [143,144]. Recent studies have utilized the results from single-EDC laboratory exposures to produce predictive population models [145,146] and have correlated EDC-perturbations in gonads or gene-expression changes with reduced reproductive performance or varying degrees of urbanization [121,147]. The study that demonstrated the clearest link between markers at lower levels of biological organization and population decline was performed over a seven-year period in an experimental lake. A population of fathead minnows exposed to an environmentally relevant concentration of ethinylestradiol experienced a population crash beginning with the disappearance of juvenile fish (age 0, <4 cm), concomitant with increased Vtg expression in males and the occurrence of intersexuality ([144] – Figure 5-2). The need for EDC studies that examine and link impacts at multiple biological scales, including the population level, has been suggested by a number of papers in the field [52,84,148,149] and such studies are becoming increasingly common. Studies should encompass end points with both high ecological significance (sex ratio, growth) and high mechanistic significance (gene expression, histopathology) [52]. The challenge for ecotoxicologists is to determine which molecular and organism level end points should

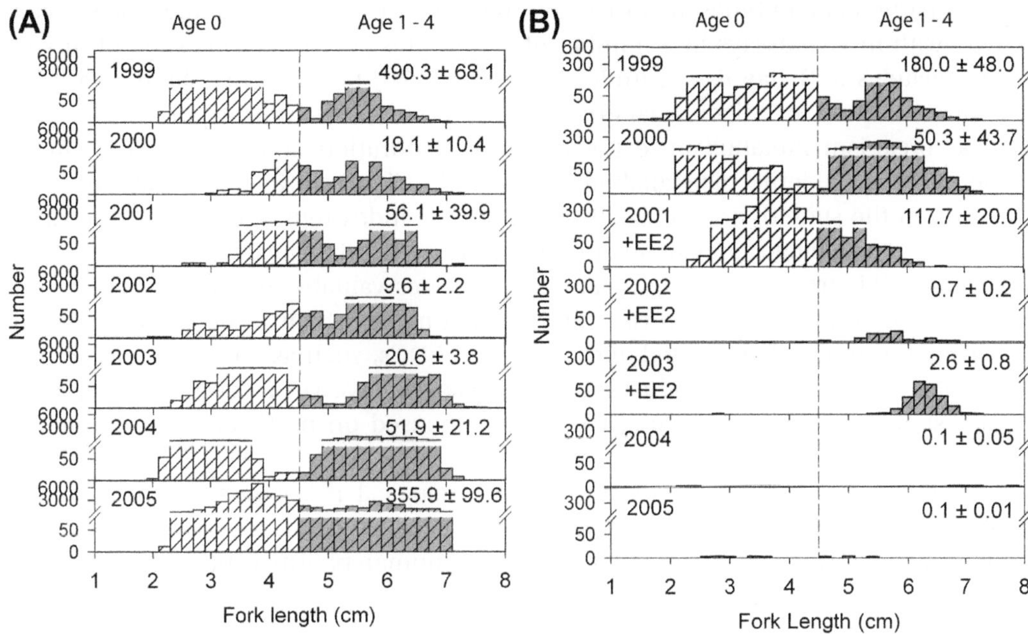

FIGURE 5-2 Length-frequency distributions of fathead minnow, which reflects the age distribution, captured in trap nets in reference Lake 442 (A) and Lake 260 (B), amended with 5–6 ng/L of ethinylestradiol in 2001–2003 during the fall, in the period 1999–2005. Distributions for each fall have been standardized to 100 trap-net days. Mean + SE daily trap-net CPUE data for adults and juveniles for the fall catches are shown in the panels. *Reprinted with permission from Kidd et al., 2007,* PNAS.

be measured in order to most accurately predict impacts at the population and even at the ecosystem levels.

5.5 Considerations for Risk Assessment

EDCs are often, but not exclusively, lipophilic compounds with a tendency to persist in the environment by binding to sediments or by accumulating in the fat tissues of organisms. EDCs that fit this description also tend to pass easily through cell membranes, binding with nuclear hormone receptors and/or otherwise modulating cellular pathways [52]. The propensity for known EDCs to have a relatively high log K_{ow} (octanol/water partition coefficient) continues to be used to inform risk assessment decisions [13], however some pharmaceuticals that can act as EDCs appear to concentrate in fish tissues regardless of lipid content [28]. Very low concentrations are known to impact fish, with predicted no effect concentrations (PNECs) for common estrogenic EDCs such as ethinylestradiol and estradiol at 0.1 and 2 ng/L, respectively [150,151].

Large-scale screening approaches, such as those adopted by the USEPA's Endocrine Disruptor Screening Program (EDSP) [13] are effective at identifying mechanisms of toxicity and prioritizing some classes of suspected EDCs for more extensive testing.

EDSP in particular combines in vitro assays that quantify receptor-binding and enzyme activity with in vivo assays in a number of vertebrates [152,153]. This suite of in vivo assays includes a 21-day fish reproduction assay [154]. Results from these assays are then prioritized via a Toxicological Priority Index (ToxPi), which incorporates elements such as organism metabolism, exposure amount/duration and individual chemical properties that underlie bioavailability (K_{OW} and cell membrane permeability) [13]. Values from the suite of assays used to generate the index represent the half-maximal activity for a particular assay divided by the lowest effective concentration [155]. The ToxPi index (Figure 5-3, [13]), for example, generates valuable information for use in assessing environmental health risks and elucidating mechanism or mode of action. However, a number of suspected and confirmed EDCs, such as particular endocrine-active pharmaceuticals and industrial contaminants, have yet to be run through the EDSP's suite of assays, as the effort is mainly focused on pesticides and herbicides [152].

Approaches tailored specifically for use in ecological risk assessment have broad similarities to EDSP in that the focus is on mechanisms of action and predictive modeling, often using known modes of action from human pharmaceutical studies, since

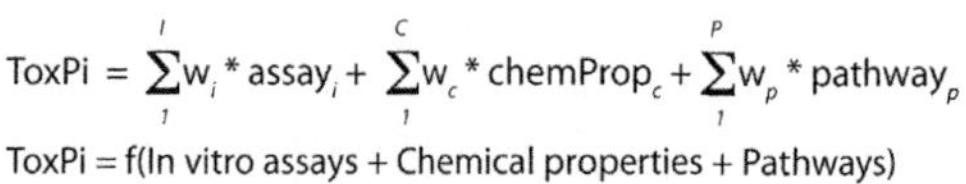

$\text{ToxPi} = \sum_{1}^{I} w_i * \text{assay}_i + \sum_{1}^{C} w_c * \text{chemProp}_c + \sum_{1}^{P} w_p * \text{pathway}_p$
ToxPi = f(In vitro assays + Chemical properties + Pathways)

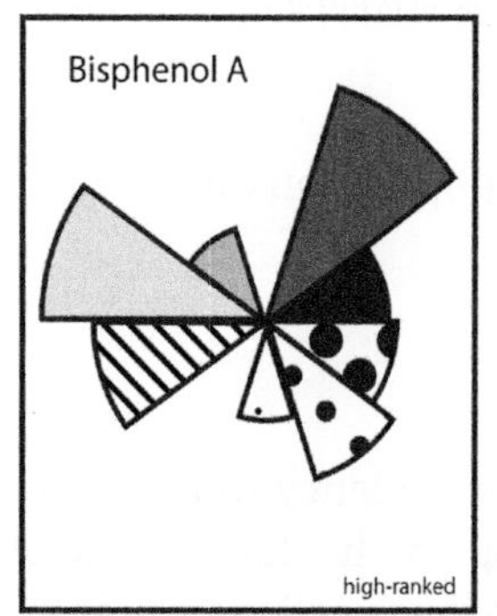

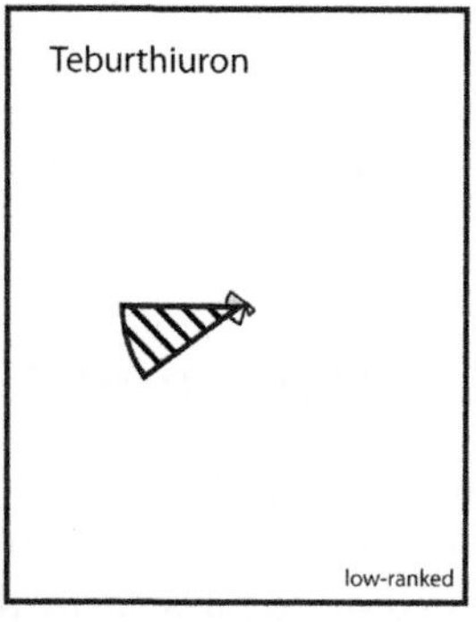

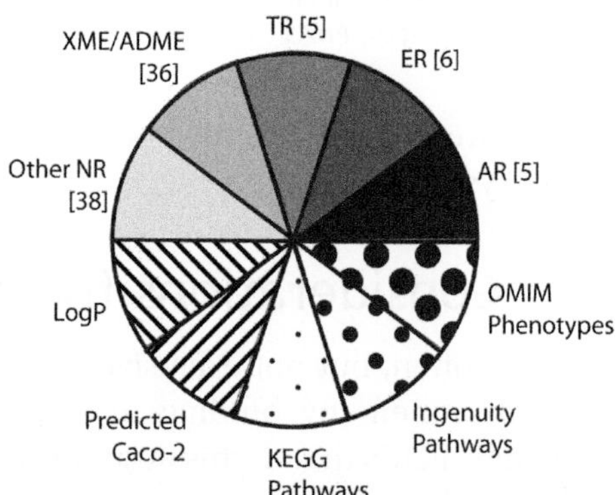

FIGURE 5-3 Definitions and notation for ToxPi. Weighted combinations of data were integrated for each chemical from multiple domains, with relative scores represented in ToxPi profiles composed of slices based on one or more components. Domains are basic data types represented by slices of a given pattern: grayscale = in vitro assay slices; stripes = chemical properties; dots = pathways. Slices represent data from related assays, properties, or pathways, including the androgen receptor (AR); estrogen receptor (ER); thyroid receptor (TR); xenobiotics-metabolizing enzymes/adsorption, distribution, metabolism & excretion (XME/ADME); other nuclear receptors (NR); relationships between phenotype and genotype (OMIM); the Kyoto Encyclopedia of Genes and Genomes (KEGG); pathways with endocrine-relevance (Ingenuity); octanol-water partition coefficent (LogP); and the cell membrane permeability assay (Caco-2). Ninety assays, two properties, and 27 pathways make up the 119 components of this endocrine ToxPi. The number of components in each slice is shown in parentheses. ToxPi profiles for BPA and tebuthiuron are shown as examples of high- and low-ranked chemicals. *Adapted and reprinted with permission from* Environmental Health Perspectives *[11].*

many pathways are highly conserved in fish [156]. Prior knowledge of endocrine toxicity mechanisms allows for extrapolation from molecular assays to higher levels of biological organization and to other species that may differ in sensitivity [156], both components of risk assessment that are inherent to ecotoxicology [157]. Incorporating these components, the newly coined "adverse outcome pathway" approach (AOP) aims to link a key toxic event at the molecular level to demographically meaningful end points (i.e., population decline, skewed sex ratio) [148]. Further enhancements to the AOP approach have been made by Garcia-Reyero and Perkins [158], via the use of reverse engineering to identify gene networks in large datasets.

Quantitative structure activity relationship (QSAR) models, which characterize the mathematical relationship between the chemical properties and biological activities of known molecules to predict the activities of unknown molecules, also have wide utility in the early stages of risk assessment [159,160]. These *in silico* methods have been developed for various classes of EDCs, such as estrogenic, androgenic and thyroid-active EDCs [161–163]; and also hold promise for the modeling of mixtures, which represent a more realistic environmental exposure regime [164]. QSAR models can be used to direct resources for laboratory assays toward the most probable mechanism(s) of toxicity. In the future, the best-informed approaches will probably combine the use of approaches such as QSAR, adverse outcome pathways, and predictive modeling.

5.6 Alternative EDC Mechanisms

The majority of EDC research has focused on reproductive impacts, particularly those caused by estrogenic compounds or to a lesser extent, androgenic compounds. Much of this research has centered on the feminization of male fish [165]. Although estrogenic and androgenic EDCs are prevalent in the aquatic environment and are of great concern, the view that research efforts should be directed toward EDCs that operate via alternative mechanisms has been put forth by several recent reviews [95,166,167]. Attention is beginning to be directed toward reproductive impacts inflicted via novel mechanisms and toward impacts on other aspects of the endocrine system, both those caused by established EDCs and by emerging EDCs. Examples of lesser-known impacts of EDCs on fish, include changes in somatic growth [45,168] and modulation of the immune system [92,169,170]. EDCs are known to disrupt pathways mediated by thyroid hormone [171], glucocorticoids [172], progestogens [173], and prostaglandins [174], among others. Additionally, as mentioned earlier, well-known estrogenic EDCs have been demonstrated to operate through diverse, cellular, signaling pathways [77,175].

5.6.1 Progestins

Progestins are components of oral contraceptives and hormone replacement treatments, are present in wastewater effluent [21], and known to interact pharmacologically with multiple steroid receptors [173,176]. In fish, endogenous progestins play an important role in oocyte maturation, sperm motility, and to function as sex pheromones [177]. Compounds typically detected in wastewater effluent such as the synthetic progestins levonorgestrel and norethindrone have recently been demonstrated to masculinize female fish, reduce fecundity, and reduce plasma-androgen concentrations in males of various species at environmentally relevant concentrations [173,177]—in fathead minnows, for example, levonorgestrel-inhibited reproduction at the environmentally relevant concentration of 0.8 ng/L and masculinized females at 3.3 ng/L [173]. Similar effects were seen in fathead minnows exposed to progesterone excreted by cattle from a CAFO [178]. This preliminary evidence suggests that progesterone and synthetic progestins found in CAFO runoff and/or wastewater effluent interfere with the androgen-signaling pathway in several species of fish via an unknown mechanism [37,173].

5.6.2 Thyroid Hormone

A number of ubiquitously distributed EDCs, such as PBDEs (flame retardants), phthalates (plasticizers), PCBs and BPA are now known to interfere with the biological actions mediated by thyroid hormone in vertebrates [171,179–181]. PBDEs and PCBs have been shown to alter the transcription of genes in the brain that regulate thyroid function [93,182], and PBDEs, PCBs and phthalates are known to bind competitively to transthyretin, a thyroid hormone transport protein [183]. PBDEs are also purported to interfere with deiodinases, enzymes that convert thyroxine (T4) to the biologically active thyroid hormone (T3) [184]. BPA, which is primarily known as an estrogenic EDC, appears to interfere with the recruitment of transcription factors to the TR [181]. Thyroid hormone signaling and reproduction are also known to be interrelated in fishes, governing the balance between stimulation of somatic growth and investment in oocyte production [185]. Therefore EDCs that disrupt thyroid action probably indirectly modulate reproductive physiology.

5.6.3 Retinoic Acid

Retinol (vitamin A) and its metabolites are vital to development, growth, differentiation, immune function, and reproduction in vertebrates [186,187]. The retinoic acid receptor (RAR) and the retinoid X receptor (RXR) act in concert with other nuclear receptors, including TR and the peroxisome proliferator-activated receptor (PPAR) [186,188]. Retinoic acid (RA) is now present in many cosmetic products and RA activity has been detected in a number of watersheds across the globe [189–192]. By forming complexes with other nuclear receptors, RA receptors can influence the expression of a large and diverse set of genes [186]. For example, organotins, which in the past had been used as marine antibiofouling agents, are known to masculinize fish and invertebrates [85]. These compounds appear to activate a heterodimer formed between RXR and PPAR, and may influence other nuclear receptor pathways and hence expression of a wide array of genes through this mechanism [193]. In another study, the administration of RA via diet to fish, attenuated the expression of genes induced by PCB-activation of the aryl hydrocarbon receptor (AhR) [187]. Considering that EDCs, which disrupt RA signaling and are AhR active (PCBs), co-occur in waterways [189], the interactions between these two classes of EDCs should be further investigated.

5.6.4 Prostaglandins

Prostaglandins are short-lived fatty acids that are involved in cellular signaling pathways modulating reproduction, inflammation and hormone regulation. Phthalates, which are used as plasticizers and are well-established EDCs, have some similarities to pharmaceuticals (nonsteroidal inflammatory drugs—NSAIDs) that inhibit the enzyme-mediated production of prostaglandins in vertebrates [174]. Additionally, many of the compounds identified as inhibitors of prostaglandin synthesis in mammalian cell lines also have confirmed estrogenic activity [1,174]. In aquatic environments downstream of wastewater outfalls, exposure to NSAIDs such as ibuprofen, is also of concern. Short-term exposure to ibuprofen in the laboratory has been shown to decrease expression of prostaglandin E2 in minnows [194]. This could have a direct impact on reproduction because some fish species use prostaglandin as a pheromone to signal spawning readiness [195,196]. Furthermore, ovulation in fish can be inhibited by antagonizing the prostaglandin receptor [197]. Medaka exposed to low concentrations of ibuprofen over a 6-week period exhibited a dose-dependent decrease in the number of weekly spawning events [23]. Chronic exposure to environmental concentrations of NSAIDs, phthalates, and potentially other yet-to-be identified EDCs could impact both mating behavior and fecundity via reduction of prostaglandin synthesis.

5.6.5 Glucocorticoids

Glucocorticoids are known to affect physiology, reproduction, and behavior in vertebrates. In teleost fishes it is known to be involved in immune and inflammatory responses, reproduction, metabolism, osmoregulation, to affect fat deposition, and to play a role in temperature-dependent sex determination [172,198–200]. The glucocorticoid cortisol was recently demonstrated to masculinize and to reduce the fecundity of female mosquitofish. Female mosquitofish exposed to elevated levels of this naturally occurring hormone developed a gonopodium (anal fin used for sperm transfer) and attempted to mate with other females [172] fathead minnows responded in a similar fashion to the synthetic cortisol dexamethasone, a frequently prescribed antiinflammatory drug that has been detected in wastewater effluent. Female fatheads experienced a decrease in fecundity and in plasma estradiol concentrations and counter-intuitively displayed an increase in the estrogen-responsive protein vitellogenin (Vtg); the development and growth of offspring was also significantly impacted [201]. Cellular interactions involving both natural and synthetic glucocorticoids are clearly complex, potentially also involving so-called orphan receptors (described below) [176], and can have a positive or negative impact on fish physiology depending on the concentration and timing of exposure [199,203]. As with other alternate EDC pathways, further research is necessary to elucidate these effects.

5.6.6 Orphan Receptor Interactions

The AHR, PXR, and the CAR are orphan receptors, meaning that a specific endogenous ligand has not been identified for each. These orphan receptors are often referred to as "xenosensors," because their role in vertebrates is to induce metabolism and excretion of the molecules they bind [204]. Well-known EDCs such as BPA are known to bind these receptors [205,206] and they are confirmed to exist in fish [202,207,208]. In Salmonids, for example, exposure to the glucocorticoid, dexamethasone, decreased PXR mRNA levels and omeprazole, a medication used for gastric reflux, altered the expression of metabolic enzymes induced by PXR binding [202]. In particular, AhR is known to exhibit cross talk with estrogen receptor-mediated responses, and coexposure with an estrogenic EDC and a known AhR agonist in fish, resulted in a reduction in the magnitude of the estrogenic response [75,76]. As a whole, downstream-signaling from these xenosensors is known to affect signaling downstream of steroid receptors [209], and the impact of EDCs on their related gene expression and metabolic influence must be included in future assessments of EDC impact.

5.6.7 Neuroendocrine Disruption

As in higher vertebrates, behaviors in fish related to reproduction, aggression, and risk-taking are mediated by neurotransmitters and hormones in the brain [26,210–213,214]. EDCs implicated in modulating these behaviors include natural and synthetic steroids, pesticides, and pharmaceuticals [130]. As described above, a number of studies have been performed on the effect of estrogenic or androgenic EDCs on reproductive behavior. However, only a few studies have evaluated how EDCs that directly target neurotransmitter release and uptake, such as antidepressants, affect both reproductive behavior and other types of behavioral responses integral to survival, such as predator avoidance and food acquisition [215–217]. For example, fathead minnows exposed to environmental concentrations of commonly used antidepressants (fluoxetine, bupropion, sertraline) had a reduced response to threat stimuli [217]. At higher doses of fluoxetine (a common serotonin-reuptake inhibitor), suppressed appetite and altered glucose metabolism in goldfish and reduced attempts to capture prey in striped bass, occurred [24,25,218]. In addition to changes in startle response and feeding behavior, fluoxetine is known to interfere with reproductive behavior [216], potentially by inhibiting the release of the hormone isotocin, which inhibits reproduction [219]. Furthermore, the enzyme that converts the amino acid tryptophan into serotonin, tryptophan hydroxylase, is also a known target of EDCs [220]. Hormones such as arginine vasotocin (AVT) are also altered in fish exposed to fluoxetine [213]. This warrants further investigation considering that AVT underlies the tendency toward aggressive behavior, and hence social rank, in many fish species [211,221]. These pharmaceuticals were designed to be effective at low doses in humans, therefore their presence at ng/L and μg/L concentrations in the aquatic environment [188] may be sufficient to disrupt behavior in fishes.

With regard to neuroendocrine hormones, recent research efforts have elucidated the action of the peptide hormone kisspeptin in non-mammalian vertebrates. Kisspeptin regulates the release of gonadotropin-releasing hormone (GnRH) from the hypothalamus and appears to play an important role in the timing of reproduction in fishes [222]. Emerging evidence demonstrates that kisspeptin's action may be altered by exposure to EDCs, particularly xenoestrogens [222,223]. Alteration in the release of GnRH, which controls the majority of reproductive functions in fishes [224], could have detrimental effects on the ability to spawn and on overall fecundity. Another class of neuroendocrine hormones, the endocannabinoids, play important roles in systemic metabolism but also influence reproduction in teleosts [225]. Recently it was shown that the xenoestrogen, nonylphenol, can modulate the transcription of cannabinoid receptors in the brain of goldfish [226]. Further research is necessary to clarify the role of endocannabinoids in teleost reproduction and the mechanisms by which they are impacted upon exposure to EDCs.

5.6.8 Insulin-like Growth Factor

The actions of insulin-like growth factors (IGFs), which have been understood in higher vertebrates for some time, are now being elucidated in fishes. IGFs play important roles in both somatic and gonadal growth, are stimulated by GH, and interact with steroid hormones in order to execute their many biological functions [94,227–230]. IGFs, which are expressed in the liver and also circulate in plasma, appear to be down-regulated in fish that are exposed to estrogenic EDCs such as nonylphenol, ethinylestradiol, and estradiol [95,231–233]. Such disruption can interfere with important biological transformations such as the morphological switch from osmoregulation in freshwater to seawater, in salmonids [232]. Furthermore, pesticides, such as the pyrethroid deltamethrin appear to interfere with GH and IGF gene expression in salmonid muscle tissue [234], with implications for impaired development and reproduction in fish exposed to pyrethroids in urban and agricultural runoff. Timing of exposure is crucial to impacts, as juvenile fishes may experience organizational changes in the IGF system that could lead to permanent alterations from the norm in both growth and reproduction [94]. As such, it is possible that IGF alterations are at least partially responsible for the reduced fecundity seen in female fishes exposed to environmental estrogens [94].

5.6.9 Immune Response

It is well established that hormones such as estrogens influence immune response in higher vertebrates [235,236]. Less information is available on fishes, but it is known that they possess the same primary immune cells as mammals—lymphocytes, granulocytes, and monocytes, as well as immune-response molecules—cytokines and immunoglobulins, and that innate immunity predominates [237–239]. Recent research strongly suggests that estrogens and androgens interact with the teleost immune system via mechanisms that are comparable those in tetrapods, including mammals [169,170,237]. For example, as occurs in mammals during pregnancy [240], fish tend to be more susceptible to disease during gametogenesis [170], and it has been shown that injections of estradiol, progesterone or 11-ketotestosterone will suppress immune function or immune-related gene expression in fishes [100,241]. Another study demonstrated that both estradiol and the estrogenic EDC nonylphenol suppress leukocytes at nM concentrations [242]. Immunotoxicity due to EDC exposure may be a more sensitive measure of EDC effects than reproductive end points. A study evaluating the toxicity of wastewater effluent found that immunotoxic effects were induced at lower concentrations than end points, indicating exposure to estrogens [243]. However, research results from experiments using mixtures are not as clear-cut, with estrogenic effluents causing differential responses depending on the potency of the effluent or the type of immune response being measured. One trial studied exposure to a potent estrogenic effluent alone or a weakly estrogenic effluent spiked with ethinylestradiol. This suppressed lymphocyte counts but elevated granulocyte and thrombocyte counts, in wild roach [86], Figure 5-4. Although more research is needed to elucidate the impact of EDCs on immunological function in fishes, evidence suggests that adding immunotoxicity to the list of considerations made during EDC risk assessment would lead to better-informed regulatory decisions.

5.6.10 Epigenetics

An emerging area of research with serious implications for environmental health is the field of epigenetics. The term "epigenetics" is defined as "the study of mitotically and/or

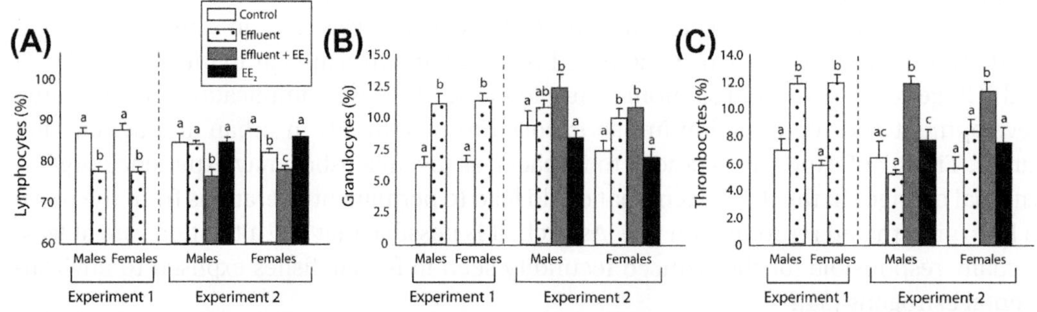

FIGURE 5-4 Immunotoxic responses in fish exposed to a potent estrogenic wastewater effluent (experiment 1), a weak estrogenic effluent with or without supplementation with ethinylestradiol or ethinylestradiol alone (experiment 2), quantified by counts of lymphocytes (A), granulocytes (B), and thrombocytes (C), shown as percentages of total leukocytes. Twelve males and 12 females were analyzed per treatment; data are shown as mean ± SE. Statistically significant differences between experimental groups for each sex are denoted by different letters ($p < 0.05$). *Adapted and reprinted with permission from* Environmental Health Perspectives, *(Filby et al., 2007).*

meiotically heritable changes in gene function that cannot be explained by changes in DNA sequence" [244]. Recent revelations demonstrate that stress and exposure to environmental chemicals, such as EDCs, can foment epigenetic change that may then be passed on to subsequent generations [245]. Epigenetic change can occur via DNA methylation, by changes to the histones around which DNA is wrapped, or through regulation by noncoding RNA [246–248]. In studies investigating the epigenomic impact of EDCs conducted with mice, phthalates and the now banned estrogenic pharmaceutical diethylstilbestrol were found to cause increased DNA methylation leading to adverse reproductive health outcomes in the offspring of exposed adults [249]. Little epigenetic research has been done on fish, but recent studies did find an increase in gonadal DNA methylation in estradiol-exposed male sticklebacks [250] and increased methylation of vitellogenin genes in ethinylestradiol-exposed zebrafish [251]. Whether epigenetic changes are passed onto the next generation has yet to be determined, but considering that this is known to occur in mammals and since the mechanisms of inheritance are generally highly conserved, it is almost certain that epigenetic inheritance occurs in fishes.

5.7 Additional Considerations

5.7.1 Complex Mixtures

Although the majority of EDC research has focused on examining the effects of single chemicals in the laboratory, in reality aquatic organisms are continuously exposed to a diverse cocktail of compounds. This is a matter of concern because the so-called "no effect" concentrations of single EDCs, can in combination cause detrimental effects [107,252]. The model of concentration-addition (CA) has been demonstrated to accurately describe the action of EDC mixtures in vitro, containing components that act via similar mechanisms (i.e., estrogenic, antiestrogenic, thyroid) [252–254]. Some studies conducted in vivo with fishes using similarly acting estrogenic EDCs also agree with the CA model [255–257]. However other research has shown that mechanistically similar EDCs in mixtures can antagonize or synergize one another, leading to a summed toxicity that is greater or less than additive, potentially because of changes in steroid metabolism triggered by EDC exposure [258–260].

When EDCs with differing mechanisms of action are combined, responses no longer adhere to the principles of CA [261]. One study in fish found that exposure to the estrogenic EDC ethinylestradiol in the presence of the androgenic EDC tributyltin (TBT) neutralized the masculinizing effects of TBT [262]. However, other studies of the combined effects of estrogens and androgens or antiestrogens have had unpredictable results [263,264]. For example, coexposure to estradiol and an antiestrogen such as tamoxifen or letrozole resulted in the reduction of the estrogen-responsive biomarker Vtg, but overall impairment of reproductive performance remained the same or worsened [264]. Results are even less clear-cut when it comes to complex environmental mixtures. For example, studies have shown simultaneous expression of the estrogen-responsive proteins or transcripts Vtg or Chg in male fish and male-biased sex ratios in the same population downstream of wastewater or pharmaceutical discharges [45,265,266]. Furthermore, one class of steroids in a complex mixture can interfere with the actions of a different class; for example, the occurrence of androgenic EDCs in wastewater influent has been associated with antagonism of progesterone and glucocorticoid receptors [267].

5.7.2 Variations in Sexual Differentiation

Testing procedures for determining the effects of EDCs on several model fishes (medaka, zebrafish, fathead minnow, stickleback) have been standardized in many respects [268–270]. However, relatively little consideration has been given to potential differences in response amongst these phylogenetically different species that vary considerably in their modes of sexual differentiation. Recent studies comparing the response of several of these model fishes to EDCs demonstrated that while sex ratio alterations occurred in all, response differed with respect to sensitivity of gonadal differentiation [271,272]. In the study comparing zebrafish and fathead minnows, zebrafish had an increased incidence of ovotestis whereas fathead minnow showed increase in undifferentiated gonads; and fathead minnow testes were delayed in maturation while zebrafish testes were not delayed [271]. As such, whether intersex or sex reversal is observed after EDC-exposure or in wild fish captured at a site receiving inputs of EDCs, may depend on the species being examined [273]. For example, fishes in which sex determination is more labile and is governed by the interaction of genes and environmental factors (i.e., temperature), may be more susceptible to alterations in sex ratio caused by environmental changes [274]. Additionally, the timing of exposure has considerable influence on the development of intersex or sex reversal; with juvenile fishes being much more sensitive to developing both conditions in response to EDC perturbations in the environment [275].

5.7.3 Adaptation

Another factor that may complicate risk-assessment decisions is the ability of fish to adapt to environmental change and interspecies variability in the aptitude to do so. It is well established that some fish species are capable of undergoing relatively rapid adaptation in response to long-term exposure to pollutants or other factors [276,277]. For example, a number of studies have demonstrated the evolved tolerance of killifish (*Fundulus heteroclitis*) to polychlorinated biphenyls (PCBs), a ubiquitous class of EDCs [275–280]. Similar adaptations to various pollutants have been observed in fathead minnows, mosquitofish and the common sole [283–285]. It is also known that this evolved tolerance is often balanced with costly trade-offs, with fish becoming less tolerant of other stressors as a result of increased resistance to a particular toxicant or suite of toxicants [286]. Furthermore, results from laboratory assays using fish specifically bred for such testing may not be representative of responses in wild populations of the same species, because of genetic variation and local adaptation [276].

5.7.4 Metabolites

With the current move toward relying on in vitro systems to determine the toxicity of contaminants (REACH, ToxCast, EDSP) [11–13,229,287], the importance of whole-organism metabolism is at risk of being overlooked in the regulatory community. A number of EDCs, including flame retardants (PBDEs), pesticides such as methoxychlor and pyrethroids, and the plasticizer BPA are converted into more endocrine-active or more toxic metabolites in vivo [48,288–293]. The metabolic activation of the methoxychlor, which was introduced as an alternative to the highly estrogenic pesticide DDT, is well understood. With the availability of microarrays, it is now known that methoxychlor metabolites can interfere with upwards of 30 different transcripts related to estrogenic and androgenic signaling pathways in the livers of fishes [290]. Pyrethroid pesticides, in particular permethrin and bifenthrin, now widely used for urban and agricultural pest management, are considered to be emerging EDCs. The endocrine activity of some pyrethroid metabolites is significantly higher than that of particular pyrethroid parent compounds [48,49], and the nature of that activity may change depending on whether an in vitro or in vivo assay is used to assess these EDCs, likely because of hepatic metabolism converting the parent compound into an estrogenic metabolite [91]. Although the use of agents (i.e., S-9) that mimic hepatic metabolism in vitro generates results that are similar to in vivo responses [294], more data on the similarities and differences between in vitro and in vivo responses to endocrine-disrupting compounds is needed, before we rely exclusively on in vitro results to make regulatory decisions. In their current state, in vitro assays are highly useful for pinpointing mechanisms but do not take into account the complicated web of pathways involved in endocrine signaling and response [209].

5.7.5 Abiotic Factors

The effects of EDCs can be exacerbated by abiotic environmental factors such as hypoxia, photoperiod and temperature [69,198,295–298], and the exposure of fish to conditions such as hypoxia may even have population-level consequences [297,299]. Studies in zebrafish and medaka demonstrate that transcription of estrogen-responsive genes in response to estradiol or nonylphenol exposure differs depending on temperature and day-length [295,296]. This has implications for observations made regarding expression of such genes in wild fish, as these responses may vary widely, based on the season during which fish were collected. It also has implications for laboratory testing, as these tests should be conducted at similar temperatures and photoperiods in order to be comparable. Another abiotic factor now recognized to be of increasing importance to endocrine disruption is hypoxia. A number of studies conducted on Atlantic croaker, a species native to the Gulf of Mexico, implicate hypoxia as a major cause of endocrine disruption.

Hypoxic events, which are on the increase globally because of anthropogenic influences [300], have been shown to masculinize this species, which is now exposed to more frequent and longer periods of low dissolved oxygen in the Gulf of Mexico [69,297]. Impacts on Atlantic croaker because of hypoxia included impairment of gonadal growth and gamete production, along with a male-skewed sex ratio. Croaker exposed to hypoxia in the laboratory had a decrease in aromatase expression, the enzyme that converts testosterone to estrogen, suggesting the potential mechanism behind masculinization [297]. Variations in abiotic factors such as temperature, photoperiod and dissolved oxygen levels should be documented, particularly in field observations of endocrine disruption, so that these conditions can be taken into account when comparing responses with other studies.

5.7.6 Non-monotonic Responses

Although the latest evaluations of EDC risk do incorporate some of the factors mentioned above that can influence EDC action [13,229], the nature of how endogenous hormones in the vertebrate endocrine system interact with their targets makes prediction of a "safe" concentration of an EDC quite difficult. For example, the method used to calculate effective concentrations for the USEPA EDSP, and hence the way risk is ultimately evaluated, assumes that response to EDCs is linear [13]. To the contrary, a number of EDC studies have observed that many endocrine-active compounds do not induce a linear or typical sigmoidal dose–response curve. Instead, exposure to a range of concentrations results in a "U"- or inverted "U"-shaped non-monotonic response because of low dose stimulation and higher dose inhibition (Figure 5-5) [45,301–304]. This is particularly evident with receptor-mediated responses that can saturate or vary depending on the concentration [304]. Computer simulations indicate that

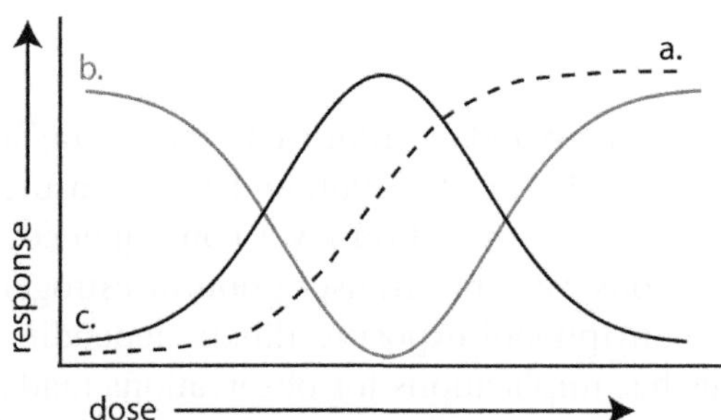

FIGURE 5-5 Unimodal or biphasic dose response versus sigmoidal dose response. a) Sigmoidal dose–response curve: demonstrative of many toxic responses, in which toxicity increases in a linear fashion, eventually reaching an asymptote. b) Biphasic or c) Unimodal dose–response curves: typical of EDCs, response is not linear due to the presence of endogenous hormones, action via different mechanisms depending on dose, and/or feedback inhibition. Responses at high doses or concentrations are often extrapolated downward for regulatory purposes, assuming a linear response, to determine a "safe" level. However, this becomes problematic with responses to EDCs that are unimodal or biphasic (both types are "non-monotonic").

a non-monotonic dose–response can also occur when receptor dimerization is disrupted. In particular, a mixed-ligand homodimer can be formed with an endogenous ligand such as estrogen and an exogenous ligand such as an EDC, impacting gene transcription and ultimately protein expression [305]. In fact, a recent high-profile review puts forth a strong justification for changing the way risk is assessed for EDCs, because the design of current toxicological assessments ignores nontypical responses, and no observable adverse effect levels would be derived based on the resulting incomplete data set [2]. This is due both to a tendency toward unimodal or non-monotonic responses and because endogenous hormones and many EDCs, including BPA, nonylphenol, the herbicide atrazine, PBDEs, PCBs, and pyrethroid pesticides, can exert action at nanomolar/micromolar concentrations, (some even as low as picomolar) that may not even be included in testing regimes [2,91].

5.8 Ecotoxicology in Risk Assessment

In 2002, the Organization for Economic Co-operation and Development (OECD) put forth a proposed plan for a tiered approach toward ecological risk assessment. This tiered plan would include an initial 14-day screening assay, fish development and reproduction testing, and a fish full life-cycle test. As of 2010, while some revisions to current testing have been made and test guidelines are available for assessing estrogenic and androgenic EDCs in fish, a final strategy toward EDC testing in wildlife is being developed by the OECD and the European Commission, but has yet to be implemented [124,306,307]. The USEPA EDSP program has formulated a similar approach, in that a fish partial life-cycle assay has been incorporated into a suite of mostly mammalian in vitro assays intended to identify whether a particular chemical interferes with estrogen, androgen or thyroid signaling [13]. Neither of these approaches integrates behavioral responses or field data, which while challenging, is necessary for an accurate evaluation of EDC impacts in fish. The complexity involved in applying biomarker responses derived from laboratory testing to assessing risk in aquatic environments is continually acknowledged by those in the field. Uncertainties exist as to whether biomarker no effect concentrations (NOECs) should be used to derive predicted no effect concentrations (PNECs), unless direct linkages can be made between lower-level responses and higher levels of biological organization such as altered development, growth, or population decline [308,309]. An alternative to deriving PNECs or threshold concentrations below which effects are unlikely to occur, is the generation of species-sensitivity distributions [157]. This method uses available response data, such as NOECs from multiple tests, to build a frequency distribution to determine the chemical concentration that is detrimental to a certain percentage of species [157,310]. Concerns arising from this approach are that the end points used to build the SSD may not be appropriate for the chemical(s) of concern or that the species used are not relevant to the ecosystem being assessed [157].

As mentioned earlier, a number of studies have made mechanistic links between gene transcription or protein expression (Vtg) and larger-scale impacts with regards to estrogenic or androgenic EDCs [45,121,144,145,147,309]. However, even when comparing the results of standard laboratory assays measuring biomarker Vtg and apical end points such as fecundity, the lowest effective concentrations are often disparate [124]. Additionally, much work remains to be done with respect to connecting the dots between changes in gene or protein expression related to alternative endocrine targets (i.e., thyroid, prostaglandins, progestins, glucocorticoids, etc.) with whole organism or population-level consequences. Endocrine disruption occurring via these pathways probably has just as much of an influence on reproductive health, for example, as disruption along estrogenic or androgenic signaling pathways does.

Another aspect crucial to accurately assessing risk to fishes is the ability to extrapolate from frequently utilized laboratory denizens such as the fathead minnow and zebrafish to other species. As mentioned above, fishes are highly plastic in regard to modes of sexual differentiation [311], so a concentration of an EDC that affects one population of fishes may not have the same impact on a phylogenetically distant species [271]. The capacity to predict response in alternative fish species lies not only in linking biomarker responses to population-level effects, but also in appropriately applied mathematical modeling approaches [143,312,313]. However knowledge gaps, such as having a better grasp on how differences in reproductive strategies influence susceptibility to EDCs, and again, how changes in gene expression influence whole organism response, must be filled before accurate models can be built [271,308,313]. Furthermore, the importance of population density and biodiversity with regard to EDC effects must be considered, particularly if the responses of a few individuals must be extrapolated to entire ecosystems [312,314,315].

5.9 Conclusions

In conclusion, although putative NOECs for estradiol and ethinylestradiol have been derived [150], much work remains to be done in determining acceptable aquatic concentrations of EDCs that disrupt thyroid, RA, progestin, prostaglandin, insulin-like growth factor and glucocorticoid signaling. Modulation of these hormones in ways that can impact immune function, growth, stress response, and behavior via mechanisms have yet to be thoroughly explored. Determining which emerging EDCs should receive the shrinking pool of research dollars in coming years will continue to challenge toxicologists, particularly those charged with discerning how toxicity-induced changes at the molecular level scale up to affect populations and ecosystems. Furthermore, in addition to assessing risk at multiple biological scales, the effects of complex environmental mixtures, differences in species sensitivity, adaptation to pollution, and the potential for epigenetic change must also be integrated into determinations of "safe" EDC concentrations. Considering the propensity of EDCs to exert effect at low doses and to exhibit non-monotonic responses, this is a task that will require increased collaboration and ingenuity amongst researchers in the field.

References

[1] Diamanti-Kandarakis E, Bourguignon J-P, Giudice LC, Hauser R, Prins GS, Soto AM, et al. Endocrine-disrupting chemicals: an endocrine society scientific statement. Endocr Rev 2009; 30:293–342.

[2] Vandenberg LN, Colborn T, Hayes TB, Heindel JJ, Jacobs DR, Lee D-H, et al. Hormones and endocrine-disrupting chemicals: low-dose effects and nonmonotonic dose responses. Endocr Rev 2012.

[3] Pait AS, Nelson JO. Endocrine disruption in fish: an assessment of recent research and results. NOAA Technical Memorandum NOS NCCOS CCMA 2002;149:48.

[4] Porte C, Janer G, Lorusso LC, Ortiz-Zarragoitia M, Cajaraville MP, Fossi MC, et al. Endocrine disruptors in marine organisms: approaches and perspectives. Comp Biochem Physiol C 2006;143:303–15.

[5] Jobling S, Williams R, Johnson A, Taylor A, Gross-Sorokin M, Nolan M, Tyler CR, van Aerle R, Santos E, Brighty G. Predicted exposures to steroid estogens in U.K. rivers correlate with widespread sexual disruption in wild fish populations. Environmental Health Perspectives 2006;114(S1):32–9.

[6] Anderson SL, Cherr GN, Morgan SG, Vines CA, Higashi RM, Bennett WA, et al. Integrating contaminant responses in indicator saltmarsh species. Mar Environ Res 2006;62:S317–21.

[7] Bogers R, De Vries-Buitenweg S, Van Gils M, Baltussen E, Hargreaves A, van de Waart B, et al. Development of chronic tests for endocrine active chemicals part 2: an extended fish early-life stage test with an androgenic chemical in the Fathead Minnow (*Pimephales promelas*). Aquat Toxicol 2006;80:119–30.

[8] Metcalfe CD, Metcalfe TL, Kiparissis Y, Koenig BG, Khan C, Hughes RJ, et al. Estrogenic potency of chemicals detected in sewage treatment plant effluents as determined by in vivo assays with Japanese Medaka (*Oryzias latipes*). Environ Toxicol Chem 2001;20:297–308.

[9] Roepke TA, Snyder MJ, Cherr GN. Estradiol and endocrine disrupting compounds adversely affect development of Sea urchin embryos at environmentally relevant concentrations. Aquat Toxicol 2005;71:155–73.

[10] Roepke TA, Chang ES, Cherr GN. Maternal exposure to estradiol and endocrine disrupting compounds alters the sensitivity of sea urchin embryos and the expression of an orphan steroid receptor. J Exp Zool 2006;305A.

[11] Judson RS, Houck KA, Kavlock RJ, Knudsen TB, Martin MT, Mortensen HM, et al. *In vitro* screening of environmental chemicals for targeted testing prioritization: the ToxCast project. Environ Health Perspect 2010;118:485–92.

[12] Wilson MP, Schwarzman MR. Toward a new U.S. chemicals policy: rebuilding the foundation to advancing new science, green chemistry, and environmental health. Environ Health Perspect 2009;117:1202–9.

[13] Reif DM, Martin MT, Tan SW, Houck KA, Judson RS, Richard AM, et al. Endocrine profiling and prioritization of environmental chemicals using ToxCast data. Environ Health Perspect 2010;118.

[14] Jobling S, Coey S, Whitmore JG, Kime DE, Van Look KJW, McAllister BG, et al. Wild intersex Roach (*Rutilus rutilus*) have reduced fertility. Biol Reprod 2002;67:515–24.

[15] Liu Z-h, Ogejo JA, Pruden A, Knowlton KF. Occurrence, fate and removal of synthetic oral contraceptives (SOCs) in the natural environment: a review. Sci Total Environ 2011;409(24):5149–61.

[16] Soares A, Guieysse B, Jefferson B, Cartmell E, Lester JN. Nonylphenol in the environment: a critical review on occurrence, fate, toxicity and treatment in wastewaters. Environ Int 2008;34:1033–49.

[17] Ma M, Rao K, Wang Z. Occurrence of estrogenic effects in sewage and industrial wastewaters in Beijing, China. Environ Pollut 2007;147:331–6.

[18] Tilton F, Benson WH, Schlenk D. Evaluation of estrogenic activity from a municipal wastewater treatment plant with predominantly domestic input. Aquat Toxicol 2002;61(3–4):211–24.

[19] Vajda AM, Barber LB, Gray JL, Lopez EM, Woodling JD, Norris DO. Reproductive disruption in fish downstream from an estrogenic wastewater effluent. Environ Sci Technol 2008;42:3407–14.

[20] Kusk KO, Krüger T, Long M, Taxvig C, Lykkesfeldt AE, Frederiksen H, et al. Endocrine potency of wastewater: contents of endocrine disrupting chemicals and effects measured by in vivo and in vitro assays. Environ Toxicol Chem 2010;30:413–26.

[21] Van Der Linden SC, Heringa MB, Man H-Y, Sonneveld E, Puijker LM, Brouwer A, et al. Detection of multiple hormonal activities in wastewater effluents and surface water, using a panel of steroid receptor CALUX bioassays. Environ Sci Technol 2008;42:5814–20.

[22] Brausch JM, Rand GM. A review of personal care products in the aquatic environment: environmental concentrations and toxicity. Chemosphere 2011;82(11):1518–32.

[23] Flippin JL, Huggett D, Foran C. Changes in the timing of reproduction following chronic exposure to ibuprofen in Japanese medaka, (*Oryzias latipes*). Aquat Toxicol 2007;81:73–8.

[24] Mennigen JA, Lado WE, Zamora JM, Duarte-Guterman P, Langlois VS, Metcalfe CD, et al. Waterborne fluoxetine disrupts the reproductive axis in sexually mature male Goldfish, (*Carassius auratus*). Aquat Toxicol 2010;100(4):354–64.

[25] Mennigen JA, Sassine J, Trudeau VL, Moon TW. Waterborne fluoxetine disrupts feeding and energy metabolism in the Goldfish (*Carassius auratus*). Aquat Toxicol 2010;100(1):128–37.

[26] Perreault HAN, Semsar K, Godwin JR. Fluoxetine treatment decreases territorial aggression in a coral reef fish. Physiol Behav 2003;79:719–24.

[27] Kolpin DW, Furlong ET, Meyer MT, Thurman EM, Zaugg SD, Barber LB, et al. Pharmaceuticals, hormones, and other organic wastewater contaminants in U.S. streams, 1999–2000: a national reconnaissance. Environ Sci Technol 2002;36:1202–11.

[28] Ramirez AJ, Brain RA, Usenko S, Mottaleb MA, O'Donnell JG, Wathen S, et al. Occurrence of pharmaceuticals and personal care products in fish: results of a national pilot study in the United States. Environ Toxicol Chem 2009;28(12):2587–97.

[29] Massarsky A, Trudeau VL, Moon TW. β-blockers as endocrine disruptors: the potential effects of human β-blockers on aquatic organisms. J Exp Zool A Ecol Genet Physiol 2011;315A(5):251–65.

[30] Anderson Td, MacRae JD. Polybrominated diphenyl ethers in fish and wastewater samples from an area of the Penobscot River in Central Maine. Chemosphere 2006;62(7):1153–60.

[31] David A, Fenet H, Gomez E. Alkylphenols in marine environments: distribution monitoring strategies and detection considerations. Mar Pollut Bull 2009;58:953–60.

[32] Slack RJ, Gronow JR, Voulvoulis N. Hazardous components of household waste. Crit Rev Environ Sci Technol 2004;34:419–45.

[33] Bonefeld-Jorgensen EC, Long M, Hofmeister MV, Vinggaard AM. Endocrine-disrupting potential of bisphenol A, bisphenol A dimethacrylate, 4-nonylphenol, and 4-octylphenol in vitro: new data and a brief review. Environ Health Perspectiv 2007;115(S-1).

[34] Diehl J, Johnson SE, Xia K, West A, Tomanek L. The distribution of 4-nonylphenol in marine organisms of North American Pacific Coast estuaries. Chemosphere 2012;87(5):490–7.

[35] Witorsch RJ, Thomas JA. Personal care products and endocrine disruption: a critical review of the literature. Crit Rev Toxicol 2010;40:1–30.

[36] Kolodziej EP, Harter T, Sedlak DL. Dairy wastewater, aquaculture, and spawning fish as sources of steroid hormones in the aquatic environment. Environ Sci Technol 2004;38(23):6377–84.

[37] Kolok AS, Sellin MK. The environmental impact of growth-promoting compounds employed by the United States beef cattle industry: history, current knowledge, and future directions. In:

Whitacre DM, editor. Reviews of environmental contamination and toxicology, vol. 195. Springer; 2008. p. 1–30.

[38] Mansell DS, Bryson RJ, Harter T, Webster JP, Kolodziej EP, Sedlak DL. Fate of endogenous steroid hormones in steer feedlots under simulated rainfall-induced runoff. Environ Sci Technol 2011.

[39] Orlando EF, Kolok AS, Binzcik GA, Gates JL, Horton MK, Lambright CS, et al. Endocrine-disrupting effects of cattle feedlot effluent on an aquatic sentinel species, the Fathead Minnow. Environ Health Perspect 2004;112:353–8.

[40] Sellin Jeffries MK, Conoan NH, Cox MB, Sangster JL, Balsiger HA, Bridges AA, et al. The anti-estrogenic activity of sediments from agriculturally intense watersheds: assessment using in vivo and in vitro assays. Aquat Toxicol 2011;105(1–2):189–98.

[41] Hartwell SI. Distribution of DDT and other persistent organic contaminants in Canyons and on the continental shelf off the Central California coast. Mar Environ Res 2008;65:199–217.

[42] McKinlay R, Plant JA, Bell JNB, Voulvoulis N. Endocrine disrupting pesticides: implications for risk assessment. Environ Int 2008;34:168–83.

[43] Metcalfe TL, Metcalfe CD, Kiparissis Y, Niimi AJ, Foran CM, Benson WH. Gonadal development and endocrine responses in Japanese Medaka (*Oryzias latipes*) exposed to *o, p′* -DDT in water or through maternal transfer. Environ Toxicol Chem 2000;19:1893–900.

[44] Werner I, Moran K. Effects of pyrethroid insecticides on aquatic organisms. In: Gan J, Spurlock F, Hendley P, Weston DP, editors. Synthetic Pyrethroids: Occurrence and Behavior in Aquatic Environments. Vol ACS Symposium Series 991. Washington D.C USA: American Chemical Society; 2008. p. 310–35.

[45] Brander SM. From 'omics to otoliths: establishing Menidia species as bioindicators of estrogenic and androgenic endocrine disruption. Dissertation. Davis, Davis: University of California; 2011.

[46] Chen H, Xiao J, Hu G, Zhou J, Xiao H, Wang X. Estrogenicity of organophosphorus and pyrethroid pesticides. Journal of Toxicology and Environmental Health, Part A 2002;65:1419–35.

[47] Sun H, Xu X-L, Xu L-C, Song L, Hong X, Chen J-F, Cui L-B, Wang X-R. Antiandrogenic activity of pyrethroid pesticides and their metabolites in a reporter gene assay. Chemosphere 2007;66:474–9.

[48] Nillos MG, Chajkowski S, Rimoldi JM, Gan J, Lavado R, Schlenk D. Stereoselective biotransformation of permethrin to estrogenic metabolites in fish. Chem Res Toxicol 2010;23(10):1568–75.

[49] Tyler CR, Beresford N, Van der Woning M, Sumpter JP, Thorpe KL. Metabolism and environmental degradation of pyrethroid insecticides produce compounds with endocrine activities. Environ Toxicol Chem 2000;19:801–9.

[50] Kinnberg K, Holbech H, Petersen GI, Bjerregaard P. Effects of the fungicide prochloraz on the sexual development of Zebrafish (*Danio rerio*). Comp Biochem Physiol C Toxicol Pharmacol 2007;145(2):165–70.

[51] Gale WL, Patino R, Maule AG. Interaction of xenobiotics with estrogen receptors alpha and beta and a putative plasma sex hormone-binding globulin from Channel Catfish (*Ictalurus punctatus*). Gen Comp Endocrinol 2004;136(3):338–45.

[52] Denslow N, Sepulveda M. Ecotoxicological effects of endocrine disrupting compounds on fish reproduction. In: The fish oocyte: from basic studies to biotechnological applications. Netherlands: Springer; 2007. p. 255–322.

[53] Germain P, Staels B, Dacquet C, Spedding M, Laudet V. Overview of nomenclature of nuclear receptors. Pharmacol Rev 2006;58:685–704.

[54] Lisse TS, Hewison M, Adams JS. Hormone response element binding proteins: novel regulators of vitamin D and estrogen signaling. Steroids 2011;76(4):331–9.

[55] McLachlan JA. Environmental signaling: what embryos and evolution teach us about endocrine disrupting chemicals. Endocr Rev 2001;22:319–41.

[56] Allen YT, Katsiadaki I, Pottinger TG, Jolly C, Matthiessen P, Mayer I, et al. Intercalibration exercise using a Stickleback endocrine disrupter screening assay. Environ Toxicol Chem 2008;27(2):404–12.

[57] Brander SM, Cole BJ, Cherr GN. An approach to detecting estrogenic endocrine disruption via choriogenin expression in an estuarine model fish species. Ecotoxicology 2012;21:1272–80.

[58] Lv X, Zhou Q, Song M, Jiang G, Shao J. Vitellogenic responses of 17 beta-estradiol and bisphenol A in male Chinese Loach (*Misgurnus anguillicaudatus*). Environ Toxicol Pharmacol 2007;24:155–9.

[59] Tabb MM, Blumberg B. New modes of action for endocrine-disrupting chemicals. Mol Endocrinol 2006;20(3):475–82.

[60] Kretschmer XC, Baldwin WS. CAR and PXR: xenosensors of endocrine disrupters? Chem. Biol Interact 2005;155(3):111–28.

[61] Baker ME. Adrenal and sex steroid receptor evolution: environmental implications. J Mol Endocrinol 2001;26(2):119–25.

[62] Filby AL, Tyler CR. Molecular characterization of estrogen receptors 1, 2a, and 2b and their tissue and ontogenic expression profiles in Fathead Minnow (*Pimephales promelas).* Biol Reprod 2005;73(4):648–62.

[63] Hawkins MB, Thornton JW, Crews D, Skipper JK, Dotte A, Thomas P. Identification of a third distinct estrogen receptor and reclassification of estrogen receptors in teleosts. Proc Nat Acad Sci 2000;97(20):10751–6.

[64] Amores A, Force A, Yan Y-L, Joly L, Amemiya C, Fritz A, et al. Zebrafish hox clusters and vertebrate genome evolution. Science 1998;282(5394):1711–4.

[65] Eick GN, Thornton JW. Evolution of steroid receptors from an estrogen-sensitive ancestral receptor. Molecular and Cellular Endocrinology 2011;334:31–8.

[66] Thornton JW. Nonmammalian nuclear receptors: evolution and endocrine disruption. Pure Appl Chem 2003;75(11–12):1827–39.

[67] Hawkins MB, Thomas P. The unusual binding properties of the third distinct teleost estrogen receptor subtype ERbetaa are accompanied by highly conserved amino acid changes in the ligand-binding domain. Endocrinology 2004;145(6):2968–77.

[68] Legler J, Zeinstra LM, Schuitemaker F, Lanser PH, Bogerd J, Brouwer A, Vethaak AD, de Voogt P, Murk AJ, van der Burg B. Comparison of in Vivo and in Vitro Reporter Gene Assays for Short-Term Screening of Estrogenic Activity. Environmental Science & Technology 2002; 36:4410–5.

[69] Thomas P, Rahman MS, Khan IA, Kummer JA. Widespread endocrine disruption and reproductive impairment in an estuarine fish population exposed to seasonal hypoxia. Proc R Soc B 2007; 274:2693–701.

[70] Thomas P. Rapid, nongenomic steroid actions initiated at the cell surface: lessons from studies with fish. Fish Physiol Biochem 2003;28:3–12.

[71] Atchison WD. Effects of toxic environmental contaminants on voltage-gated calcium channel function: from past to present. J Bioenerg Biomembr 2003;35:507–32.

[72] Nadal A, Alonso-Magdalena P, Ripoll C, Fuentes E. Disentangling the molecular mechanisms of action of endogenous and environmental estrogens. Pflügers Archiv European Journal of Physiology 2005;449:335–43.

[73] Janer G, Porte C. Sex steroids and potential mechanisms of non-genomic endocrine disruption in invertebrates. Ecotoxicology 2007;16:145–60.

[74] Bemanian V, Male R, Goksoyr A. The aryl hydrocarbon receptor-mediated disruption of vitellogenin synthesis in the fish liver: cross-talk between AHR- and ER alpha-signalling pathways. Comp Hepatol 2004;3:14.

[75] Palumbo AJ, Koivunen M, Tjeerdema RS. Optimization and validation of a California Halibut environmental estrogen bioassay using a heterologous ELISA. Sci Total Environ 2009;407:953–61.

[76] Palumbo AJ, Denison MS, Doroshov SI, Tjeerdema RS. Reduction of vitellogenin synthesis by an aryl hydrocarbon receptor agonist in the White Sturgeon (*Acipenser transmontamus*). Environ Toxicol Chem 2009;28:1749–55.

[77] Silva E, Kabil A, Kortenkamp A. Cross-talk between non-genomic and genomic signaling pathways—distinct effect profiles of environmental estrogens. Toxicol Appl Pharmacol 2010;245(2):160–70.

[78] Igarashi Y, Okuno Y, Goto S, Kanehisa M. Common features in substrates of multidrug resistance transporters. Genome Informatics 2003;14:601–2.

[79] Rao US, Fine RL, Scarborough GA. Antiestrogens and steroid hormones: substrates of the human P-glycoprotein. Biochem Pharmacol 1994;48(2):287–92.

[80] Epel D, Luckenbach T, Stevenson CN, MacManus-Spencer LA, Hamdoun A, Smital AT. Efflux transporters: newly appreciated roles in protection against pollutants. Environ Sci Technol 2008;42(11):3914–20.

[81] Hamdoun A, Epel D. Embryo stability and vulnerability in an always changing world. Proc Nat Acad Sci 2007;104(6):1745–50.

[82] James MO. Steroid catabolism in marine and freshwater fish. J Steroid Biochem Mol Biol 2011;127(3–5):167–75.

[83] Kleinow KM, Hummelke GC, Zhang Y, Uppu P, Baillif C. Inhibition of P-glycoprotein transport: a mechanism for endocrine disruption in the Channel Catfish? Mar Environ Res 2004;58(2–5):205–8.

[84] Rempel MA, Schlenk D. Effects of environmental estrogens and antiandrogens on endocrine function, gene regulation, and health in fish. In: International Review of Cell and Molecular Biology, Int Rev Cell Mol Biol, vol. 267; 2008. p. 207–252.

[85] Thibaut R, Porte C. Effects of endocrine disrupters on sex steroid synthesis and metabolism pathways in fish. J Steroid Biochem Mol Biol 2004;92(5):485–94.

[86] Filby AL, Neuparth T, Thorpe KL, Owen R, Galloway TS, Tyler CR. Health impacts of estrogens in the environment, considering complex mixture effects. Environ Health Perspect 2007a;115:1704–10.

[87] Seki M, Fujishima S, Nozaka T, Maeda M, Kobayashi K. Comparison of response to 17β-estradiol and 17β-trenbolone among three small fish species. Environ Toxicol Chem 2006; 25(10):2742–52.

[88] Beggel S, Connon R, Werner I, Geist J. Changes in gene transcription and whole organism responses in larval Fathead Minnow (*Pimephales promelas*) following short-term exposure to the synthetic pyrethroid bifenthrin. Aquat Toxicol 2011;105(1–2):180–8.

[89] Jin M, Zhang Y, Ye J, Huang C, Zhao M, Liu W. Dual enantioselective effect of the insecticide bifenthrin on locomotor behavior and development in embryonic–larval Zebrafish. Environ Toxicol Chem 2010;29(7):1561–7.

[90] Zhao M, Chen F, Wang C, Zhang Q, Gan J, Liu W. Integrative assessment of enantioselectivity in endocrine disruption and immunotoxicity of synthetic pyrethroids. Environ Pollut 2010;158(5): 1968–73.

[91] Brander SM, He G, Smalling KL, Denison MS, Cherr GN. The in vivo estrogenic and in vitro anti-estrogenic activity of permethrin and bifenthrin. Environmental Toxicology and Chemistry 2012;31(12):2848–55.

[92] Filby AL, Thorpe KL, Maack G, Tyler CR. Gene expression profiles revealing the mechanisms of anti-androgen and estrogen-induced feminization in fish. Aquat Toxicol 2007b;81:219–31.

[93] Lema SC, Dickey JT, Schultz IR, Swanson P. Dietary exposure to 2,2′,4,4′-tetrabromodiphenyl ether (PBDE-47) alters thyroid status and thyroid hormone-regulated gene transcription in the pituitary and brain. Environ Health Perspect 2008;116:1694–9.

[94] Reinecke M. Insulin-like growth factors and fish reproduction. Biol Reprod 2010a;82:656–61.

[95] Reinecke M. Influences of the environment on the endocrine and paracrine fish growth hormone-insulin-like growth factor-I system. J Fish Biol 2010b;76(6):1233–54.

[96] Baker ME, Ruggeri B, Sprague J, Eckhardt C, Lapira J, Wick I, et al. Analysis of endocrine disruption in Southern California coastal fish using an aquatic multi-species microarray. Environ Health Perspect 2008:40. 10.1289/ehp. 11627.

[97] Connon RE, Geist J, Pfeiff J, Loguinov AV, D'Abronzo LS, Wintz H, et al. Linking mechanistic and behavioral responses to sublethal esfenvalerate exposure in the endangered Delta Smelt; (*Hypomesus transpacificus*) (Fam. Osmeridae). BMC Genomics 2009:18. http://dx.doi.org/10.1186/1471-2164-10-608.

[98] Denslow ND, Garcia-Reyero N, Barber DS. Fish 'n' chips: the use of microarrays for aquatic toxicology. Mol BioSyst 2007;3(3):172–7.

[99] Wang R-L, Bencic D, Villeneuve DL, Ankley GT, Lazorchak J, Edwards S. A transcriptomics-based biological framework for studying mechanisms of endocrine disruption in small fish species. Aquat Toxicol 2010;98(3):230–44.

[100] Williams TD, Diab AM, George SG, Sabine V, Chipman JK. Gene expression responses of European Flounder (Platichthys flesus) to 17-[beta] estradiol. Toxicol Lett 2007;168(3):236–48.

[101] Hook SE. Promise and progress in environmental genomics: a status report on the applications of gene expression-based microarray studies in ecologically relevant fish species. J Fish Biol 2010;77(9):1999–2022.

[102] Sellin Jeffries MK, Mehinto AC, Carter BJ, Denslow ND, Kolok AS. Taking microarrays to the field: differential hepatic gene expression of caged Fathead Minnows from Nebraska watersheds. Environ Sci Technol 2012;46(3):1877–85.

[103] Costa V, Angelini C, De Feis I, Ciccodicola A. Uncovering the complexity of transcriptomes with RNA-Seq. Journal of biomedicine & biotechnology 2010;2010:853916.

[104] Oleksiak MF, Karchner SI, Jenny MJ, Franks DG, Welch DBM, Hahn ME. Transcriptomic assessment of resistant to effects of an aryl hydrocarbon receptor (AhR) agonist in embryos of Atlantic Killifish (*Fundulus heteroclitus*) from a marine superfund site. BMC Genomics 2011;12(263).

[105] Escher BI, Bramaz N, Eggen RIL, Richter M. In vitro assessment of modes of toxic action of pharmaceuticals in aquatic life. Environ Sci Technol 2005;39:3090–100.

[106] Routledge EJ, Sumpter JP. Estrogenic activity of surfactants and some of their degradation products assessed using a recombinant yeast screen. Environ Toxicol Chem 1996;15(3):241–8.

[107] Silva E, Rajapakse N, Kortenkamp A. Something from "nothing"—eight weak estrogenic chemicals combined at concentrations below NOECs produce significant mixture effects. Environ Sci Technol 2002;36:1751–6.

[108] Sohoni P, Sumpter JP. Several environmental oestrogens are also anti-androgens. J Endocrinol 1998;158(3):327–39.

[109] Leusch FDL, de Jager C, Levi Y, Lim R, Puijker L, Sacher F, et al. Comparison of five in vitro bioassays to measure estrogenic activity in environmental waters. Environ Sci Technol 2010;44(10):3853–60.

[110] Rogers JM, Denison MS. Recombinant cell bioassays for endocrine disruptors: development of a stably transfected human ovarian cell line for the detection of estrogenic and anti-estrogenic chemicals. In Vitro and Molecular Toxicology 2000;13:67–82.

[111] Giudice BD, Young TM. Mobilization of endocrine disrupting chemicals and estrogenic activity in simulated rainfall runoff from land-applied biosolids. Environ Toxicol Chem 2010;10:2220–8.

[112] Cox J, Mann M. Is proteomics the new genomics? Cell 2007;130:395–8.

[113] Arukwe A, Goksoyr A. Eggshell and egg yolk proteins in fish: hepatic proteins for the next generation: oogenetic, population, and evolutionary implications of endocrine disruption. Comparative Hepatology 2003;2.

[114] Holbech H, Andersen L, Petersen GI, Korgaard B, Pedersen KL, Bjerregaard P. Development of an ELISA for vitellogenin in whole body homogenate of zebrafish (Danio rerio). Comparative Biochemistry and Physiology, Part C 2001;130:119–31.

[115] Bjorkblom C, Hogfors E, Salste L, Bergelin E, Olsson P-E, Katsiadaki I, et al. Estrogenic and androgenic effects of municipal wastewater effluent on reproductive endpoint biomarkers in Three-spined Stickleback (*Gasterosteus aculeatus*). Environ Toxicol Chem 2009;28:1063–71.

[116] Sanchez W, Goin C, Brion F, Olsson P-E, Goksoyer A, Porcher J-M. A new ELISA for the Three-spined Stickleback (*Gasterosteus aculeatus* L.) spiggin, using antibodies against synthetic peptide. Comp Biochem Physiol C 2008;147:129–37.

[117] Sanchez BC, Ralston-Hooper K, Sepúlveda MS. Review of recent proteomic applications in aquatic toxicology. Environ Toxicol Chem 2010;30(2):274–82.

[118] Tomanek L. Environmental proteomics: changes in the proteome of marine organisms in response to environmental stress, pollutants, infection, symbiosis, and development. Ann Rev Mar Sci 2011;3(1):373–99.

[119] Uleberg K-E, Larssen E, Oysaed KB, Maeland M, Bjornstad-Hjelle A. Ecotoxicology goes MudPIT? Mar Environ Res 2010;69(Suppl. 1):S34–6.

[120] Fent K, Sumpter JP. Progress and promises in toxicogenomics in aquatic toxicology: is technical innovation driving scientific innovation? Aquat Toxicol 2011;105:25–39.

[121] Harris CA, Hamilton PB, Runnalls TJ, Vinciotti V, Henshaw A, Hodgson D, et al. The consequences of feminization in breeding groups of wild fish. Environ Health Perspect 2011;119(3):306–11.

[122] Parrott JL, Wood CS, Boutot P, Dunn S. Changes in growth and secondary sex characteristics of Fathead Minnows exposed to bleached sulfite mill effluent. Environ Toxicol Chem 2003;22(12):2908–15.

[123] Scholz S, Mayer I. Molecular biomarkers of endocrine disruption in small model fish. Mol Cell Endocrinol 2008;293:57–70.

[124] Dang Z, Li K, Yin H, Hakkert B, Vermeire T. End point sensitivity in fish endocrine disruption assays: regulatory implications. Toxicol Lett 2011;202(1):36–46.

[125] Blazer VS. Histopathological assessment of gonadal tissue in wild fishes. Fish Physiol Biochem 2002;26(1):85–101.

[126] Leon A, Teh SJ, Hall LC, Teh FC. Androgen disruption of early development in Qurt strain Medaka (*Oryzias latipes*). Aquat Toxicol 2007;82(3):195–203.

[127] Velasco-Santamarìa YM, Bjerregaard P, Korsgaard B. Gonadal alterations in male Eelpout (*Zoarces viviparus*) exposed to ethinylestradiol and trenbolone separately or in combination. Mar Environ Res 2010;69(Suppl. 1):S67–9.

[128] Tetreault GR, Bennett CJ, Shires K, Knight B, Servos MR, McMaster ME. Intersex and reproductive impairment of wild fish exposed to multiple municipal wastewater discharges. Aquat Toxicol 2011;104(3–4):278–90.

[129] Williams RJ, Keller VDJ, Johnson AC, Young AR, Holmes MGR, Wells C, et al. A national risk assessment for intersex in fish arising from steroid estrogens. Environ Toxicol Chem 2009;28:220–30.

[130] Clotfelter ED, Bell AM, Levering KR. The role of animal behavior in the study of endocrine-disrupting chemicals. Anim Behav 2004;68:665–76.

[131] Hellou J. Behavioural ecotoxicology, an "early warning" signal to assess environmental quality. Environ Sci Pollut Res 2011;18(1):1–11.

[132] Brian JV, Augley JJ, Braithwaite VA. Endocrine disrupting effects on the nesting behavior of male Three-spined Stickleback (*Gasterosteus aculeatus* L). J Fish Biol 2006;68:1883–90.

[133] Dzieweczynski TL. Short-term exposure to an endocrine disruptor affects behavioral consistency in male Three-spine Stickleback. Aquat Toxicol 2011;105:681–7.

[134] Garcia-Reyero N l, Lavelle CM, Escalon BL, Martinovic D, Kroll KJ, Sorensen PW, et al. Behavioral and genomic impacts of a wastewater effluent on the Fathead Minnow. Aquat Toxicol 2010;101(1): 38–48.

[135] Lavelle C, Sorensen PW. Behavioral responses of adult male and female Fathead Minnows to a model estrogenic effluent and its effects on exposure regime and reproductive success. Aquat Toxicol 2011;101(3–4):521–8.

[136] Salierno JD, Kane AS. 17 alpha-ethinylestradiol alters reproductive behaviors, circulating hormones, and sexual morphology in male Fathead Minnows (*Pimephales promelas*). Environ Toxicol Chem 2009;28:953–61.

[137] Schoenfuss HL, Bartell SE, Bistodeau TB, Cediel RA, Grove KJ, Zintek L, et al. Impairment of reproductive potential of male Fathead Minnows by environmentally relevant exposures to 4-nonylphenol. Aquat Toxicol 2008;86:91–8.

[138] Coe TS, Hamilton PB, Hodgson D, Paull GC, Stevens JR, Sumner K, et al. An environmental estrogen alters reproductive hierarchies, disrupting sexual selection in group-spawning fish. Environ Sci Technol 2008;42(13):5020–5.

[139] Coe TS, Hamilton PB, Hodgson D, Paull GC, Tyler CR. Parentage outcomes in response to estrogen exposure are modified by social grouping in Zebrafish. Environ Sci Technol 2009;43(21): 8400–5.

[140] Colman JR, Baldwin D, Johnson LL, Scholz NL. Effects of the synthetic estrogen, 17α-ethinylestradiol, on aggression and courtship behavior in male Zebrafish (*Danio rerio*). Aquat Toxicol 2009;91(4):346–54.

[141] Partridge C, Boettcher A, Jones AG. Short-term exposure to a synthetic estrogen disrupts mating dynamics in a Pipefish. Horm Behav 2010;58(5):800–7.

[142] Reyhanian N, Volkova K, Hallgren S, Bollner T, Olsson P-E, Olsen H, et al. 17[alpha]-Ethinyl estradiol affects anxiety and shoaling behavior in adult male Zebrafish (*Danio rerio*). Aquat Toxicol 2011;105(1–2):41–8.

[143] Gurney WSC. Modeling the demographic effects of endocrine disruptors. Environ Health Perspect 2006;114(S1):122–6.

[144] Kidd KA, Blanchfield PJ, Mills KH, P PV, Evans RE, Lazorchak JM, Flick RW. Collapse of a fish population after exposure to a synthetic estrogen. Proceedings of the National Academy of Science 2007;104:8897–901.

[145] Miller DH, Jensen KM, Villeneuve DL, Kahl MD, Makynen EA, Durhan EJ, et al. Linkage of biochemical responses to population-level effects: a case study with vitellogenin in the Fathead Minnow (*Pimephales promelas*). Environ Toxicol Chem 2007;26:521–7.

[146] Ankley GT, Miller DH, Jensen KM, Villeneuve DL, Martinovic D. Relationship of plasma sex steroid concentrations in female fathead minnows to reproductive success and population status. Aquatic Toxicology 2008;88:69–74.

[147] Crago J, Corsi SR, Weber D, Bannerman R, Klaper R. Linking biomarkers to reproductive success of caged Fathead Minnows in streams with increasing urbanization. Chemosphere 2011;82(11):1669–74.

[148] Ankley GT, Bennett RS, Erickson RJ, Hoff DJ, Hornung MW, Johnson RD, et al. Adverse outcome pathways: a conceptual framework to support ecotoxicology research and risk assessment. Environ Toxicol Chem 2010;29(3):730–41.

[149] Hinton DE, Kullman SW, Hardman RC, Volz DC, Chen P-J, Carney M, Bencic DC. Resolving mechanisms of toxicity while pursuing ecotoxicological relevance? Marine Pollution Bulletin 2005;51:635–48.

[150] Caldwell DJ, Mastrocco F, Anderson PD, Länge R, Sumpter JP. Predicted-no-effect concentrations for the steroid estrogens estrone, 17β-estradiol, estriol, and 17α-ethinylestradiol. Environ Toxicol Chem 2012; http://dx.doi.org/10.1002/etc.1825.

[151] Caldwell DJ, Mastrocco F, Hutchinson TH, Lange R, Heijerick D, Janssen C, et al. Derivation of an aquatic predicted no-effect concentration for the synthetic hormone, 17 alpha-ethinyestradiol. Environ Sci Technol 2008;42:7046–54.

[152] U.S. EPA (U.S. Environmental Protection Agency). Draft information collection request (ICR): Tier 1 screening of certain chemicals under the endocrine disruptor screening program (EDSP). Available from: http://www.regulations.gov/search/Regs/home.html#documentDetail?R=0900006480375f4e; 2007 [accessed 27.03.12].

[153] U.S. EPA (Environmental Protection Agency). Final list of initial pesticide active ingredients and pesticide inert ingredients to be screened under the Federal Food, Drug, and Cosmetic Act. Available from: http://www.epa.gov/scipoly/oscpendo/pubs/final_list_frn_041509.pdf; 2009 [accessed 05.03.12].

[154] Villeneuve DL, Garcia-Reyero Nl, Martinovic D, Mueller ND, Cavallin JE, Durhan EJ, Makynen EA, Jensen KM, Kahl MD, Blake LS, Perkins EJ, Ankley GT. I. Effects of a dopamine receptor antagonist on fathead minnow, Pimephales promelas, reproduction. Ecotoxicology and Environmental Safety 2010;73:472–7.

[155] U.S. EPA (Environmental Protection Agency). ToxCast™ Data Sets & Published Research. Available from: http://epa.gov/ncct/toxcast/data_sets.html; 2010b [accessed 04.03.12].

[156] Ankley GT, Bencic DC, Breen MS, Collette TW, Conolly RB, Denslow ND, et al. Endocrine disrupting chemicals in fish: developing exposure indicators and predictive models of effects based on mechanism of action. Aquat Toxicol 2009;92:168–78.

[157] Calow P, Forbes VE. Peer reviewed: does ecotoxicology inform ecological risk assessment? Environ Sci Technol 2003;37(7):146A–51A.

[158] Garcia-Reyero N, Perkins EJ. Systems biology: leading the revolution in ecotoxicology. Environ Toxicol Chem 2011;30(2):265–73.

[159] Bradbury SP, Feijtel TCJ, Leeuwen CJV. Peer reviewed: meeting the scientific needs of ecological risk assessment in a regulatory context. Environ Sci Technol 2004;38(23):463A–70A.

[160] Kruhlak NL, Benz RD, Zhou H, Colatsky TJ. (Q)SAR modeling and safety assessment in regulatory review. Clin Pharmacol Ther 2012;91(3):529–34.

[161] Liu H, Papa E, Gramatica P. QSAR prediction of estrogen activity for a large set of diverse chemicals under the guidance of OECD principles. Chem Res Toxicol 2006;19(11):1540–8.

[162] Tamura H, Ishimoto Y, Fujikawa T, Aoyama H, Yoshikawa H, Akamatsu M. Structural basis for androgen receptor agonists and antagonists: interaction of SPEED 98-listed chemicals and related compounds with the androgen receptor based on an in vitro reporter gene assay and 3D-QSAR. Bioorg Med Chem 2006;14:7160–74.

[163] Yang W, Shen S, Mu L, Yu H. Structure–activity relationship study on the binding of PBDEs with thyroxine transport proteins. Environ Toxicol Chem 2011;30(11):2431–9.

[164] Altenburger R, Nendza M, Schuurmann G. Mixture toxicity and its modeling by quantitative structure-activity relationships. Environ Toxicol Chem 2003;22:1900–15.

[165] Schlenk D. Are steroids really the cause for fish feminization? A mini-review of in vitro and in vivo guided TIES. Mar Pollut Bull 2008;57:250–4.

[166] Guillette Jr LJ. Endocrine disrupting contaminants—beyond the dogma. Environ Health Perspect 2006;114(S1):9–12.

[167] Hotchkiss AK, Rider CV, Bylstone CR, Wilson VS, Hartig PC, Ankley GT, et al. Fifteen years after "Wingspread" – environmental endocrine disruptors and human and wildlife health: where we are today and where we need to go. Toxicol Sci 2008;105:235–59.

[168] Pottinger TG, Cook A, Jurgens MD, Rhodes G, Katsiadaki I, Balaam JL, et al. Effects of sewage effluent remediation on body size, somatic RNA: DNA ratio, and markers of chemical exposure in Three-spined Sticklebacks. Environ Int 2010;37(1):158–69.

[169] Casanova-Nakayama A, Wenger M, Burki R, Eppler E, Krasnov A, Segner H. Endocrine disrupting compounds: can they target the immune system of fish? Mar Pollut Bull 2011; 63(5–12):412–6.

[170] Milla S, Depiereux S, Kestemont P. The effects of estrogenic and androgenic endocrine disruptors on the immune system of fish: a review. Ecotoxicology 2011:1–15.

[171] Carr JA, Patino R. The hypothalamus-pituitary-thyroid axis in teleosts and amphibians: endocrine disruption and its consequences to natural populations. Gen Comp Endocrinol 2011;170(2): 299–312.

[172] Knapp R, Marsh-Matthews E, Vo L, Rosencrans S. Stress hormone masculinizes female morphology and behaviour. Biology Letters 2010;7:150–2.

[173] Zeilinger J, Steger-Hartmann T, Maser E, Goller S, Vonk R, Lange R. Effects of synthetic gestagens on fish reproduction. Environmental Toxicology and Chemistry 2009;28:2663–70.

[174] Kristensen DM, Skalkam ML, Audouze K, Lesne L, Desdoits-Lethimonier C, Frederiksen H, et al. Many putative endocrine disruptors inhibit prostaglandin synthesis. Environ Health Perspect 2011;119(4).

[175] Huang C, Wu S, Zhang X, Chang H, Zhao Y, Giesy JP, et al. Modulation of estrogen synthesis through activation of protein kinase A in H295R cells by extracts of estuary sediments. Environ Toxicol Chem 2011;30(12):2793–801.

[176] Sitruk-Ware R. Pharmacological profile of progestins. Maturitas 2004;47:277–83.

[177] Paulos P, Runnalls TJ, Nallani G, La Point T, Scott AP, Sumpter JP, Huggett DB. Reproductive responses in fathead minnow and Japanese medaka following exposure to a synthetic progestin, Norethindrone. Aquatic Toxicology 2010;99:256–62.

[178] DeQuattro ZA, Peissig EJ, Antkiewicz DS, Lundgren EJ, Hedman CJ, Hemming JDC, et al. Effects of progesterone on reproduction and embryonic development in the Fathead Minnow (Pimephales promelas). Environmental Toxicology and Chemistry 2012;31:851–6.

[179] Brar NK, Waggoner C, Reyes JA, Fairey R, Kelley KM. Evidence for thyroid endocrine disruption in wild fish in San Francisco Bay, California, USA. Relationships to contaminant exposures. Aquat Toxicol 2009;96(3):203–15.

[180] Legler J. New insights into the endocrine disrupting effects of brominated flame retardants. Chemosphere 2008;73:216–22.

[181] Sheng Z-G, Tang Y, Liu Y-X, Yuan Y, Zhao B-Q, Chao X-J, et al. Low concentrations of bisphenol A suppress thyroid hormone receptor transcription through a nongenomic mechanism. Toxicol Appl Pharmacol 2012;259(1):133–42.

[182] Yu L, Deng J, Shi X, Liu C, Yu K, Zhou B. Exposure to DE-71 alters thyroid hormone levels and gene transcription in the hypothalamic-pituitary-thyroid axis of Zebrafish larvae. Aquat Toxicol 2010;97(3):226–33.

[183] Boas M, Feldt-Rasmussen U, Skakkebaek NE, Main KM. Environmental chemicals and thyroid function. European Journal of Endocrinology 2006;154:599–611.

[184] Noyes PD, Hinton DE, Stapleton HM. Accumulation and debromination of decabromodiphenyl ether (BDE-209) in juvenile Fathead Minnows (*Pimephales promelas*) induces thyroid disruption and liver alterations. Toxicol Sci 2011;122(2):265–74.

[185] Habibi HR, Nelson ER, Allan ERO. New insights into thyroid hormone function and modulation of reproduction in Goldfish. Gen Comp Endocrinol 2012;175(1):19–26.

[186] Vilhais-Neto GC, Pourquie O. Retinoic acid. Curr Biol 2008;18:R191–2.

[187] Arukwe A, Nordbo B. Hepatic biotransformation responses in Atlantic salmon exposed to retinoic acids and 3,3,4,4-tetrachlorobiphenyl (PCB conger 77). Comparative Biochemistry and Physiology, Part C 2008;147:470–82.

[188] Hakansson H, Esteban J, Halldin K, Herlin M, Heimeier R, Thornqvist PO. Retinoid receptors and metabolism in endocrine disruption. Toxicol Lett 2010;196(Suppl. 1):S12.

[189] Allinson M, Shiraishi F, Salzman S, Allinson G. In vitro assessment of retinoic acid and aryl hydrocarbon receptor activity of treated effluent from 39 wastewater-treatment plants in Victoria, Australia. Arch Environ Contam Toxicol 2011;61(4):539–46.

[190] Inoue D, Nakama K, Sawada K, Watanabe T, Takagi M, Sei K, et al. Contamination with retinoic acid receptor agonists in two rivers in the Kinki region of Japan. Water Res 2010;44(8): 2409–18.

[191] Sawada K, Inoue D, Wada Y, Sei K, Nakanishi T, Ike M. Detection of retinoic acid receptor agonistic activity and identification of causative compounds in municipal wastewater treatment plants in Japan. Environ Toxicol Chem 2011;31(2):307–15.

[192] Zhen H, Wu X, Hu J, Xiao Y, Yang M, Hirotsuji J, et al. Identification of retinoic acid receptor agonists in sewage treatment plants. Environ Sci Technol 2009;43(17):6611–6.

[193] le Maire A, Grimaldi M, Roecklin D, Dagnino S, Vivat-Hannah V, Balaguer P, et al. Activation of RXR-PPAR heterodimers by organotin environmental endocrine disruptors. EMBO Rep 2009;10(4): 367–73.

[194] Bhandari K, Venables B. Ibuprofen bioconcentration and prostaglandin E2 levels in the Bluntnose Minnow (Pimephales notatus). Comp Biochem Physiol C Toxicol Pharmacol 2010;153(2):251–7.

[195] Appelt CW, Sorensen PW. Female goldfish signal spawning readiness by altering when and where they release a urinary pheromone. Anim Behav 2007;74:1329–38.

[196] Moore A, Lower N. The impact of two pesticides on olfactory-mediated endocrine function in mature male Atlantic Salmon (*Salmo salar* L.) parr. Comp Biochem Physiol B 2001;129: 269–76.

[197] Fujimori C, Ogiwara K, Hagiwara A, Rajapakse S, Kimura A, Takahashi T. Expression of cyclooxygenase-2 and prostaglandin receptor EP4b mRNA in the ovary of the medaka fish, (*Oryzias latipes*): possible involvement in ovulation. Mol Cell Endocrinol 2011;332(1–2): 67–77.

[198] Hattori RS, Fernandino JI, Kishii A, Kimura H, Kinno T, Oura M, et al. Cortisol-induced masculinization: does thermal stress affect gonadal fate in Pejerrey, a teleost fish with temperature-dependent sex determination? PLoS One 2009;4(8):e6548.

[199] Milla S, Wang N, Mandiki SNM, Kestemont P. Corticosteroids: friends or foes of teleost fish reproduction? Comp Biochem Physiol A Mol Integr Physiol 2009;153(3):242–51.

[200] Yamaguchi T, Yoshinaga N, Yazawa T, Gen K, Kitano T. Cortisol is involved in temperature-dependent sex determination in the Japanese Flounder. Endocrinology 2010;151(8):3900–8.

[201] LaLone CA, Villeneuve DL, Olmstead AW, Medlock EK, Kahl MD, Jensen KM, et al. Effects of a glucocorticoid receptor agonist, dexamethasone, on Fathead Minnow reproduction, growth, and development. Environ Toxicol Chem 2012;31(3):611–22.

[202] Wassmur B, Grans J, Kling P, Celander MC. Interactions of pharmaceuticals and other xenobiotics on hepatic pregnane X receptor and cytochrome P450 3A signaling pathway in Rainbow Trout (*Oncorhynchus mykiss*). Aquat Toxicol 2010;100(1):91–100.

[203] Kugathas S, Sumpter JP. Synthetic glucocorticoids in the environment: first results on their potential impacts on fish. Environ Sci Technol 2011;45(6):2377–83.

[204] Rüegg J, Penttinen-Damdimopoulou P, Mäkelä S, Pongratz I, Gustafsson J-Å, Luch A. Receptors mediating toxicity and their involvement in endocrine disruption. In: Luch A, editor. Molecular, clinical and environmental toxicology, vol. 1. Birkhäuser Basel; 2009. p. 289–323.

[205] Sui Y, Ai N, Park S-H, Rios-Pilier J, Perkins JT, Welsh WJ, et al. Bisphenol A and its analogues activate human pregnane X receptor. Environ Health Perspect 2012;120(3):399–405.

[206] Wolf CR, Elcombe C, Rode A, Scheer N, Bower C, Henderson C. Role of CAR and PXR in xenobiotic toxicity. Toxicol Lett 2011;205(Suppl. 1):S13.

[207] Billiard SM, Timme-Laragy AR, Wassenberg DM, Cockman C, Di Giulio RT. The role of the aryl hydrocarbon receptor pathway in mediating synergistic developmental toxicity of polycyclic aromatic hydrocarbons to Zebrafish. Toxicol Sci 2006;92:526–36.

[208] Hahn ME, Karchner SI, Evans BR, Franks DG, Merson RR, Lapseritis JM. Unexpected diversity of aryl hydrocarbon receptors in non-mammalian vertebrates: insights from comparative genomics. J Exp Zool 2006;305A:693–706.

[209] Pascussi J-M, Gerbal-Chaloin S, Duret C, Daujat-Chavanieu M, Vilarem M-J, Maurel P. The tangle of nuclear receptors that controls xenobiotic metabolism and transport: cross talk and consequences. Annu Rev Pharmacol Toxicol 2008;48(1):1–32.

[210] Burnard D, Gozlan RE, Griffiths SW. The role of pheromones in freshwater fishes. J Fish Biol 2008;73:1–16.

[211] Lema SC. Identification of multiple vasotocin receptor cDNAs in teleost fish: sequences, phylogentic analysis, sites of expression, and regulation in the hypothalamus and gill in response to hyperosmotic challenge. Mol Cell Endocrinol 2010;312:215–30.

[212] Lord L-D, Bond J, Thompson RR. Rapid steroid influences on visually guided sexual behavior in male Goldfish. Horm Behav 2009;56(5):519–26.

[213] Semsar K, Perreault HAN, Godwin JR. Fluoxetine-treated male Wrasses exhibit low AVT expression. Brain Res 2004;1029:141–7.

[214] Forlano PM, Bass AH. Neural and hormonal mechanisms of reproductive-related arousal in fishes. Horm Behav 2011;59(5):616–29.

[215] Mennigen JA, Stroud P, Zamora JM, Moon TW, Trudeau VL. Pharmaceuticals as neuroendocrine disruptors: lessons learned from fish on Prozac. J Toxicol Environ Health B 2011;14(5–7): 387–412.

[216] Mennigen JA, Martyniuk CJ, Crump K, Xiong H, Zhao E, Popesku J, et al. Effects of fluoxetine on the reproductive axis of female Goldfish (*Carassius auratus*). Physiol Genom 2008;35(3):273–82.

[217] Painter MM, Buerkley MA, Julius ML, Vajda AM, Norris DO, Barber LB, et al. Antidepressants at environmentally relevant concentrations affect predator avoidance behavior of larval Fathead Minnows (*Pimephales promelas*). Environ Toxicol Chem 2009;28(12):2677–84.

[218] Gaworecki KM, Klaine SJ. Behavioral and biochemical responses of hybrid striped Bass during and after fluoxetine exposure. Aquat Toxicol 2008;88(4):207–13.

[219] Popesku JT, Martyniuk CJ, Mennigen J, Xiong H, Zhang D, Xia X, et al. The Goldfish (*Carassius auratus*) as a model for neuroendocrine signaling. Mol Cell Endocrinol 2008;293(1–2):43–56.

[220] Rahman MS, Khan IA, Thomas P. Tryptophan hydroxylase: a target for neuroendocrine disruption. J Toxicol Environ Health B 2011;14(5–7):473–94.

[221] Iwata E, Nagai Y, Sasaki H. Social rank modulates brain arginine vasotocin immunoreactivity in false Clown Anemonefish (*Amphiprion ocellaris*). Fish Physiol Biochem 2010;36(3):337–45.

[222] Tena-Sempere M, Felip A, Gomez A, Zanuy S, Carrillo M. Comparative insights of the kisspeptin/kisspeptin receptor system: lessons from non-mammalian vertebrates. Gen Comp Endocrinol 2011;175(2):234–43.

[223] Navarro VM, Sanchez-Garrido MA, Castellano JM, Roa J, Garcia-Galiano D, Pineda R, et al. Persistent impairment of hypothalamic KiSS-1 system after exposures to estrogenic compounds at critical periods of brain sex differentiation. Endocrinology 2009;150(5):2359–67.

[224] Chen C-C, Fernald RD. GnRH and GnRH receptors: distribution, function and evolution. J Fish Biol 2008;73:1099–120.

[225] Cottone E, Guastalla A, Mackie K, Franzoni MF. Endocannabinoids affect the reproductive functions in teleosts and amphibians. Mol Cell Endocrinol 2008;286S:S41–5.

[226] Pomatto V, Palermo F, Mosconi G, Cottone E, Cocci P, Nabissi M, et al. Xenoestrogens elicit a modulation of endocannabinoid system and estrogen receptors in 4NP treated Goldfish, (*Carassius auratus*). Gen Comp Endocrinol 2011;174(1):30–5.

[227] Chen W, Wang Y, Li W, Lin H. Insulin-like growth factor binding protein-2 (IGFBP-2) in Orange-spotted Grouper, (*Epinephelus coioides*): molecular characterization, expression profiles and regulation by 17[beta]-estradiol in ovary. Comp Biochem Physiol B Biochem Mol Biol 2010;157(4):336–42.

[228] Kling P, Jonsson E, Nilsen TO, Einarsdottir IE, Ronnestad I, Stefansson SO, et al. The role of growth hormone in growth, lipid homeostasis, energy utilization and partitioning in Rainbow Trout: interactions with leptin, ghrelin and insulin-like growth factor I. Gen Comp Endocrinol 2012;175(1):153–62.

[229] LeBlanc G, Kullman SW, Norris DO, Baldwin WS, Kloas W, Greally JM. Draft detailed review paper: state of the science on novel in vitro and in vivo screening and testing methods and endpoints for evaluating endocrine disruptors. 2011.

[230] Norbeck LA, Sheridan MA. An in vitro model for evaluating peripheral regulation of growth in fish: effects of 17[beta]-estradiol and testosterone on the expression of growth hormone receptors, insulin-like growth factors, and insulin-like growth factor type 1 receptors in Rainbow Trout (Oncorhynchus mykiss). Gen Comp Endocrinol 2011;173(2):270–80.

[231] Arsenault JTM, Fairchild WL, MacLatchy DL, Burridge L, Haya K, Brown SB. Effects of water-borne 4-nonylphenol and 17 beta-estradiol exposure during parr-smolt transformation on growth and plasma IGF-I of Atlantic Salmon (*Salmo salar* L.). Aquat Toxicol 2004;66:255–65.

[232] McCormick SD, O'Dea MF, Moeckel AM, Lerner DT, Bjornsson BT. Endocrine disruption of parr-smolt transformation and seawater tolerance of Atlantic Salmon by 4-nonylphenol and 17 beta-estradiol. Gen Comp Endocrinol 2005;142:280–8.

[233] Shved N, Berishvili G, Baroiller J-F, Segner H, Reinecke M. Environmentally relevant concentrations of 17alpha-ethinylestradiol (EE2) interfere with the growth hormone (GH)/insulin-like growth factor (IGF)-1 system in developing bony fish. Toxicol Sci 2010;106(1):93–102.

[234] Aksakal EM, Ceyhun SB, Erdogan O, Ekinci D. Acute and long-term genotoxicity of deltamethrin to insulin-like growth factors and growth hormone in Rainbow Trout. Comp Biochem Physiol C Toxicol Pharmacol 2010;152(4):451–5.

[235] Straub RH. The complex role of estrogens in inflammation. Endocr Rev 2007;28(5):521–74.

[236] Verthelyi D. Sex hormones as immunomodulators in health and disease. Int Immunopharmacol 2001;1:983–93.

[237] Aoki T, Takano T, Santos MD, Kondo H, Hirono I. Molecular innate immunity in teleost fish: review and future perspectives. In: Tsukamoto K, Kawamura T, Takeuchi T, BeardJr. TD,

Kaiser MD, editors. Fisheries for global welfare and environment, 5th World Fisheries Congress. Tokyo: TERRAPUB 2008; 2008. p. 263–76.

[238] Plouffe DA, Hanington PC, Walsh JG, Wilson EC, Belosevic M. Comparison of select innate immune mechanisms of fish and mammals. Xenotransplantation 2005;12(4):266–77.

[239] Lieschke GJ, Trede NS. Fish immunology. Curr Biol 2009;19(16):R678–82.

[240] Weinberg ED. Pregnancy-associated depression of cell-mediated immunity. Rev Infect Dis 1984;6(6):814–31.

[241] Watanuki H, Yamaguchi T, Sakai M. Suppression in function of phagocytic cells in Common Carp (*Cyprinus carpio L*). injected with estradiol, progesterone or 11-ketotestosterone. Comp Biochem Physiol C Toxicol Pharmacol 2002;132(4):407–13.

[242] Shelley LK, Ross PS, Kennedy CJ. The effects of an in vitro exposure to 17 beta-estradiol and nonylphenol on Rainbow Trout (*Oncorhynchus mykiss*) peripheral blood leukocytes. Comp Biochem Physiol C Toxicol Pharmacol 2011;155(3):440–6.

[243] Liney KE, Hagger JA, Tyler CR, Depledge MH, Galloway TS, Jobling S. Health effects in fish of long-term exposure to effluents from wastewater treatment works. Environ Health Perspect 2006;114(S1):81–9.

[244] Riggs AD, Martienssen RA, Russo VEA. Introduction. In: Russo VEA, Martienssen RA, Riggs AD, editors. Epigenetics. Cold Spring Harbor, New York: Spring Harbor Laboratory Press; 1996. p. 1–4.

[245] Seong K-H, Li D, Shimizu H, Nakamura R, Ishii S. Inheritance of stress-induced, ATF-2-dependent epigenetic change. Cell 2011;145(7):1049–61.

[246] Head JA, Dolinoy DC, Basu N. Epigenetics for ecotoxicologists. Environ Toxicol Chem 2012;31(2):221–7.

[247] Lema C, Cunningham MJ. MicroRNAs and their implications in toxicological research. Toxicol Lett 2010;198(2):100–5.

[248] Vandegehuchte M, Janssen C. Epigenetics and its implications for ecotoxicology. Ecotoxicology 2011;20(3):607–24.

[249] Newbold RR, Padilla-Banks E, Jefferson WN. Adverse Effects of the Model Environmental Estrogen Diethylstilbestrol Are Transmitted to Subsequent Generations. Endocrinology 2006;147:s11–7.

[250] Aniagu SO, Williams TD, Allen Y, Katsiadaki I, Chipman JK. Global genomic methylation levels in the liver and gonads of the three-spine stickleback (Gasterosteus aculeatus) after exposure to hexabromocyclododecane and 17-Œ= oestradiol. Environment International 2008; 34:310–7.

[251] Stromqvist M, Tooke N, Brunstrom B. DNA methylation levels in the 5' flanking region of the vitellogenin I gene in liver and brain of adult Zebrafish (*Danio rerio*)—sex and tissue differences and effects of 17 alpha-ethinylestradiol exposure. Aquat Toxicol 2010;98(3):275–81.

[252] Kortenkamp A. Low dose mixture effects of endocrine disrupters: implications for risk assessment and epidemiology. Int J Androl 2008;31(2):233–40.

[253] Rajapakse N, Silva E, Kortenkamp A. Combining xenoestrogens at levels below individual no-observed-effect concentrations dramatically enhances steroid hormone action. Environ Health Perspect 2002;110:917–21.

[254] Silva E, Rajapakse N, Scholze M, Backhaus T, Ermler S, Kortenkamp A. Joint Effects of Heterogeneous Estrogenic Chemicals in the E-Screen: Exploring the Applicability of Concentration Addition. Toxicological Sciences 2011;122:383–94.

[255] Brian JV, Harris CA, Scholze M, Backhaus T, Booy P, Lamoree MH, et al. Accurate prediction of the response of a freshwater fish to a mixture of estrogenic chemicals. Environ Health Perspect 2005;113:721–8.

[256] Correia AD, Freitas S, Scholze M, Goncalves JF, Booij P, Lamoree MH, et al. Mixtures of estrogenic chemicals enhance vitellogenic response in Sea Bass. Environ Health Perspect 2007; 115(S-1).

[257] Thorpe KL, Cummings RI, Hutchinson TH, Scholze M, Brighty G, Sumpter JP, et al. Relative potencies and combination effects of steroidal estrogens in fish. Environ Sci Technol 2003;37: 1142–9.

[258] Charles GD, Gennings C, Zacharewski TR, Gollapudi BB, Carney EW. Assessment of interactions of diverse ternary mixtures in an estrogen receptor alpha reporter assay. Toxicol Appl Pharmacol 2002;180(1):11–21.

[259] Charles GD, Gennings C, Tornesi B, Kan HL, Zacharewski TR, Bhaskar Gollapudi B, et al. Analysis of the interaction of phytoestrogens and synthetic chemicals: an in vitro/in vivo comparison. Toxicol Appl Pharmacol 2007;218(3):280–8.

[260] Chen J, Ahn KC, Gee NA, Ahmed MI, Duleba AJ, Zhao L, et al. Triclocarban enhances testosterone action: a new type of endocrine disruptor? Endocrinology 2008;149:1173–9.

[261] Spurgeon DJ, Jones OAH, Dorne J-LCM, Svendsen C, Swain S, Sturzenbaum SR. Systems toxicology approaches for understanding the joint effects of environmental chemical mixtures. Science of The Total Environment 2010;408:3725–34.

[262] Santos MM, Micael J, Carvalho AP, Morabito R, Booy P, Massanisso P, et al. Estrogens counteract the masculinizing effect of tributyltin in Zebrafish. Comp Biochem Physiol C Toxicol Pharmacol 2006;142(1–12):151–5.

[263] Sarria MP, Santos MM, Reis-Henriques MA, Vieira NM, Monteiro NM. The unpredictable effects of mixtures of androgenic and estrogenic chemicals on fish early life. Environ Int 2011;37(2): 418–24.

[264] Sun L, Zha J, Wang Z. Effects of binary mixtures of estrogen and antiestrogens on Japanese Medaka (*Oryzias latipes*). Aquat Toxicol 2009;93(1):83–9.

[265] Rempel MA, Reyes J, Steinert S, Hwang W, Armstrong J, Sakamoto K, et al. Evaluation of relationships between reproductive metrics, gender and vitellogenin expression in demersal Flatfish collected near the municipal wastewater outfall of Orange County, California, USA. Aquat Toxicol 2006;77(3):241–9.

[266] Sanchez W, Sremski W, Piccini B, Palluel O, Maillot-Marechal E, Betoulle S, et al. Adverse effects in wild fish living downstream from pharmaceutical manufacture discharges. Environ Int 2011;37(8): 1342–8.

[267] Bellet V, Hernandez-Raquet G, Dagnino S, Seree L, Pardon P, Bancon-Montiny C, et al. Occurrence of androgens in sewage treatment plants influents is associated with antagonist activities on other steroid receptors. Water Res 2012;46(6):1912–22. Research 2012, 46, (6), 1912–1922.

[268] Ankley G, Johnson RD. Small fish models for identifying and assessing the effects of endocrine-disrupting chemicals. ILAR 2004;45(4):469–83.

[269] Scholz S, Mayer I. Molecular biomarkers of endocrine disruption in small model fish. Mol Cell Endocrinol 2008;293:57–70.

[270] Segner H. Zebrafish (Danio rerio) as a model organism for investigating endocrine disruption. Comparative Biochemistry and Physiology, Part C 2009;149:187–95.

[271] Thorpe KL, Marca Pereira ML, Schiffer H, Burkhardt-Holm P, Weber K, Wheeler JR. Mode of sexual differentiation and its influence on the relative sensitivity of the Fathead Minnow and Zebrafish in the fish sexual development test. Aquat Toxicol 2011;105:412–20.

[272] Holbech H, Kinnberg K, Petersen GI, Jackson P, Hylland K, Norrgren L, Bjerregaard P. Detection of endocrine disrupters: Evaluation of a Fish Sexual Development Test (FSDT). Comparative Biochemistry and Physiology Part C: Toxicology & Pharmacology 2006;144:57–66.

[273] Devlin RH, Nagahama Y. Sex determination and sex differentiation in fish: an overview of genetic, physiological, and environmental influences. Aquaculture 2002;208:191–364.

[274] Strüssmann CA, Conover DO, Somoza GM, Miranda LA. Implications of climate change for the reproductive capacity and survival of new world Silversides (family *Atherinopsidae*). J Fish Biol 2010;77(8):1818–34.

[275] Peters REM, Courtenay SC, Hewitt LM, MacLatchy DL. Effects of 17[alpha]-ethynylestradiol on early-life development, sex differentiation and vitellogenin induction in Mummichog (*Fundulus heteroclitus*). Mar Environ Res 2009;69(3):178–86.

[276] Brown AR, Hosken DJ, Balloux F, Bickley LK, LePage G, Owen SF, et al. Genetic variation, inbreeding and chemical exposure – combined effects in wildlife and critical considerations for ecotoxicology. Philos Trans R Soc B Biol Sci 2009;364(1534):3377–90.

[277] Whitehead A. Comparative genomics in ecological physiology: toward a more nuanced understanding of acclimation and adaptation. J Exp Biol 2012;215(6):884–91.

[278] Greytak SR, Tarrant AM, Nacci D, Hahn ME, Callard GV. Estrogen responses in Killifish (*Fundulus heteroclitus*) from polluted and unpolluted environments are site- and gene-specific. Aquat Toxicol 2010;99(2):291–9.

[279] Nacci DE, Champlin D, Coiro L, McKinney R, Jayaraman S. Predicting the occurrence of genetic adaptation to dioxinlike compounds in populations of the estuarine fish *Fundulus heteroclitus*. Environ Toxicol Chem 2002;21:1525–32.

[280] Nacci D, Champlin D, Jayaraman S. Adaptation of the estuarine fish *Fundulus heteroclitus* (Atlantic Killifish) to polychlorinated biphenyls (PCBs). Estuaries and Coasts 2010;33(4):853–64.

[281] Nacci D, Huber M, Champlin D, Jayaraman S, Cohen S, Gauger E, et al. Evolution of tolerance to PCBs and susceptibility to a bacterial pathogen (*Vibrio harveyi*) in Atlantic Killifish (*Fundulus heteroclitus*) from New Bedford (MA, USA) harbor. Environ Pollut 2009;157: 857–64.

[282] Whitehead A, Triant DA, Champlin D, Nacci D. Comparative transcriptomics implicates mechanisms of evolved pollution tolerance in a Killifish population. Mol Ecol 2010;19(23):5186–203.

[283] Schlueter MA, Guttman SI, Oris JT, Bailer AJ. Differential survival of Fathead Minnows, (*Pimephales promelas*), as affected by copper exposure, prior population stress, and allozyme genotypes. Environ Toxicol Chem 1997;16(5):939–47.

[284] Theodorakis CW. Integration of genotoxic and population genetic endpoints in biomonitoring and risk assessment. Ecotoxicology 2001;10(4):245–56.

[285] Guinand B, Durieux EDH, Dupuy C, Cerqueira F, Begout M-L. Phenotypic and genetic differentiation in young-of-the-year common sole (Solea solea) at differentially contaminated nursery grounds. Marine Environmental Research 2011;71:195–206.

[286] Meyer JN, Di Giulio RT. Heritable adaptation and fitness costs in Killifish (*Fundulus heteroclitus*) inhabiting a polluted estuary. Ecol Appl 2003;13:490–503.

[287] Weisbrod AV, Sahi J, Segner H, James MO, Nichols J, Schultz I, et al. The state of in vitro science for use in bioaccumulation assessments for fish. Environ Toxicol Chem 2009;28(1):86–96.

[288] Akgul Y, Derk RC, Meighan T, Rao KMK, Murono EP. The methoxychlor metabolite, HPTE, inhibits rat luteal cell progesterone production. Reprod Toxicol 2011;32(1):77–84.

[289] James MO, Stuchal LD, Nyagode BA. Glucuronidation and sulfonation, in vitro, of the major endocrine-active metabolites of methoxychlor in the Channel Catfish, (*Ictalurus punctatus*), and induction following treatment with 3-methylcholanthrene. Aquat Toxicol 2008;86(2):227–38.

[290] Martyniuk CJ, Spade DJ, Blum JL, Kroll KJ, Denslow ND. Methoxychlor affects multiple hormone signaling pathways in the Largemouth Bass (*Micropterus salmoides*) liver. Aquat Toxicol 2011; 101(3–4):483–92.

[291] McCarthy AR, Thomson BM, Shaw IC, Abell AD. Estrogenicity of pyrethroid metabolites. J Environ Monitor 2006;8:197–202.

[292] McCormick JM, Es TV, Cooper KR, White LA, Haaggblom MM. Microbially mediated O-methylation of bisphenol A: results in metabolites with increased toxicity to the developing Zebrafish (*Danio rerio*) embryo. Environ Sci Technol 2011;45(15):6567–74.

[293] Mercado-Feliciano M, Bigsby RM. Hydroxylated metabolites of the polybrominated diphenyl ether mixture DE-71 are weak estrogen receptor-alpha ligands. Environ Health Perspect 2008;116:1315–21.

[294] Taxvig C, Olesen PT, Nellemann C. Use of external metabolizing systems when testing for endocrine disruption in the T-screen assay. Toxicol Appl Pharmacol 2011;250(3):263–9.

[295] Jin Y, Chen R, Sun L, Liu W, Fu Z. Photoperiod and temperature influence endocrine disruptive chemical-mediated effects in male adult Zebrafish. Aquat Toxicol 2009;92(1):38–43.

[296] Jin Y, Shu L, Huang F, Cao L, Sun L, Fu Z. Environmental cues influence EDC-mediated endocrine disruption effects in different developmental stages of Japanese Medaka (*Oryzias latipes*). Aquat Toxicol 2011;101(1):254–60.

[297] Thomas P, Rahman MS. Extensive reproductive disruption, ovarian masculinization and aromatase suppression in Atlantic Croaker in the Northern Gulf of Mexico hypoxic zone. Proc R Soc B: Biol Sci 2011.

[298] Wu RSS, Richards Jeffrey G, Farrell AP, Brauner CJ. Chapter 3 effects of hypoxia on fish reproduction and development. In: Fish physiology, vol. 27. Academic Press; 2009;79–141.

[299] Rose KA, Adamack AT, Murphy CA, Sable SE, Kolesar SE, Craig JK, et al. Does hypoxia have population-level effects on coastal fish? Musings from the virtual world. J Exp Mar Biol Ecol 2009; 381(Suppl.):S188–203.

[300] Diaz RJ, Rosenberg R. Spreading dead zones and consequences for marine ecosystems. Science 2008;321(5891):926–9.

[301] Alworth LC, Howdeshell KL, Ruhlen RL, Day JK, Lubahn DB, Huang TH-M, et al. Uterine responsiveness to estradiol and DNA methylation are altered by fetal exposure to diethylstilbestrol and methoxychlor in CD-1 mice: effects of low versus high doses. Toxicol Appl Pharmacol 2002; 183:10–22.

[302] Calabrese EJ. Estrogen and related compounds: biphasic dose responses. Crit Rev Toxicol 2001; 31(4–5):503–15.

[303] Welshons WV, Nagel SC, vom Saal FS. Large effects from small exposures. III. Endocrine mechanisms mediating effects of bisphenol A at levels of human exposure. Endocrinology (Suppl. 6):S56–69, http://dx.doi.org/10.1210/en.2005-1159, 2006;147.

[304] Welshons WV, Thayer KA, Judy BM, Taylor JA, Curran EM, vom Saal FS. Large effects from small exposures. I. Mechanisms for endocrine-disrupting chemicals with estrogenic activity. Environ Health Perspect 2003;111(8):994–1006.

[305] Li L, Andersen ME, Heber S, Zhang Q. Non-monotonic dose-response relationship in steroid hormone receptor-mediated gene expression. Journal of Molecular Endocrinology 2007;38:569–85.

[306] Knacker T, Boettcher M, Frische T, Rufli H, Stolzenberg H-C, Teigeler M, et al. Environmental effect assessment for sexual endocrine-disrupting chemicals: fish testing strategy. Integr Environ Assess Manage 2010;6(4):653–62.

[307] Hutchinson TH, Pickford DB. Ecological risk assessment and testing for endocrine disruption in the aquatic environment. Toxicology 2002;181–2:383–7.

[308] Bonnineau C, Moeller A, Barata C, Bonet B, Proia L, Sans-Piché F, et al. Advances in the multi-biomarker approach for risk assessment in aquatic ecosystems. In: Emerging and priority pollutants in rivers, vol. 19. Berlin/Heidelberg: Springer; 2012. p. 147–179.

[309] Hutchinson TH, Ankley GT, Segner H, Tyler CR. Screening and testing for endocrine disruption in fish – biomarkers as "signposts," not "traffic lights," in risk assessment. Environ Health Perspect 2006;114(S1):106–14.

[310] Van Leeuwen CJ, Hermes JLM. Risk assessment of chemicals: an introduction. Norwell, MA: Kluwer Academic Publishers; 1995;374.

[311] Godwin J. Neuroendocrinology of sexual plasticity in teleost fishes. Front Neuroendocrinol 2010; 31(2):203–16.

[312] Forbes VE, Calow P, Sibly RM. The extrapolation problem and how population modeling can help. Environ Toxicol Chem 2008;27:1987–94.

[313] Celander MC, Goldstone JV, Denslow ND, Iguchi T, Kille P, Meyerhoff RD, et al. Species extrapolation for the 21st century. Environ Toxicol Chem 2010;30(1):52–63.

[314] Burton GA, De Zwart D, Diamond J, Dyer S, Kapo KE, Liess M, et al. Making ecosystem reality checks the status quo. Environ Toxicol Chem 2012;31(3):459–68.

[315] de Vries P, Smit MGD, van Dalfsen JA, De Laender F, Karman CC. Consequences of stressor-induced changes in species assemblage for biodiversity indicators. Environ Toxicol Chem 2010;29(8):1868–76.

[316] Filby AL, Paull GC, Searle F, Ortiz-Zarragoitia M, Tyler CR. Environmental estrogen-induced alterations of male aggression and dominance hierarchies in fish: a mechanistic analysis. Environ Sci Technol 2012;46(6):3472–9.

[317] Fernandes D, Schnell S, Porte C. Can pharmaceuticals interfere with the synthesis of active androgens in male fish? An in vitro study. Mar Pollut Bull 2011;62(10):2250–3.

[318] Schultz MM, Painter MM, Bartell SE, Logue A, Furlong ET, Werner SL, et al. Selective uptake and biological consequences of environmentally relevant antidepressant pharmaceutical exposures on male Fathead Minnows. Aquat Toxicol 2011;104(1–2):38–47.

[319] Han S, Choi K, Kim J, Ji K, Kim S, Ahn B, et al. Endocrine disruption and consequences of chronic exposure to ibuprofen in Japanese Medaka (*Oryzias latipes*) and freshwater Cladocerans *Daphnia magna* and *Moina macrocopa*. Aquat Toxicol 2010;98(3):256–64.

[320] Chung E, Genco MC, Megrelis L, Ruderman JV. Effects of bisphenol A and triclocarban on brain-specific expression of aromatase in early Zebrafish embryos. Proc Nat Acad Sci 2011;108(43): 17732–7.

[321] Vandenberg LN, Maffini MV, Sonnenschein C, Rubin BS, Soto AM. Bisphenol-A and the great divide: a review of controversies in the field of endocrine disruption. Endocr Rev 2009;30:75–95.

[322] Le Page Y, Vosges M l, Servili A, Brion F, Kah O. Neuroendocrine effects of endocrine disruptors in teleost fish. J Toxicol Environ Health B 2011;14(5–7):370–86.

[323] Xia J, Niu C, Pei X. Effects of chronic exposure to nonylphenol on locomotor activity and social behavior in Zebrafish (*Danio rerio*). J Environ Sci 2010;22(9):1435–40.

[324] Genovese G, Da CuÒa R, Towle DW, Maggese MC, Lo Nostro F. Early expression of zona pellucida proteins under octylphenol exposure in *Cichlasoma dimerus* (*Perciformes, Cichlidae*). Aquat Toxicol 2010;101(1):175–85.

[325] Lema SC, R SI, Scholz NL, Incardona JP, Swanson P. Neural defects and cardiac arrhythmia in fish larvae following embryonic exposure to 2,2',4,4'-tetrabromodiphenyl ether (PBDE 47). Aquat Toxicol 2007;82:296–307.

[326] Morthorst JE, Holbech H, Bjerregaard P. Trenbolone causes irreversible masculinization of Zebrafish at environmentally relevant concentrations. Aquat Toxicol 2010;98(4):336–43.

[327] Skolness SY, Durhan EJ, Garcia-Reyero N, Jensen KM, Kahl MD, Makynen EA, et al. Effects of a short-term exposure to the fungicide prochloraz on endocrine function and gene expression in female Fathead Minnows (*Pimephales promelas*). Aquat Toxicol 2011;103(3–4):170–8.

[328] Martinovic D, Blake LS, Durhan EJ, Greene KJ, Kahl MD, Jensen KM, et al. Reproductive toxicity of vinclozolin in the Fathead Minnow: confirming an anti-androgenic mode of action. Environ Toxicol Chem 2008;27:478–88.

[329] Baudo R, Muntau H. Lesser known in-place pollutants and diffuse source problems. In: Baudo R, Giesy J, Muntau H, editors. Sediments chemistry and toxicity of in-place pollutants. Boston, MA: Lewis Publishers, Inc.; 1990. p. 405.

[330] Brink K, Jansen van Vuren J, Bornman R. The lack of endocrine disrupting effects in Catfish (*Clarias gariepinus*) from a DDT sprayed area. Ecotoxicol Environ Saf 2012;79(0):256–63.

[331] Chang H, Hu J, Shao B. Occurrence of natural and synthetic glucocorticoids in sewage treatment plants and receiving river waters. Environ Sci Technol 2007;41(10):3462–8.

[332] Guguen-Guillouzo C, Guillouzo A, Maurel P. General review on in vitro hepatocyte models and their applications. In: Hepatocytes, vol. 640. Humana Press; 2010. p. 1–40.

[333] Jobling S, Casey D, Rodgers-Gray T, Oehlmann J, Schulte-Oehlmann U. Pawlowski, comparative responses of molluscs and fish to environmental estrogens and estrogenic effluent. Aquat Toxicol 2003;65:205–20.

[334] Tena-Sempere M, Felip A, Gomez A, Zanuy S, Carrillo M. Comparative insights of the kisspeptin/kisspeptin receptor system: lessons from non-mammalian vertebrates. Gen Comp Endocrinol 2011;175(2):234–43.

[335] Thomas P, Dressing G, Pang Y, Berg H, Tubbs C, Benninghoff A, et al. Progestin, estrogen and androgen G-protein coupled receptors in fish gonads. Steroids 2006;71:310–6.

[336] Villeneuve DL, Knoebl I, Kahl MD, Jensen KM, Hammermeister DE, Greene KJ, et al. Relationship between brain and ovary aromatase activity and isoform-specific aromatase mRNA expression in the Fathead Minnow (*Pimephales promelas*). Aquat Toxicol 2006;76(3–4):353–68.

[337] William HC, Jason RR. Community responses to contaminants: using basic ecological principles to predict ecotoxicological effects. Environ Toxicol Chem 2009;28(9):1789–800.

[338] Whyte SK. The innate immune response of Finfish-a review of current knowledge. Fish Shellfish Immunol 2007;23(6):1127–51.

[339] Jaensson A, Scott AP, Moore A, Kylin H, Olsen KH. Effects of a pyrethroid pesticide on endocrine responses to female odors and reproductive behavior in male parr of brown trout (Salmo trutta L.). Aquatic Toxicology 2007;81:1–9.

[340] Fan W, Yanase T, Morinaga H, Gondo S, Okabe T, Nomura M, Komatsu T, Morohashi K-I, Hayes TB, Takayanagi R, Nawata H. Atrazine-induced aromatase expression is SF-1 dependent: implications for endocrine disruption in wildlife and reproductive cancers in humans. Environmental Health Perspectives 2007;115:720–7.

[341] Flippin JL, Huggett D, Foran C. Changes in the timing of reproduction following chronic exposure to ibuprofen in Japanese medaka, Oryzias latipes. Aquatic Toxicology 2007;81:73–8.

[342] Yamamoto H, Tamura I, Hirata Y, Kato J, Kagota K, Katsuki S, Yamamoto A, Kagami Y, Tatarazako N. Aquatic toxicity and ecological risk assessment of seven parabens: Individual and additive approach. Science of The Total Environment 2011;410-411:102–11.

[343] Samiksha AR, Robert AA. Triclosan has endocrine-disrupting effects in male western mosquitofish, Gambusia affinis. Environmental Toxicology and Chemistry 2010;2010(29):1287–91.

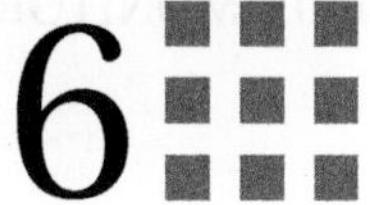

Water Quality Monitoring and Environmental Risk Assessment in a Developing Coastal Region, Southeastern North Carolina

Lawrence B. Cahoon*, Michael A. Mallin†

*DEPARTMENT OF BIOLOGY AND MARINE BIOLOGY, UNC WILMINGTON, WILMINGTON, NC, USA

†CENTER FOR MARINE SCIENCE, UNC WILMINGTON, WILMINGTON, NC, USA

CHAPTER OUTLINE

Monitoring Water Quality, http://dx.doi.org/10.1016/B978-0-444-59395-5.00006-6

6.1 Introduction

Environmental monitoring may have several goals: determining compliance with discharge and ambient standards, identifying areas with persistent problems for remediation efforts, establishing cause and effect relationships, and providing information for modeling and management tools, such as total maximum daily loads, among others. Monitoring programs need not be extensive in time and space to serve some of these goals, such as compliance determinations for point source discharges, but long-term, spatially comprehensive monitoring programs provide additional power to address larger scale questions about environmental hazards, particularly with regard to sets of parameters that describe different aspects of those hazards. Such large-scale approaches lend themselves to more formal *risk assessments*, which can help focus the attention of managers, policy makers, the regulated community, and the public at large on the most important problems.

Risk assessment protocols include four steps: (1) identification of a hazard; (2) identification of the population(s) exposed to that hazard; (3) establishment of dose-response or other quantitative cause and effect relationships for the hazard and for exposed population(s); (4) overall assessment of the risk posed by the specific hazard to the exposed population. Environmental hazards can be diverse, encompassing a wide array of pollutants, direct and indirect impacts, and interactive effects. In an environmental context, one must consider both risks to *human users* (who generally create, expose themselves to, and manage those risks) as well as risks to *ecologic integrity*—the ability of an ecosystem to support its "normal" suite of ecosystem services from an anthropocentric perspective, and to support its inherent ecological properties, such as stability, resilience, and biodiversity, among others. For instance, academic researchers have produced strong field and laboratory evidence that implicate key concentrations of various metals and organic toxins that cause negative effects upon benthic organisms in freshwater and estuarine sediments [1,2]. Cause and effect relationships for environmental hazards are sometimes sufficiently well established to justify regulations and standards, so that more informative risk assessments can be conducted [3].

This chapter develops risk assessment metrics for four environmental parameters in aquatic ecosystems (dissolved oxygen (DO), chlorophyll *a*, fecal coliform bacteria, and turbidity) for which regulatory standards have been established and that have been monitored extensively in time and space in southeastern North Carolina. The aims are twofold: to compare relative risks to human and ecosystem health among differing Coastal Plain aquatic habitats and to compare the risks posed by nonattainment of standards for different water-quality parameters. DO is important for aerobic aquatic life forms and has been implicated as a major factor in fish kills and shellfish losses [4–6]. The United States Environment Protection Agency (USEPA) and most states have adopted a general minimum standard of 5 mg/L (ppm), with exceptions for certain habitats (higher for "trout waters" and lower for "swamp waters") [7]. *Chlorophyll a* is a measure of phytoplankton biomass, and therefore a diagnostic for algae blooms, which can (1) cause problems with low DO through creation of a biochemical oxygen demand (BOD) upon bloom death and decay [8,9]; (2) impact taste and odor in drinking water supplies [10]; (3) release microalgal toxins in certain settings [11]; and (4) provide a usable food source for mixotrophic toxic algae and protozoa [12]. Algal blooms also can have a substantial human economic impact based on toxic effects, nuisance implications, and other problems [13]. USEPA and some states have set a maximum standard of 40 μg/L for chlorophyll *a* in surface waters [7]. Note, however, that this applies only to the planktonic (suspended) plant biomass, not to filamentous algae, associated with surfaces, benthic microalgae, or aquatic macrophytes, for which no standards have been promulgated as there are no approved standard methods for sampling and/or quantifying them. (We note that some states, such as Florida, Ohio, and Wisconsin are investigating the initiation of benthic microalgal standards). It is notable that both DO and chlorophyll *a* are commonly used standards for estimating the degree of eutrophication of water bodies [14]. *Fecal coliform* bacteria, and more recently *Escherichia coli*, are indicators of fecal microbial contamination from warm-blooded animals, including humans, and as such are strongly associated with exposure to water-borne pathogens through drinking, eating contaminated aquatic organisms, or body contact. USEPA, the United States Food and Drug Administration (US FDA) and most states set maximal fecal coliform bacteria standards as zero, 14, or 200 colony-forming units (CFU) per 100 ml for drinking water, shellfish waters, and recreational waters, respectively. (Note that the latter two values are applied as geometric means of multiple samples taken under specified conditions, with separate provisions for single instantaneous values) [7,15]. Fecal enterococcus bacteria are used more generally now as fecal pollution indicators in salt water environments [16], owing to their relatively higher persistence in that more stressful environment [17,18]. *Turbidity* is a measure of the light-scattering properties of suspended particulates, which is strongly indicative of fine-grained sediment loads but also addresses the ability of suspended matter, including phytoplankton, to attenuate light flux. Suspended sediment interferes with shellfish filter-feeding [19]; alters bottom habitat to the detriment of native organisms [20]; adsorbs pollutants by a variety of physical and chemical means [21] resulting in suspended sediment and turbidity concentrations being strongly correlated with pollutants, especially fecal bacteria, phosphorus, and BOD [22–24]. Attenuation of light through turbidity restricts growth of aquatic plants [25,26] and reduces visibility and prey-capture success in fish [27,28]. USEPA and many states have set standards for turbidity as 25 nephelometric turbidity units (NTU) for seawater and 50 NTU for freshwater [7], reflecting the effects of salinity on particle flocculation and settlement.

Data used in this analysis were derived from two large-scale regional monitoring programs in southeastern North Carolina. The South Brunswick Water and Sewer Authority (SBWSA) was formed in 1994 and was charged with developing a comprehensive regional program for managing wastewater and storm water in southwest Brunswick County, a rapidly developing coastal area that included the incorporated towns of Calabash and Sunset Beach as well as adjoining portions of the unincorporated county (Figure 6-1). Water-quality monitoring by University of North Carolina Wilmington

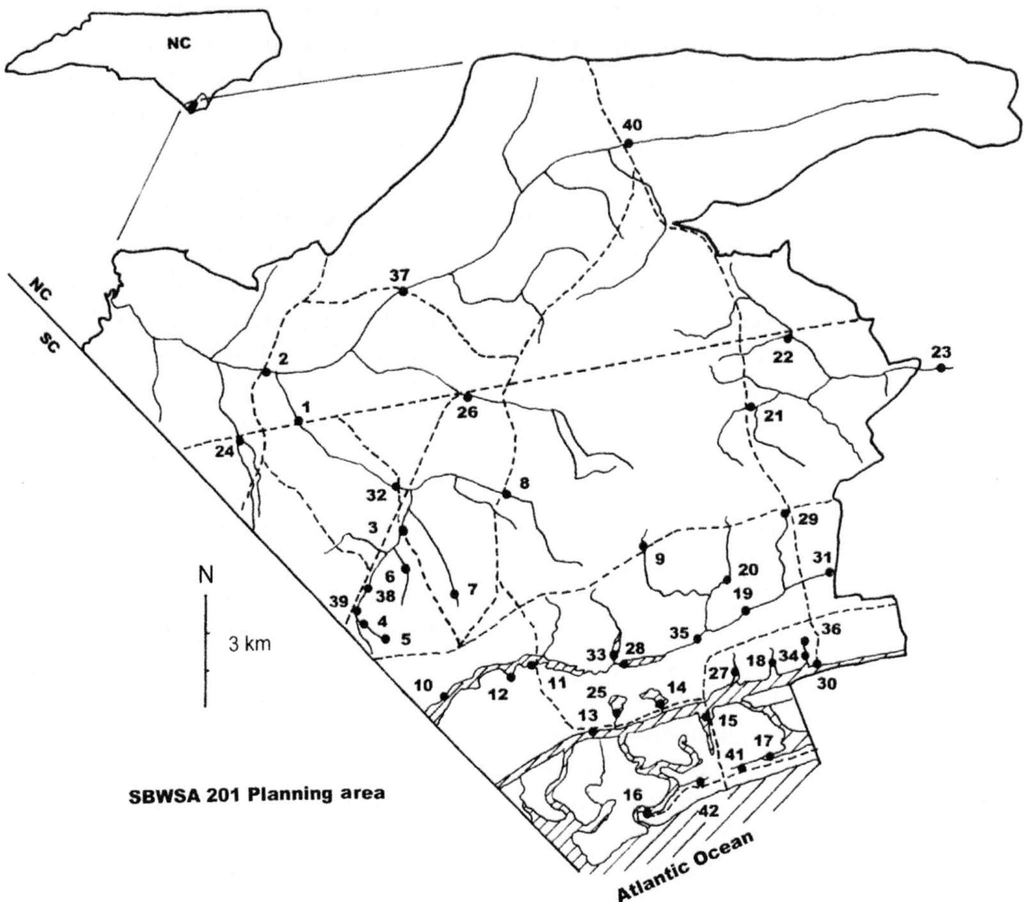

FIGURE 6-1 Map of SBWSA sampling sites (ML), numbered in sequence of designation, in southwest Brunswick County, North Carolina, within the boundary of the 201 Planning Area (NC/SC line and heavy solid line). Lighter solid lines are perennial watercourses, dashed lines are major roads, and cross hatched areas are open surface waters. Incorporated towns of Calabash, Carolina Shores, and Sunset Beach include ML 10–12; ML 3–7, 38, and 39; and ML 14–18, 25, 27, 30, 34, 36, 41, and 42, respectively.

researchers under contract to SBWSA began in late 1996 and continued through mid-2003, when SBWSA was dissolved by Brunswick County. The Lower Cape Fear River Program (LCFRP) was developed in response to a perceived need by a large number (~20) of industrial and municipal dischargers to meet the monitoring requirements of their National Pollutant Discharge Elimination System (NPDES) permits by combining efforts and contracting with UNC Wilmington researchers to conduct field sampling and some measurements, with others sub-contracted to state-certified commercial laboratories. The lower Cape Fear basin included portions of the Cape Fear River below Elizabethtown in Bladen County, as well as two major tributaries, the Black and Northeast Cape Fear rivers, and the Cape Fear River estuary (Figure 6-2). Monitoring began in 1995 and is continuing.

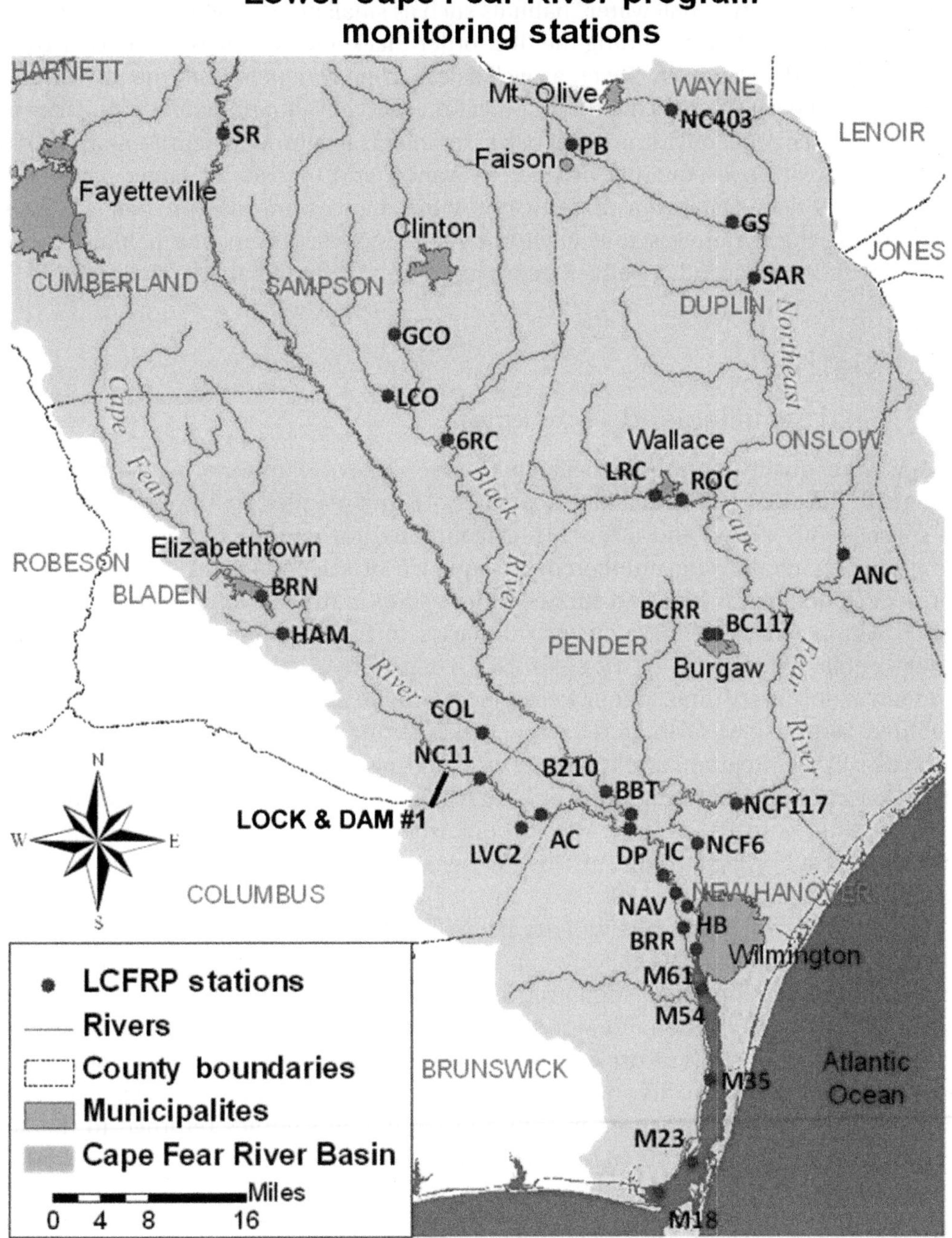

FIGURE 6-2 Map of the Lower Cape Fear River system and the LCFRP sampling stations. (For color version of this figure, the reader is referred to the online version of this book.)

These two monitoring programs have yielded data on a wide variety of aquatic habitats, ranging from "small water" habitats such as small ponds and first-order streams to larger bodies of water, including the fifth-order Black and Northeast Cape Fear rivers and the sixth-order Cape Fear River, as well as estuarine waters in both the SBWSA area and the Cape Fear Estuary. This geographical diversity of aquatic habitats provides an opportunity to conduct environmental risk assessments in a broadly representative set of water bodies, with consequently broader relevance, and to look for larger patterns of environmental risks. Data from these two monitoring programs also provide an opportunity to test the hypothesis that environmental risks vary from one habitat type to another, with some hazards relatively more important in some habitats than others.

6.2 Methods

6.2.1 Field Sampling and Techniques

SBWSA: Water quality monitoring began in October 1996 at 22 monitoring locations (ML) throughout the SBWSA 201 Area (approximately 41 square miles), with twenty additional ML subsequently added and a few ML discontinued for various reasons (Figure 6-1). Monitoring locations were numbered in sequence of selection, and were selected to represent all drainage basins and surface water bodies in the sampling area. All ML were located within waters of the state or tributaries to these waters, and were accessed through public rights of way or with permission of private owners. Each ML was sampled on a set schedule every three to four weeks, always in the midmorning hours, 0900 to 1200 local time. Sampling was conducted once weekly during the first one and a half and last three years of the program, and twice per week during the middle period. Sampling was conducted at geographic subgroups of 7–10 ML each time, allowing sufficient time for return of time-sensitive samples (fecal coliforms) within specified time limits [29]. Samples were all collected from just below the surface at each ML following specified protocols for each parameter [29].

Values of the four water quality parameters examined in this study were measured either in situ or from samples collected and returned to the laboratory for each sampling location and time. All protocols followed standard methods and QA/QC as specified by APHA (1998 or earlier); the laboratory was state certified to conduct these and other measurements for the SBWSA project. DO was measured and logged in situ using a YSI 85 multi-parameter water quality meter, which was calibrated before each daily sampling trip. Chlorophyll *a* was measured in triplicate for 200 ml samples returned to the lab, filtered through glass fiber filters, analyzed fluorometrically following the methods of Welschmeyer (1994) [30] and USEPA (1997) [31], and reported as μg/L. Fecal coliform analyses employed the membrane filtration method (MFC, method 9222D [29]). We filtered, incubated, and counted five filtered subsamples (3×10 ml, 1×1 ml, and 1×100 ml) from each sample, reporting a mean value $\pm$ one std dev. for the three 10-ml subsamples as CFU/100 ml when values were within specified colony count ranges, and otherwise used values from single filters as appropriate. Turbidity was measured as NTU on a single separate water sample using a DRT-15CE nephelometer (Fisher Scientific) that was calibrated regularly with manufacturer's formazin standards.

LCFRP: The main stem of the Cape Fear River is formed by the merging of the Haw and the Deep rivers in Chatham County in the North Carolina Piedmont. The main stem of the river has been altered by the construction of several dams and water control structures. In the coastal plain, the river is joined by two major tributaries, the Black and the Northeast Cape Fear rivers (Figure 6-2). These fifth-order blackwater streams drain extensive riverine swamp forests and add organic colored dissolved organic matter (DOM) to the main stem. The watershed (about 9149 square miles) is the most heavily industrialized in North Carolina with 244 NPDES permitted wastewater discharges with a total permitted flow of approximately 425 million gallons per day, and (as of 2000) over 1.83 million people residing in the basin, and approximately 4,200,000 head of swine housed in concentrated animal feeding operations, or CAFOs [32]. Approximately 24% of the land use in the watershed is devoted to agriculture and livestock production [32], with livestock production dominated by swine and poultry operations. Thus, the watershed receives considerable point and nonpoint source loading of pollutants. However, the estuary is a well-flushed system, with flushing time ranging from 1 to 22 days with a median flushing time of about seven days, much shorter than the other large N.C. estuaries to the north [33].

Water quality is monitored by the Aquatic Ecology Laboratory of the University of North Carolina Wilmington Center for Marine Science. Nine stations in the Cape Fear estuary (from Navassa to Southport) and one station in the Northeast Cape Fear estuary are sampled by boat (Figure 6-2). Riverine stations sampled by boat include NC11, AC, DP, IC, and BBT (Figure 6-2). NC11 is located upstream of any major point source discharges in the lower river and estuary system, and is considered to be representative of water quality entering the lower system (we note that the City of Wilmington and portions of Brunswick County get their drinking water from the river just upstream of Lock and Dam #1). Station BBT is located on the Black River between Thoroughfare (a stream connecting the Cape Fear and Black rivers) and the main stem Cape Fear, and is influenced by both rivers. We consider B210 and NCF117 to represent water quality entering the lower Black and Northeast Cape Fear rivers, respectively. Data have also been collected at stream and river stations throughout the watersheds of the Cape Fear, the Northeast Cape Fear, and the Black rivers. (Figure 6-2 [9,34]).

6.2.2 Techniques

All samples and field parameters collected for the estuarine stations of the Cape Fear River (NAV down through M18) were gathered on an ebb tide. This was done so that the data better represented the river water flowing downstream through the system rather than the tidal influx of coastal ocean water. Sample collection and analyses were conducted according to the procedures in the Lower Cape Fear River Program Quality Assurance/Quality Control (QA/QC) manual, and included periodic inspection by representatives from the North Carolina Division of Water Quality to verify University of North Carolina Wilmington (UNCW) field and laboratory procedures.

Field parameters (DO, water temperature, pH, turbidity, salinity, conductivity) were measured at each LCFRP site using a YSI 6920 (or 6820) multi-parameter water quality sonde displayed on a YSI 650 MDS. Each parameter was measured with individual probes on the sonde. At stations sampled by boat, physical parameters were measured at 0.1 m, the middle of the water column, and at the bottom (up to 12 m). Occasionally, high flow prohibited the sonde from reaching the actual bottom and measurements were taken as deep as possible. At the terrestrially sampled stations the physical parameters were measured at a depth of 0.1 m. The Aquatic Ecology Laboratory at UNCW is state-certified by the N.C. Division of Water Quality to perform field parameter measurements.

Fecal coliform bacteria were analyzed at a state-certified laboratory contracted by the LCFRP. Samples were collected approximately 0.1 m below the surface in sterile plastic bottles provided by the contract laboratory and placed on ice for no more than 6 h before analysis. Chlorophyll *a* was measured for the LCFRP by UNCW personnel as above, using the methods described in Refs. [30,31]. The Aquatic Ecology Laboratory at the UNCW Center for Marine Science is state-certified by the N.C. Division of Water Quality for the analysis of chlorophyll *a*.

6.2.3 Habitat Groupings

Aquatic habitats in the SBWSA area were grouped into three categories. "Estuary" sites were all those located in aquatic habitats behind Sunset Beach or in the Calabash River and they had measurable (>1 part per trillion [ppt]) salinity most of the time. "Pond" sites were in freshwater ponds, some natural but most man-made. "Stream" sites were all in flowing freshwater bodies, most natural and some man-made or altered. Aquatic habitats in the LCFRP area were grouped into five categories by location and stream order. Estuary sites were all those sites in the Cape Fear River estuary where salinity routinely exceeded 1–2 ppt, from approximately Station NAV downstream to the ocean (Figure 6-2), covering an approximate reach of 25 miles in the river and broader estuary. "Cape Fear Main" sites were located in the main stem of the Cape Fear River between Elizabethtown and the estuary proper. "NECFR/Black Main" sites were located in the main channels of the Northeast Cape Fear and Black rivers, respectively. "NECFR Upper" sites were all located within the tributaries, i.e., lower order streams, of the Northeast Cape Fear river. Similarly, "Black Upper" sites were all located within tributaries to the Black River.

6.2.4 Data Analysis

Values of each measured parameter were compared to relevant regulatory water quality standards, and the frequencies of values failing to meet those standards were calculated for each habitat category. For DO, an instantaneous, i.e., single time of measurement, value of 5 mg/L was used in estuary-monitoring locations. Following guidance from NC

DENR the instantaneous standard for swamp waters, 4 mg/L was used for the other freshwater monitoring locations. Values of chlorophyll *a* above 40 μg/L were considered to fail the standard in all habitat categories. Values of fecal coliform bacteria above 200 CFU/100 ml were always considered as failing to meet the standard for human body contact. Values above 14 CFU/100 ml were considered to fail the standard in Estuary waters, where shellfishing is the appropriate designated use. The drinking water standard of 0 (zero) CFU/100 ml was considered irrelevant, as none of these waters are used untreated as drinking water sources. Turbidity values above 25 NTU in "estuary" sites and above 50 NTU in all other sites were considered to fail to meet the standards. The percentages of values that failed to meet North Carolina (NC) standards were calculated and used directly in subsequent calculations of risk assessment.

6.3 Results

6.3.1 Dissolved Oxygen

Average values for DO concentrations were generally lower in the small waters sampled by the SBWSA program than in the larger water bodies sampled by the LCFRP (Table 6-1). Significant portions of the DO data sets for all groupings were below the respective state

Table 6-1 Groupings and Average Values of Water Quality Parameters of Monitoring Locations (ML, Shown in Maps, Figures 6-1 and 6-2) in the SBWSA 201 Planning Area, (1996–2003) and in the Lower Cape Fear River Program Sampling Area, (1995–2010)

	ML	DO	Chla	Fecal	Turb
SBWSA Group					
Estuary: N = 507	10, 11, 13, 15, 16, 17, 30, 41, 42	4.96 (2.36)	13.1 (23.8)	174 (498)	14.5 (33.0)
Pond: N = 763	12, 14, 18, 21, 22, 25, 27, 28, 33, 34	5.09 (2.43)	11.0 (16.9)	160 (413)	14.0 (40.0)
Stream: N = 1165	1, 2, 3, 4, 5, 6, 7, 8, 9, 19, 20, 24, 26, 31, 32, 35, 37, 38, 39, 40	5.52 (2.47)	14.4 (64.7)	1120 (18900)	19.6 (47.9)
Overall: N = 2435	All	5.25 (2.43)	10.5 (36.4)	611 (9240)	16.7 (42.3)
LCFRP Group					
Estuary: N = 1830	SPD, M18, M23, M35, M42, M54, BRR, M61, HB, NAV	7.27 (2.15)	5.03 (5.19)	45.5 (155)	16.7 (13.5)
CFR Main: N = 910	IC, DP, AC, LVC, NC11	7.52 (2.56)	4.00 (5.35)	60.7 (174)	20.1 (19.7)
NECFR/Black Main: N = 552	B210, NCF6, NCF117	6.47 (2.38)	1.77 (3.05)	82.0 (293)	7.7 (7.7)
NECFR Upper: N = 1583	BC117, BCRR, ANC, ROC, LRC, SAR, GS, NC403, PB	6.65 (3.21)	7.67 (25.2)	523 (2730)	10.7 (15.9)
Black Upper: N = 890	COL, 6RC, LCO, GCO, SR	6.83 (2.80)	3.22 (13.2)	167 (710)	5.0 (6.7)
Overall: N = 5765		6.99 (2.63)	5.00 (11.7)	201 (964)	12.7 (8.8)

N = Number of Samples (Locations × Times) Taken for Each ML Grouping. Data are Means (Std Dev.)

standards of 5 or 4 mg/L, varying from 7.1% of the Cape Fear main data to 52.9% of the SBWSA estuary data (Table 6-2). Much of the extensive seasonal variability in DO concentrations is related to temperature, which controls oxygen solubility (Cahoon et al., in prep.), so the differences among habitat types probably reflect habitat characteristics. The shallower, small water habitats probably have relatively greater groundwater impact, less aeration effect by wind and wave action, and higher respiratory consumption of oxygen by sediments and benthic organisms [35]. Blackwater fluvial systems drain swamps, which feature shallow water that has considerable contact with the organically-rich sediments of the riparian swamp floor. Much microbial respiration occurs in this situation, reducing the DO in the swamp water [36]. The Cape Fear estuary proper, especially the lower estuary, is buffered against severe bottom-water hypoxia through strong currents and gravitational mixing [37].

6.3.2 Chlorophyll *a*

Average values of chlorophyll *a* were generally higher in the small water habitats of the SBWSA area than in the larger water bodies sampled by the LCFRP (Table 6-1). All average values were well below the NC standard of 40 μg/L, and so the percentages of values exceeding the state standard were also quite low, varying from 0% in the main channels of the Northeast Cape Fear and Black rivers to about 7% in the Stream habitats of the SBWSA area. Phytoplankton require a combination of nutrient loading, water clarity, and

Table 6-2 Percent of Values for Regulated Water Quality Parameters That Fail to Meet Relevant Standards in Habitat Groups in the SBWSA 201 Planning Area, (1996–2003) and in the Lower Cape Fear River Program Sampling Area, (1995–2010), Respectively

	ML	DO	Chla	Fecal	Turb
SBWSA Group					
Estuary: N = 507	10, 11, 13, 15, 16, 17, 30, 41, 42	52.9	5.9	14.9/66.4	9.9
Pond: N = 763	12, 14, 18, 21, 22, 25, 27, 28, 33, 34	36.1	3.5	17.3	2.9
Stream: N = 1165	1, 2, 3, 4, 5, 6, 7, 8, 9, 19, 20, 24, 26, 31, 32, 35, 37, 38, 39, 40	28.5	7.0	25.0	6.2
Overall: N = 2435	All	36.0	5.4	20.5	5.9
LCFRP Group					
Estuary: N = 1830	SPD, M18, M23, M35, M42, M54, BRR, M61, HB, NAV	14.8	0.2	3.1/56.9	17.9
CFR Main: N = 1266	IC, DP, AC, LVC, NC11	7.1	0.4	12.0	5.5
NECFR/Black Main: N = 552	B210, NCF6, NCF117	12.7	0.0	2.4	0.4
NECFR Upper: N = 1583	BC117, BCRR, ANC, ROC, GS, NC403, PB, SAR, LRC	21.2	3.6	30.9	2.5
Black Upper: N = 890	COL, 6RC, LCO, GCO, SR	15.1	1.6	11.8	0.2
Overall: N = 6121	All	15.1	1.3	13.4	7.3

The First Value for Fecal is %>200 CFU/100 ml and the Second Value is for %>14 CFU/100 ml

residence time to create the blooms that values above the state standard indicate, a combination that appears only infrequently satisfied in these habitats.

6.3.3 Fecal Coliform Bacteria

Average values of fecal coliform bacteria concentrations were highly variable among habitat groupings, ranging over almost an order of magnitude in each sampling area and by over an order of magnitude among habitat groups (Table 6-1). The standard deviations within habitat groups were similarly large, always exceeding the means by substantial multiples. Similarly, the proportions of fecal coliform values exceeding relevant standards were variable and in some cases quite high, with values exceeding the shellfishing standard of 14 CFU/100 ml in estuary sites in both areas over 56% of the time (Table 6-2) and exceeding the human body contact standard (200 CFU/100 ml) from between 2.4% in the main channels of the Northeast Cape Fear and Black rivers to 30.9% in the tributaries of the Northeast Cape Fear River. Such high variability and high frequency of excessive values probably reflect the multiplicity of sources of fecal contamination, notably including high concentrations of CAFOs in the upper Northeast Cape Fear watershed in particular [38], as well as abundant and densely sited septic systems in much of the SBWSA area [39].

6.3.4 Turbidity

Average values of turbidity were generally similar, given wide variation among values over time, but were notably lower in the tributaries to the Black River (Table 6-1). There were generally higher frequencies of excessive turbidity in the Estuary habitats of both sampling areas, owing to the lower turbidity standard in those waters. Otherwise, the frequency of excessive turbidity values was generally well below 10% (Table 6-2). Generally low turbidity values make sense in this setting; on the Coastal Plain the slope of headwaters streams averages about 0.1% (or a 1 m drop per km) and the slope of middle order streams averages a mere 0.02% [40], making for slow-flowing waters and little potential for erosion. Additionally, there are widespread riparian wetlands in the region, and sediment and particle trapping in these wetlands is a factor in low turbidity [23].

6.4 Discussion: Environmental Risk Analysis

The hazards posed by non-compliant values of environmental parameters clearly depend on the at-risk population. Therefore, hazards to humans must be considered in a different light from hazards to ecosystem integrity. Considering hazards to humans first, two kinds of problems may be defined: (1) direct threats to human health and (2) more indirect threats to human uses of ecosystem services, such as recreational uses, passive enjoyment of the waterscape, and, perhaps, property values as affected by impairment of water quality. Hazards to ecosystem integrity are broader and, to some degree, more difficult to separate into cleanly defined categories. Moreover, some responses to non-compliant conditions might be perceived as positive, e.g., stimulation of primary production by nutrient loading that causes higher phytoplankton biomass (chlorophyll *a*) might ultimately support more total ecosystem production, such as fishery productivity, at least in some circumstances. Consequently, hazards to ecosystem integrity might consider (a) direct mortality threats to aquatic flora and fauna, (b) sublethal stressors to aquatic flora and fauna, (c) effects on primary production, (d) effects on secondary production, (e) effects on species of interest to humans (fishery and recreational wildlife), including indirect effects, and (f) effects on overall system resilience.

Each of the water-quality parameters considered in this analysis may now be evaluated in terms of the above hazard categories. For the purposes of this examination, the magnitude of impact of each potential threat by each parameter is rated as "high" (well known as a major effect), "medium" (effects known only under extreme conditions), and "low" (no major effects known). For example, low DO is known to be a primary cause of fish kills, and so would be rated as "high" with regard to its hazard potential for "effects on species of interest to humans." Similarly, most violations of the fecal coliform standards are relatively low in absolute magnitude and pose essentially no direct mortality threat to aquatic flora and fauna, and would be rated as "low."

Table 6-3 shows one such matrix of hazards and impact magnitude by parameter; there may well be circumstances elsewhere in which different judgments would be

Table 6-3 Hazard and Impact Matrix for Noncompliance with Regulatory Standards for Dissolved Oxygen (DO), Chlorophyll *a* (CHla), Fecal Coliform Bacteria (FCB; >200 CFU/100 ml = Human Body Contact Standard; >14 CFU/100 ml = Shellfish Consumption Standard); and Turbidity (Turb), Based on Experience in the SBWSA and LCFRP Monitoring Programs

	Parameter				
			FCB		
Hazard	**DO**	**Chla**	**>200**	**>14**	**Turb**
Human impacts					
Health	Low	Med	High	High	Low
Contact	Low	Med	High	High	Med
Non-contact sport	Med	Med	Med	Low	Low
Aesthetics	Med	Med	Low	Low	High
Property values	Med	Med	Med	Med	Med
Human impact score Σ =	8	10	11	10	8
Ecologic integrity					
Mortality	High	Med	Med	Low	High
Stressor	High	Med	Low	Low	High
Primary production	Low	High	Med	Low	Med
Secondary production	High	Med	Low	Low	High
Human interest	High	Med	Med	High	Med
Resilience	High	Med	Low	Low	Med
Ecologic integrity score Σ =	16	13	9	8	16
Overall impact score Σ =	24	23	20	18	24

Impacts of Noncompliance are Rated as High (=3), Med (=2), and Low (=1)

offered, but the judgments offered here reflect our long experience with these two aquatic ecosystems. The primary mechanism by which noncompliant (low) values of DO affect aquatic ecosystems is by killing or stressing aquatic organisms, hence the "high" impact ratings for most ecosystem effects (the exception being that primary producers, as oxygen producers, would be less affected) and "low" impact rating for human health and contact impacts but "medium" impact rating for those human uses tied more to the welfare of aquatic organisms. High values of chlorophyll *a*, which signal algae blooms and the occasionally more harmful effects of toxic algae blooms, were assigned "medium" impact ratings for most hazard categories, the exception being an obviously "high" rating for primary production, when high phytoplankton biomass might also correlate with high primary-producer biomass in general, a strong eutrophication signal. Values of fecal coliform bacteria above the human body contact standard of 200 CFU/100 ml obviously have a "high" impact potential for situations involving human contact with, or consumption of, aquatic resources. To the degree to which high fecal coliform bacteria levels may correlate with sewage spills or other inputs of wastes, animal or human, with their associated pathogens, nutrients, and other chemicals, the impact of high fecal coliforms might be rated as "medium" in contexts aside from human body contact or shellfish consumption. When fecal coliform bacteria levels exceed the shellfishing standard of 14 CFU/100 ml, there is a clear threat to human health that would engender the only commonly encountered warnings widely posted in the environment to warn of hazards: shellfish closure signs (Figure 6-3). These generally lower levels of fecal coliform bacteria create little impact on ecosystem function, except for human perception of ecosystem integrity. Studies have shown that when people are aware of elevated fecal bacterial counts in adjacent waterways, residential land values can be negatively impacted, posing a financial burden to humans [41].

FIGURE 6-3 Shellfishing closure sign in estuarine portion of SBWSA sampling area behind Sunset Beach. (For color version of this figure, the reader is referred to the online version of this book.)

As mentioned above, noncompliant levels of turbidity in and of themselves and as strong correlates with excessive sedimentation create several kinds of impacts, including making water appear muddy (Figure 6-4); adsorbing other pollutants; reducing visibility of hazards to humans and prey to consumers; reducing light flux to aquatic plants; and even smothering or otherwise interfering with benthic organisms. Consequently noncompliant turbidity values were generally rated as having "high" or "medium" impacts.

The generation of an overall environmental risk assessment requires that values for the frequency of noncompliant levels of the several parameters estimated in this exercise in Table 6-2 must now be integrated with the hazard-impact matrix in Table 6-3, to allow identification of the most pressing water quality problems for human uses and ecologic integrity among the different aquatic habitats of the two regions evaluated here. Simple multiplication of the frequencies in Table 6-2 by the hazard-matrix scores in Table 6-3 yields relative values that permit rankings of relative risks by parameter and habitat type

FIGURE 6-4 High turbidity event in a stream in the SBSWA monitoring area. (For color version of this figure, the reader is referred to the online version of this book.)

and comparisons of risks to humans directly and to ecosystem integrity (Table 6-4). This approach can be applied in other settings as well, presuming the existence of suitable data sets and informed judgments about impact potentials.

Several features of this overall risk assessment emerge immediately. First, low DO is estimated to pose the greatest overall threat to both direct human uses and to ecologic integrity in both monitoring areas and in almost all habitat types. This estimate derives from the widespread occurrence of noncompliant DO values and the impacts of low DO values on the ability of these habitats to fully support aquatic life. Undoubtedly some portion of low DO values may be attributable to temperature effects on the solubility of oxygen in the warmer months, and to naturally high loadings of BOD in this organic-rich

Table 6-4 Risk Assessment Matrix

	DO	Chla	Fecal	Turb
SBWSA Group				
Estuary				
Human impact	423	60	164/664	79
Ecologic integrity impact	846	77	134/531	159
Pond				
Human impact	289	35	190	23
Ecologic integrity impact	578	45	156	47
Stream				
Human impact	228	70	275	50
Ecologic integrity impact	456	91	225	100
Overall				
Human impact	288	55	225	48
Ecologic integrity impact	576	71	184	95
LCRFP Group				
Estuary				
Human impact	118	2	34/569	143
Ecologic integrity impact	237	2	28/455	286
CFR Main				
Human impact	57	4	132	44
Ecologic integrity impact	114	6	108	89
NECFR/Black Main				
Human impact	102	0	27	3
Ecologic integrity impact	204	0	22	6
NECFR Upper				
Human impact	170	36	340	21
Ecologic integrity impact	340	46	277	40
Black Upper				
Human Impact	121	16	130	2
Ecologic integrity impact	242	20	106	4
Overall				
Human impact	121	13	147	59
Ecologic integrity impact	242	17	121	117

setting of coastal plain habitats. Shallower waters may also experience relatively high oxygen consumption by sediments and benthos [36], an argument supported by the relative rankings of the various habitat types in terms of average DO values (Table 6-1). A human contribution can also be found in the broad and frequent occurrence of low DO values, however, as large scale loadings of nutrients across the landscapes of these two monitoring areas drive substantially increased autochthonous and allochthonous organic production that drive oxygen demand in these aquatic habitats [34,42].

The assessment of low DO as a substantial risk to human uses is based more on the loss of associated ecosystem services than on any direct impacts on human health, *per se*, although low DO arising from cultural eutrophication may coincide with incidences of sewage pollution, major runoff events that also load contaminants, and harmful algae blooms. Fish kills, although not common throughout the monitoring areas, have occurred during the respective monitoring periods, and in the large majority of cases were attributable to low DO [43,44]. Loss of fish resources has a significant impact on human uses and signals a lack of ecologic integrity that affects human valuations of these habitats. This valuation effect may be local and respond to specific issues in a particular body of water, however, we speculate that the broader sense of environmental degradation conveyed by such headline events as fish kills may lower public perception of the value of these ecosystems and facilitate other degrading uses, in a sense giving permission to abuse "less-than-pristine" ecosystems.

Second, fecal coliform contamination clearly poses major risks for human users of aquatic resources in this region, signaling pollution by human and/or animal wastes that can carry pathogens as well as nutrients that can support growth of other harmful microbes. Widespread closures of estuarine waters to shellfishing and occasional closures of surface waters to swimming indicate the magnitude of this risk and the seriousness with which it is viewed by the public and environmental managers [23]. Nevertheless, the frequency with which noncompliant values of fecal coliform bacteria were observed in both monitoring programs indicates that preventive practices and compliance with established rules are still inadequate and require more effort. Significant emphasis in recent years on control of storm water runoff as a major nonpoint source of fecal contamination has led to newer and tighter regulations, but the perceived relative magnitude of the human health risk has not yet been sufficient to "close the deal."

Third, turbidity poses relatively lower risk overall than either low DO or fecal coliform contamination in these aquatic habitats, but with a few interesting variants. Foremost among these is the relatively high risk estimated for estuarine systems, which reflects the much lower turbidity standard for estuarine waters (25 NTU) in comparison with freshwater habitats (50 NTU), which is emplaced to protect submerged aquatic vegetation and filter-feeding bivalves. To the degree to which turbidity reflects sedimentation as a problem, these data suggest that estuarine areas still receive excess sediment loading, despite an aggressive effort to control sediment runoff by a variety of measures: silt fencing and ground cover on disturbed soils, widespread use of sedimentation basins (both wet and dry ponds), setback rules, etc. The turbidity data may be biased to some extent toward the lightest, smallest, and most easily suspended materials that are managed with the most difficulty by these management tactics. Observations of sediment composition in local estuarine habitats indicate a strong bias toward very fine particulates as major components of terrigenous sediment loading [45]. The high affinity of various pollutants for these finest particulates [21], however, leads to the troublesome conclusion that current rules manage the least problematic components of the overall sedimentation problem.

Fourth, the risk assessment results for chlorophyll *a* generally indicate less concern about this aspect of water quality, with slightly more risk associated with small water habitats than in the more extensive surface waters of the lower Cape Fear system. With the exception of any major algal blooms that actually produce toxins, the direct risks for human use are minor in these habitats. However, small water habitats elsewhere in the southeast US that receive nutrient loading (such as golf course ponds and suburban wet-detention ponds) have harbored toxic algal blooms that cause fish kills [46]. Some summer cyanobacterial blooms were observed during 2009–2011 in both the main stem of the Cape Fear River and in the Northeast Cape Fear River. The reasons for their recent appearance are currently under investigation and may involve high summer temperatures, low flow, and upstream nutrient loading. These blooms have on some occasions caused taste and odor problems to water-treatment facilities and their customers drawing water from the Cape Fear River, which is a nuisance and a human economic issue, although no problems with algal toxins have been reported so far. One must recognize, however, that chlorophyll *a* biomass represents only a fraction of the total primary producer biomass in these aquatic ecosystems, and thus only a fraction of the total biological cycling of nutrients, dissolved gases, food production, and habitat structure. In some habitats native vegetation, i.e., *Spartina* in salt marshes, represents a much larger portion of total primary-producer biomass. In other cases, notably golf course ponds in the SBWSA area, filamentous algae, and aquatic macrophytes were much more prevalent in quantities clearly beyond what would be expected without human impacts [47]. Consequently, chlorophyll *a* measures understate the environmental risks of excessive primary production.

Fifth, across habitat types, i.e., by summing risk assessment scores across rows in Table 6-4, one can discern several additional patterns. The small water habitats monitored in the SBWSA area generally show higher risk-assessment values than the broader surface waters of the Cape Fear River system, because the relative effects of terrigenous inputs would be much higher for the former. When shellfishing is considered as human use, the estuary habitats clearly present the highest risks of compromised integrity in both areas. Interestingly, the tributaries to the Northeast Cape Fear River exhibit much higher risk scores than the tributaries to the Black River, although the two watersheds lie adjacent to each other and both support relatively low human populations and high populations of domestic animals in CAFOs. We note there are three small waste water treatment facilities influencing Northeast Cape Fear system streams, and these facilities frequently have excessive pollution issues in addition to nonpoint source CAFO runoff issues in the area.

Finally, this assessment necessarily makes a variety of assumptions about the representative nature of the data sets, the valuations selected for assessing hazards and impacts in Table 6-3, and the perceptions of different user groups. Moreover, the degree to which external factors, such as climate variability, control the various parameters assessed in this analysis might change over a longer time frame and requires consideration. Major hurricanes, a severe El Niño rainy season in 1998, and severe droughts in 2002 and 2007 had substantial effects detectable in one or both data sets [43]. Future climate variation, whether associated with oscillators like the El Niño Southern Oscillation or progressive climate warming, may change patterns in important and unforeseen ways. Humans are also major actors in this developing coastal region. Increasing water withdrawals, continuing construction of impermeable surfaces, and changes in waste-treatment practices may all have important consequences. This approach is, therefore an assessment based on recent monitoring data, local knowledge and professional judgment, and remains subject to change.

6.5 Summary and Conclusions

Environmental risks to human uses of these two Coastal Plain areas are generally moderate across different uses, but higher when exposure to fecal pollution is an issue. Risks to ecologic integrity are higher, however, from frequent noncompliance with relevant standards for DO, turbidity and chlorophyll *a*. This dichotomy in risk assessment indicates a potential for conflict among resource allocations for environmental remediation. These risk assessments indicate both the vulnerabilities and needs for more effective regional environmental management approaches.

Acknowledgments

This research was supported by the South Brunswick Water and Sewer Authority and the Lower Cape Fear River Program. We thank S. Ensign, J. Hales, E. Henry, J. Johnson, S. Loucaides, J. Merritt, M. McIver, J. Nearhoof, D. Parsons, K. Rowland, C. Shank, E. Steffy, B. Sykes, J. Tombro, B. Toothman, M. Tzannis and H. Wells for numerous contributions.

References

[1] Long ER, McDonald DD, Smith SL, Calder FD. Incidence of adverse biological effects within ranges of chemical concentrations in marine and estuarine sediments. Environ Manag 1995;19:81–97.

[2] MacDonald DD, Ingersoll CG, Berger TA. Development and evaluation of consensus-based sediment quality guidelines for freshwater ecosystems. Arch Environ Contam Toxicol 2000;39:20–31.

[3] USEPA. Guidance for assessing chemical contaminant data for use in fish advisories. EPA-823-B-00–007. In: Fish sampling and analysis, vol. 1. Washington, D.C: United States Environmental Protection Agency, Office of Water; 2000.

[4] Diaz RJ, Rosenberg R. Marine benthic hypoxia: a review of its ecological effects and the behavioral responses of benthic macrofauna. Oceanogr Mar Biol Annu Rev 1995;33:245–303.

[5] Diaz RJ, Rosenberg R. Spreading dead zones and consequences for marine ecosystems. Science 2008;321:926–9.

[6] Riisgard HU, Andersen P, Hoffman E. From fish to jellyfish in the eutrophied Limfjorden (Denmark). Estuaries and Coasts, http://dx.doi.org/10.1007/s12237-012-9480-4; 2012.

[7] NCDENR. Classification and water quality standards applicable to surface waters and wetlands of north Carolina. Administrative Code Section: 15A NCAC 2B. 0100. Raleigh, N.C: North Carolina Department of Environment and Natural Resources, Division of Water Quality; 1999.

[8] Boesch DF, Brinsfield RB, Magnien RE. Chesapeake Bay eutrophication: scientific understanding, ecosystem restoration, and challenges for agriculture. J Environ Qual 2001;30:303–20.

[9] Mallin MA, Johnson VL, Ensign SH, MacPherson TA. Factors contributing to hypoxia in rivers, lakes and streams. Limnol Oceanogr 2006;51:690–701.

[10] Burkholder JM. Cyanobacteria. In: Bitton G, editor. Encyclopedia of environmental microbiology. New York: Wiley Publishers; 2002. p. 952–82.

[11] Burkholder JM. Implications of harmful microalgal and heterotrophic dinoflagellates in management of sustainable marine fisheries. Ecol Appl 1998;8:S37–62.

[12] Burkholder JM, Glibert PM, Skelton HM. Mixotrophy, a major mode of nutrition for harmful algal species in eutrophic waters. Harmful Algae 2008;8:77–93.

[13] Hoagland P, Anderson DM, Kaoru Y, White AW. The economic effects of harmful algal blooms in the United States: estimates, assessment efforts and information needs. Estuaries 2002;25:819–37.

[14] Bricker SB, Clement CG, Pirhalla DE, Orlando SP, Farrow DRG. National estuarine eutrophication assessment: effects of nutrient enrichment in the nation's estuaries. NOAA, National Ocean Service, Special Projects Office and the National Centers for Coastal Ocean Science; 1999.

[15] USFDA. Sanitation of shellfish growing areas. National Shellfish Sanitation Program Manual of Operations, Part 1. Washington, D.C: United States Department of Health and Human Services, Food and Drug Administration, Office of Seafood; 1995.

[16] NCDEH. North Carolina recreational water quality monitoring program, Division of Environmental Health, Shellfish Sanitation and Recreational Water Quality Section website, http://www.deh.enr.state.nc.us/shellfish/index.html; 2004.

[17] Dufour AP. Bacterial indicators of recreational water quality. Can J Public Health 1984;75:49–55.

[18] Hanes NB, Fragala R. Effect of seawater concentrations on survival of indicator bacteria. J Water Pollut Control Fed 1967;39:97–104.

[19] Rothschild BJ, Ault JS, Goulletquer P, Heral M. Decline of the Chesapeake Bay oyster population: a century of habitat destruction and overfishing. Mar Ecol Prog Ser 1994;111:29–39.

[20] Waters TF. Sediment in streams: sources, biological effects, and control. Am Fish Soc Monogr 1995;7.

[21] Olsen CR, Cutshall NH, Larsen IL. Pollutant-particle associations and dynamics in coastal marine environments: a review. Mar Chem 1982;11:501–33.

[22] Baudardt J, Grabulos J, Barussean J-P, Lebaron P. *Salmonella* spp. and fecal coliform loads in coastal waters from a point vs. nonpoint source of pollutants. J Environ Qual 2000;29:241–50.

[23] Mallin MA, Ensign SH, McIver MR, Shank GC, Fowler PK. Demographic, landscape, and meteorological factors controlling the microbial pollution of coastal waters. Hydrobiol 2001;460: 185–93.

[24] Mallin MA, Johnson VL, Ensign SH. Comparative impacts of storm water runoff on water quality of an urban, a suburban, and a rural stream. Environ Monit Assess 2009;159:475–91.

[25] Dennison WC, Orth RJ, Moore KA, Stevenson JC, Carter V, Kollar S, et al. Assessing water quality with submersed aquatic vegetation. BioSci 1993;43:86–93.

[26] Li X, Weller DE, Gallegos GL, Jordan TE, Kim H-C. Effects of watershed and estuarine characteristics on the abundance of submerged aquatic vegetation in Chesapeake Bay subestuaries. Est Coasts 2007;30:840–54.

[27] Berg L, Northcote TG. Changes in territorial, gill-flaring, and feeding behavior in juvenile Coho Salmon (*Onchorhynchus kisutch*) following short-term pulses of suspended sediments. Can J Fish Aquat Sci 1985;42:1410–7.

[28] Gardner MB. Effects of turbidity on feeding rates and selectivity of bluegills. Trans Am Fish Soc 1981;110:446–50.

[29] APHA. In: Greenberg AE, editor. Standard methods for the examination of water and wastewater. Washington, D.C.: American Public Health Association; 1998.

[30] Welschmeyer NA. Fluorometric analysis of chlorophyll *a* in the presence of chlorophyll *b* and phaeopigments. Limnol Oceanogr 1994;39:1985–92.

[31] USEPA. Methods for the aetermination of chemical substances in marine and estuarine environmental matrices. EPA/600/R-97/072. 2nd ed. Cincinnati, Ohio: National Exposure Research Laboratory, Office of Research and Development, U.S. Environmental Protection Agency; 1997.

[32] NCDENR. Cape Fear River Basinwide Water Quality Plan. Raleigh, NC: North Carolina Department of Environment and Natural Resources, Division of Water Quality/Planning; 2005. p. 27699–1617.

[33] Ensign SH, Halls JN, Mallin MA. Application of digital bathymetry data in an analysis of flushing times of two North Carolina estuaries. Comp Geosci 2004;30:501–11.

[34] Mallin MA, McIver MR, Ensign SH, Cahoon LB. Photosynthetic and heterotrophic impacts of nutrient loading to blackwater streams. Ecol Appl 2004;14:823–38.

[35] MacPherson TA, Mallin MA, Cahoon LB. Biochemical and sediment oxygen demand: patterns of oxygen depletion in tidal creeks. Hydrobiol 2007;586:235–48.

[36] Meyer JL. A blackwater perspective on riverine ecosystems. BioSci 1990;40:643–51.

[37] Lin J, Xie L, Pietrafesa LJ, Shen J, Mallin MA, Durako MJ. Dissolved oxygen stratification in two micro-tidal partially-mixed estuaries. Est Coastal Shelf Sci 2006;70:423–37.

[38] Mallin MA, Cahoon LB. Industrialized animal production—a major source of nutrient and microbial pollution to aquatic ecosystems. Popul Environ 2003;24:369–85.

[39] Cahoon LB, Hales JC, Carey ES, Loucaides S, Rowland KR, Nearhoof JE. Shellfish closures in southwest Brunswick County, North Carolina: septic tanks vs. storm-water runoff as fecal coliform sources. J Coast Res 2006;22:319–27.

[40] Smock LA, Gilinsky E. Coastal Plain blackwater streams. In: Hackney CT, Adams SM, Martin WH, editors. Biodiversity of the Southeastern United States. New York, NY: John Wiley and Sons, Inc; 1992.

[41] Leggett CG, Bockstael NE. Evidence of the effects of water quality on residential land prices. J Environ Econ Manag 2000;39:121–44.

[42] Cahoon LB, Mikucki JA, Mallin MA. Nutrient imports to the Cape Fear and Neuse River basins to support animal production. Environ Sci Technol 1999a;33:410–5.

[43] Mallin MA, Posey MH, McIver MR, Parsons DC, Ensign SH, Alphin TD. Impacts and recovery from multiple hurricanes in a Piedmont-Coastal Plain river system. BioSci 2002;52:999–1010.

[44] Mallin MA, Cahoon LB, Toothman BR, Parsons DC, McIver MR, Ortwine ML, et al. Impacts of a raw sewage spill on water and sediment quality in an urbanized estuary. Mar Pollut Bull 2007;54:81–8.

[45] Cahoon LB, Nearhoof JE, Tilton CL. Sediment grain size effect on benthic microalgal biomass in shallow aquatic ecosystems. Estuaries 1999b;22:735–41.

[46] Lewitus AJ, Schmidt LB, Mason LJ, Kempton JW, Wilde SB, Wolny JL, et al. Harmful algal blooms in South Carolina residential and golf course ponds. Popul Environ 2003;24:387–413.

[47] Cahoon LB, Owen DA, Creech EL, Hackney JH, Waters TW, Townsend EC. Eutrophication of Eastern Lake, Sunset Beach, North Carolina. J Elisha Mitchell Sci Soc 1990;106:1–6.

Analytical Measurements to Improve Nonpoint Pollution Assessments in Indiana's Lake Michigan Watershed

Julie R. Peller*, Erin P. Argyilan†, Jeremiah J. Cox*, Nicole Grabos†

*DEPARTMENT OF CHEMISTRY, INDIANA UNIVERSITY NORTHWEST, GARY, IN, USA

†DEPARTMENT OF GEOSCIENCES, INDIANA UNIVERSITY NORTHWEST, GARY, IN, USA

CHAPTER OUTLINE

7.1 Introduction

Lake Michigan's southern shoreline borders approximately 59 miles of Northwest Indiana, rendering Indiana an important Great Lakes state. A large industrial base has occupied the southern Lake Michigan shoreline, from the Illinois border of the lake to the Michigan border, since the turn of the twentieth century. This is due to the beneficial functions provided by the lake. When both Milwaukee, Wisconsin, and Chicago, Illinois are considered, the southern Lake Michigan shoreline is home to nearly 8 million people and considered heavily urbanized. Many municipal wastewater treatment plants discharge treated wastewater into the lake and combined sewer overflow (CSO) and sanitary sewer overflow (SSO) events increase municipal waste pollution. It is estimated 24 billion gallons of untreated sewage and storm water flow into the Great Lakes each year [1,2]. Over the past few decades, municipal waste has increased in complexity while wastewater treatment plants have aged, further stressing freshwater bodies with chemicals of emerging

Monitoring Water Quality, http://dx.doi.org/10.1016/B978-0-444-59395-5.00007-8

concern, a problem recognized worldwide [3–5] and highlighted in a recent United States Environment Protection Agency (USEPA) literature review report [6].

7.2 Background

The essential protection of natural bodies of water from industrial and wastewater point sources of pollution was established in 1972 with the enactment of the USEPA's Clean Water Act. Unfortunately, a significant amount of severe environmental degradation had taken place for decades in and around the Great Lakes, which is evident from the Great Lakes Areas of Concern (AOC) program, established by the USEPA to address severely degraded geographical areas in the Great Lakes basin [7,8]. In Indiana, the Grand Calumet River, which flows into Lake Michigan, remains on the AOC list, a clear example of the results of years of high levels of point and nonpoint source pollution [9]. While point-source pollution regulations are continually challenged and updated using the most current scientific knowledge, important attention has also turned to the effects from nonpoint sources. The less understood and often more complex nonpoint sources of pollution include atmospheric pollutant deposition, urbanization, leaking septic systems, and certain agricultural practices, among others.

In efforts to restore and protect waters from nonpoint source pollutants, the USEPA and other government agencies have adopted the watershed management approach, whereby water quality issues are addressed within a geographic area delineated by watershed boundaries [10]. The design and implementation of watershed management plans, including the responsibility for stream monitoring, often involves collaborations initiated by environmental nonprofit, organizations, municipalities subject to MS-4 regulations, or community groups. The general goals of a watershed management plan typically include decreasing nonpoint source pollution and nutrient inputs, improving biotic communities and the instream habitat, reducing pathogen (*Escherichia coli*) concentrations, increasing stakeholder education and involvement, and facilitating the implementation of best-management practices. The determination of the nonpoint sources and impacts of watershed pollutants, however, is complex and requires strategic management planning in collaboration with research science.

The Little Calumet-Galien watershed discharges to the heavily modified channel of the Little Calumet River which enters directly in to Lake Michigan at Burns Harbor, Porter County, Indiana (Figure 7-1). The water impairment categories listed for the Little Calumet-Galien watershed are pathogens, nutrients, and sediments. According to the 2010 USEPA Indiana Watershed Monitoring and Assessment Report [11], of the 67% of rivers and streams assessed in the state, approximately 69% are considered impaired, due mostly to pathogens, mercury, or polychlorinated biphenyl (PCB) contamination. Only 30% of the assessed miles support full-body contact recreational use. Of the lakes, reservoirs, and ponds in Indiana that were assessed, only 4% qualify as good, leaving 96% considered impaired, due mainly to mercury and polychlorinated biphenyl (PCBs). While the impairments are clearly measurable, the sources of pollution are mostly unknown.

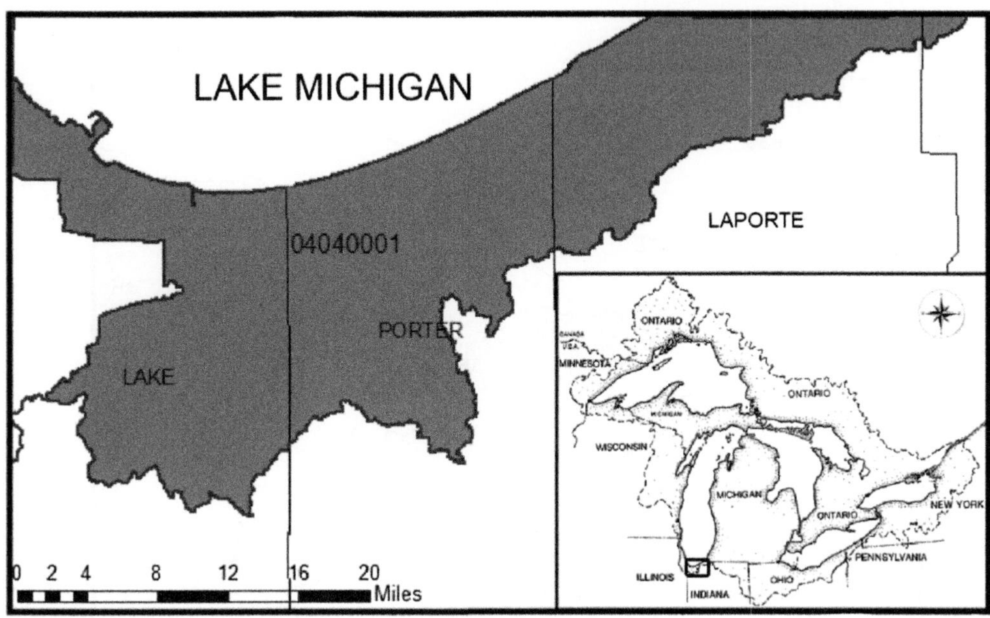

FIGURE 7-1 Map of the Little Calumet-Galien watershed (HUC04040001), located in Lake, Porter and Laporte counties of northwest Indiana and its relation to the Great Lakes (inset figure). (For color version of this figure, the reader is referred to the online version of this book.)

For the Indiana rivers and streams category, the majority of the sources are unknown. However, it is known that the prominent sources are associated with agriculture or sewage discharge. Alarmingly, although almost all of Indiana's 59 miles of Lake Michigan shoreline outside the Indiana Harbor fully support aquatic life, few of the shoreline waters are considered acceptable for full-body contact recreational use.

Early stages of the nonpoint source pollution study reported here are focused on the Salt Creek watershed in northern Indiana, which ultimately drains into Lake Michigan (Figure 7-2). Salt Creek and its tributaries flow into the Little Calumet River. Both waterways are considered impaired by the Indiana Department of Environmental Management (IDEM) Office of Water Quality, specifically because of excessive *E. coli* levels and impaired biotic communities. The Salt Creek Watershed Plan (SCWMP) was initiated in 2006 and implemented in 2008 by the Save the Dunes Conservation Fund (http://savedunes.org/) with collaboration and funding from IDEM. The IDEM-approved watershed management plan for Salt Creek requires monthly measurements of general chemistry parameters (dissolved oxygen, pH, temperature, and conductivity), flow, turbidity, and *E. coli* from May to October. Additional parameters periodically measured, but not required, include total suspended solids (TSS) and various ions: nitrate, nitrite, ammonium, total phosphorous, phosphate, and dissolved phosphorous. Biological parameters that are monitored monthly are macroinvertebrates and their habitats. Despite the diligence of monitoring, the resources for necessary research and evaluation of the specific causes of

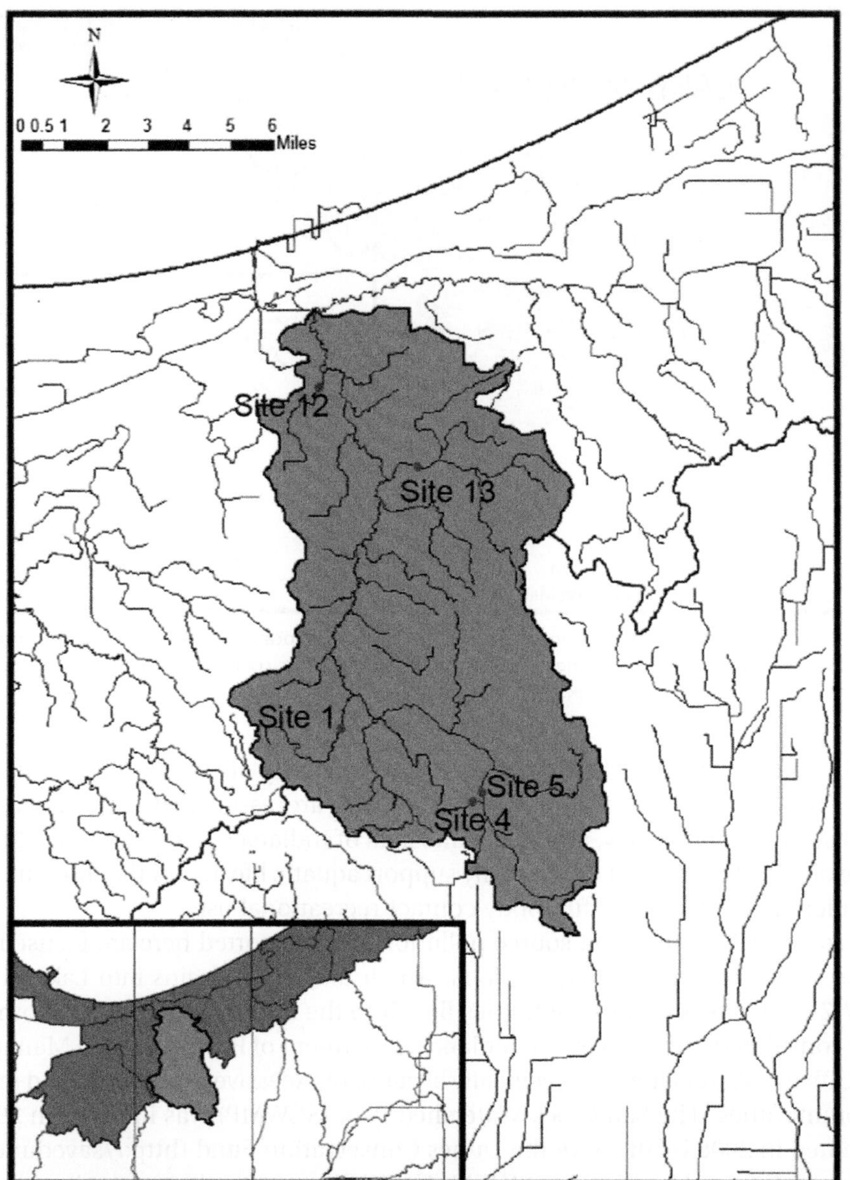

FIGURE 7-2 Map of the Salt Creek watershed (HUC04040001050) located in Porter County showing the locations of the five sampling sites (1, 4, 5, 12, 13). (For color version of this figure, the reader is referred to the online version of this book.)

water quality impairments are often limited. Moreover, the Salt Creek watershed continues to undergo land use changes, with the conversion of agriculture land to housing and the alteration of open land to a new hospital. This makes the integration of the watershed-management group with research science even more vital [12].

Targeted chemical and physical measurements and assessments are required to ascertain the sources and effects of nonpoint pollution on waterways that ultimately discharge into Lake Michigan. This pilot study was designed to utilize the strengths of the watershed management component in combination with research science measurements and analyses. Initial planning, sampling, and analyses occurred from summer 2010 through fall 2011 in the Salt Creek watershed, Porter County, Indiana and included certain parameters along the Lake Michigan shoreline during fall 2011 (Figure 7-2). The timing of the study involved sampling during the winter months (November–April), which are not monitored under the SCWMP. This provides data on the effects of winter processes, including road salting and snowmelt. The results offer insights into the possible impacts of projected climate-change weather patterns, which suggest more dynamic hydrologic processes in the winter season resulting from warmer air temperatures in the Great Lakes region [13,14].

The long-term analytical analyses plan focuses on the most probable pollutants present according to the land use in the specific area of the watershed. For example, waterways surrounded by agricultural practices should be monitored mainly for fertilizers and pesticides. The analytical scope of the study can be grouped into (1) common ions, such as chloride, sulfate, and nitrate; (2) pesticides; (3) polycyclic aromatic hydrocarbons (PAHs) and other hydrocarbons, indicators of transportation and industrial air pollutants, which will partition into the sediments; (4) fluorescent whitening agents (FWAs), indicators of household contamination; (5) total organic carbon (TOC), a general parameter potentially indicating sewage leakage or excess yard/livestock waste; and (6) TSS. The first year's data collection and analyses, presented here, focused on FWA studies, the loads of suspended solids and dissolved ions, and the seasonal variations of these parameters. It is the first step toward gauging some of the major sources of nonpoint pollution in Indiana's Lake Michigan watershed and shows the added understanding gained from research science involvement in watershed management plans.

7.3 Research Plan

7.3.1 Study Sites

The Salt Creek watershed (HUC 04040001050) is located in Porter County, Indiana within the Little Calumet-Galien watershed which is the only watershed in Indiana that discharges to Lake Michigan (Figure 7-1). The Salt Creek watershed is one of six subwatersheds that make up the Little Calumet-Galien watershed (Figure 7-2). The Salt Creek drainage area encompasses 49,573 acres and includes the large communities of Portage and Valparaiso, which are two of the fastest growing communities in northwest Indiana. According to the 2010 US Census, the population of Portage was 36,828, a 9.9% increase since 2000; the population of Valparaiso was 31,730, a 15.7% increase over the same period. The various land uses of the Salt Creek watershed are classified as recreational, industrial, commercial, agricultural, urban, forest, grassland, and residential

(SCWMP 2008). In the past 10 years, there has been significant conversion of agricultural land to residential use. For the first year of the study, five sites within the Salt Creek watershed were chosen for the integrated watershed testing and water collections from the 23 sites established as part of the SCWMP. Sites were selected on the basis of their different land uses and expected human impacts and are summarized in Table 7-1. Several samples were also collected along the Lake Michigan shoreline (Portage and Chesterton, IN) during fall 2011 and from two tributaries, Burns Ditch, and the Little Calumet River (west branch in Portage, IN).

7.3.2 Water Collections and Analyses

Instream measurements for temperature, pH, dissolved oxygen, and conductivity were determined in the Salt Creek streams using a Hach Quanta Datasonde, which was calibrated prior to sampling. Total discharge was determined at the time of sampling by segmenting the cross sectional area along permanent transects established at each site. Mean velocity was determined for each segment at 0.6 depth and multiplied by the segment width. Total discharge was calculated as the sum of the total discharges for individual stream segments. Samples for TSS analyses were collected from the center of the stream at one half the water depth and processed according to USEPA Method 160.2.

All water samples collected for chemical analytical analyses were collected in brown glass bottles to preclude plasticizers from the high-performance liquid chromatography (HPLC) fluorescence analyses and the water was transported in coolers. Within hours of the sampling, the water samples were vacuum filtered in a dark area of the lab using 0.2 um filter paper; the filtered water was refrigerated and stored in the dark to avoid photochemical changes.

For the analysis of FWAs, two liters of filtered water were passed through a solid phase extraction disk (ENVI-Disk™, purchased from Supelco Analytical) using a vacuum filtration assembly. The disks were utilized as follows: 5 mL of methanol was poured through the filter followed by 5 mL of water and then the solution of interest. Finally, 10 mL of methanol was used to recover the membrane-extracted solute. The methanol extracts were then analyzed using a Beckman System Gold HPLC fitted with a Supelco Discovery® C18 Column (5 μm, 250 mm × 4.6 mm) and a JASCO FP 1520 Fluorescence Detector to separate and quantify the fluorescent compounds. The excitation/emission wavelengths for all samples were 350 and 430 nm. The mobile phase for the chromatography utilized two solvents: acetonitrile and 0.45% aqueous tetrabutylammonium hydrogen sulfate. The isocratic solvent flow consisted of 15% acetonitrile and 85% aqueous ammonium salt at a flow rate of 1 mL/min. Photine CBUS (benzenesulfonic acid, 2,2′-([1,1′-biphenyl]-4,4′-diyldi-2,1-ethenediyl) bis-disodium salt (CAS-No. 27,344-41-8), a common FWA and verified in our lab in to be present in treated wastewater effluent, was donated by Keystone Chemicals and used as the standard for comparison.

Ion chromatography was performed with a Waters HPLC system equipped with an IC-Pak™ Anion HR column and a Waters conductivity detector. Concentrated sodium

Table 7-1 Sampling Sites, Research Science Assessments Planned for the Long Term Nonpoint Pollution Studies, and Ranges for the General Measurement Parameters

	Site 1	**Site 4**	**Site 5**	**Site 12**	**Site 13**
Site descriptions	Agriculture. Treated WW effluent from affluent community. Thermal pollution.	Agriculture. Nursery upstream. Low flow and volume.	Highway department road salt storage. High wetland density. Large number of septic systems.	Several feet from busy county road. History of high conductivity. Many outfall pipes.	Residential. Agriculture. Downstream from new hospital (construction underway 2010–12).
Analytical measurements					
Pesticides/ fertilizers	◆	◆			◆
Fluorescent whitening agents	◆		◆	◆	◆
Anions	◆	◆	◆	◆	◆
PAHs/hydrocarbons			◆	◆	
TOC	◆	◆	◆	◆	◆
Total suspended solids	◆	◆	◆	◆	◆
Temperature range (°C)	6.81–15.78	–	0.26–11.74	−0.06–10.88	−0.10–13.66
Conductivity range (nnhos/cm)	663–909	–	450–758	892–2710	484–954
pH range	7.06–7.91	–	7.14–8.10	7.19–7.85	7.21–8.57
Dissolved O_2 range (mg/L)	6.97–8.47	–	6.54–11.05	4.02–10.88	6.10–10.64

borate gluconate was diluted with water and mixed with n-butanol and acetonitrile, as specified by the Waters care and use manual. A stock solution consisting of fluoride (1 ppm), chloride (2 ppm), nitrite (4 ppm), bromide (4 ppm), nitrate (4 ppm), phosphate (6 ppm) and sulfate (4 ppm) was prepared and run prior to all the sample analyses. Additional standards were prepared for the ions that were quantified: nitrate, nitrite, chloride and sulfate.

7.4 Results and Discussion

Five stream sites in the Salt Creek Watershed were studied from the fall of 2010 through spring 2011, months when stream monitoring was not required by the watershed management plan. Throughout this time, the on-site measurements of temperature, pH, dissolved oxygen and conductivity (ranges shown in Table 7-1 for Sites 1, 5, 12 and 13), and flow were collected. Table 7-2 shows an example of the stream parameter data collected from Site 13 (Figure 7-2) in relation to local weather. The measured values were relatively consistent throughout the months of changing weather. The dissolved oxygen varied from 6.10 mg/L in September (temperature = 19.3 °C) to 10.64 mg/L in February, demonstrating the expected relationship to temperature. Air-saturated water at 25 °C and sea level has an oxygen content of 8.6 mg/L, compared to the increased solubility of oxygen, 14.6 mg/L, at 0.0 °C. Conductivity and pH varied from 484 to 954 mmhos/cm and 7.21 to 8.57, respectively. The temperature range for Site 1, shown in Table 7-1, is worth noting. This site is downstream from a community wastewater treatment facility, which is a source of thermal pollution. Different from the other sites, the temperature never dropped below 6.8 °C.

FWAs, also known as optical brighteners (OBs), are chemical compounds added to most laundry detergent formulations and other common products, such as household cleaners, toilet and other papers, fabrics, and plastics. Laundry activities produce a significant percentage of household wastewater and the presence of anthropogenic chemicals, such as FWAs, in natural bodies of water may be linked to improper sewage water disposal, such as leaking septic systems [15]. FWAs and other commonly-used compounds have been detected in many sewage influents and effluents and, therefore, can be utilized as markers of human contamination [16,17]. During our months of sampling, 2 L water samples were collected 6 inches below the water surface in large brown bottles and concentrated to analyze for the presence of FWAs. In our unreported work on the detection and measurement of FWAs, the method described in the experimental section was developed. Several treated wastewater effluent samples were tested with this method and the presence of Photine CBUS, one of the most highly utilized FWAs, was verified and quantified.

Over the first few months, many water samples from the Salt Creek watershed were tested for FWAs and none tested positive. We expected to detect FWAs from Site 1, because of the presence of an upstream community sewage treatment plant, and possibly at the other sites where septic fields are present. The fluorescent compounds undergo

Table 7-2 Water Collection and On-site Measurement Dates, Weather Conditions, and Data for Site 13 in the Salt Creek Watershed, Porter County, Indiana

Week	Date	Weather Description (Temp in °F)	Dissolved Oxygen, mg/L	pH	Temp. °C	Conductivity, mmhos/cm	Discharge, L/s
1	September 10, 2010	Bright and sunny, very little rain past week	9.88	8.23	16.5	842	70.1
2	September 17, 2010	Sunny with little rain, low 70s					–
3	September 24, 2010	Rain in the morning, cloudy and windy	6.10	7.95	19.3	852	85.7
4	October 8, 2010	Breezy, partly cloudy	8.40	8.08	11.1	848	83.8
5	October 15, 2010	Overcast to sunny, high 50s	7.72	8.03	12.5	888	89.8
6	October 22, 2010	Sunny, no precipitation last week, high 50s	9.52	8.57	7.7	861	72.6
7	November 5, 2010	Precipitation previous day, flurries/ sleet, upper 30s					
8	November 19, 2010	Cold and windy, no precipitation	10.49	8.28	4.6	881	91.6
9	December 10, 2010	1–2-inch snow, ice over most sites, cold and sunny	7.52	7.91		–	135.5
10	February 11, 2011	Heavy snowfall: over 1-inch snow in areas, freezing	10.64	7.41	−0.1	954	
11	February 15, 2011	Tons of snow melting! Warm					321.4
12	February 18, 2011	Most snow melted, cold	9.42	7.21	4.7	484	2412.0
13	March 4, 2011	About 0.5-inch rain, high 50s	10.25	7.76	8.1	730	422.4
14	April 1, 2011	0.1-inch rain/sleet in early morning, 40s					
NC	June 29, 2010		5.99	7.64	22.4	603	
NC	July 21, 2010		7.32	7.95	23.7	818	
NC	August 26, 2011		8.92	8.26	19.9	833	

NC = no water collected.

photolytic isomerization when exposed to ultraviolet light, and are rendered nonfluorescent. In the shallow streams (depth $<$ 8 inch), it is highly probable that these compounds would be exposed to sunlight and photolyze, and, therefore, the shallow streams were not sampled. But three of the five sampling sites contained water at a depth greater than 8 inches and sampling was performed below this depth. As a result of the unexpected absence of the FWAs, we conducted a simple experiment to establish the residence time, or persistence, of FWAs in these stream waters.

A 10-gallon fish tank was filled with 7 L of water and 2 inches of sediment collected from the Salt Creek watershed. The tank of stream water was kept in the dark and fitted with an aerator to simulate natural conditions. A baseline water sample was taken to ensure that the FWA was not present. The tank water was then spiked with the Photine CBUS for a concentration of 7 ppm. A sample was taken after 1 h; the concentration of the FWA decreased to 0.5 ppm and after a few hours was no longer detected. While a sediment extraction was not performed to verify the presence of the CBUS in the sediment, the results of this simple experiment imply that Photine CBUS preferentially partitions on the sediment, as is often the case with compounds with large nonpolar structures [18]. Although this FWA contains sulfonate groups (Figure 7-3) that offer some solubility in water, it is predominantly a hydrocarbon structure and will preferably partition into a more nonpolar medium, such as suspended particulates, sediments or fibers [19]. To determine the presence of FWAs in watershed streams, sediments need to be collected, extracted, and analyzed. This sediment-extraction procedure is now part of the 2011–2012 nonpoint-source study and should provide information on the possibility of leaking sewage waters in the watershed, in addition to potential links to high bacteria counts.

Dissolved anions in the filtered water samples were separated and analyzed by ion chromatography using an IC-Pak™ Anion HR Column and a Waters Conductivity Detector. Chloride and sulfate ions were detected in every sample, while nitrate, nitrite, fluoride, and phosphate were intermittently detected and often were below the detection limits of this method. When the detected chloride ion concentrations from four of the sampling sites are graphed over the course of the sampling weeks, concentration fluctuations suggest a response to seasonal stresses, with higher chloride concentrations in the winter months (Figure 7-4). This is probably associated with the management of

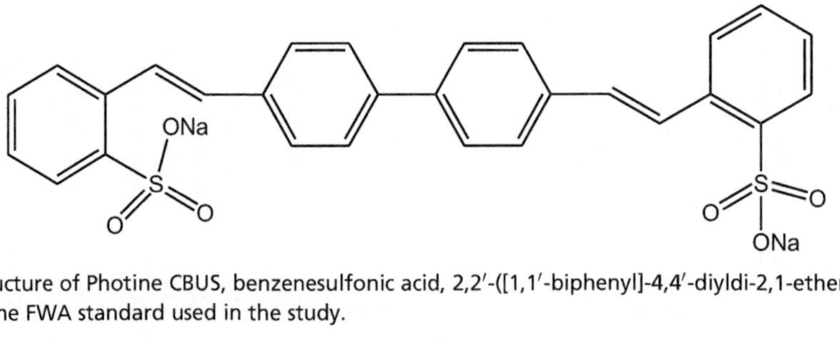

FIGURE 7-3 Structure of Photine CBUS, benzenesulfonic acid, 2,2′-([1,1′-biphenyl]-4,4′-diyldi-2,1-ethenediyl)bis,-disodium salt, the FWA standard used in the study.

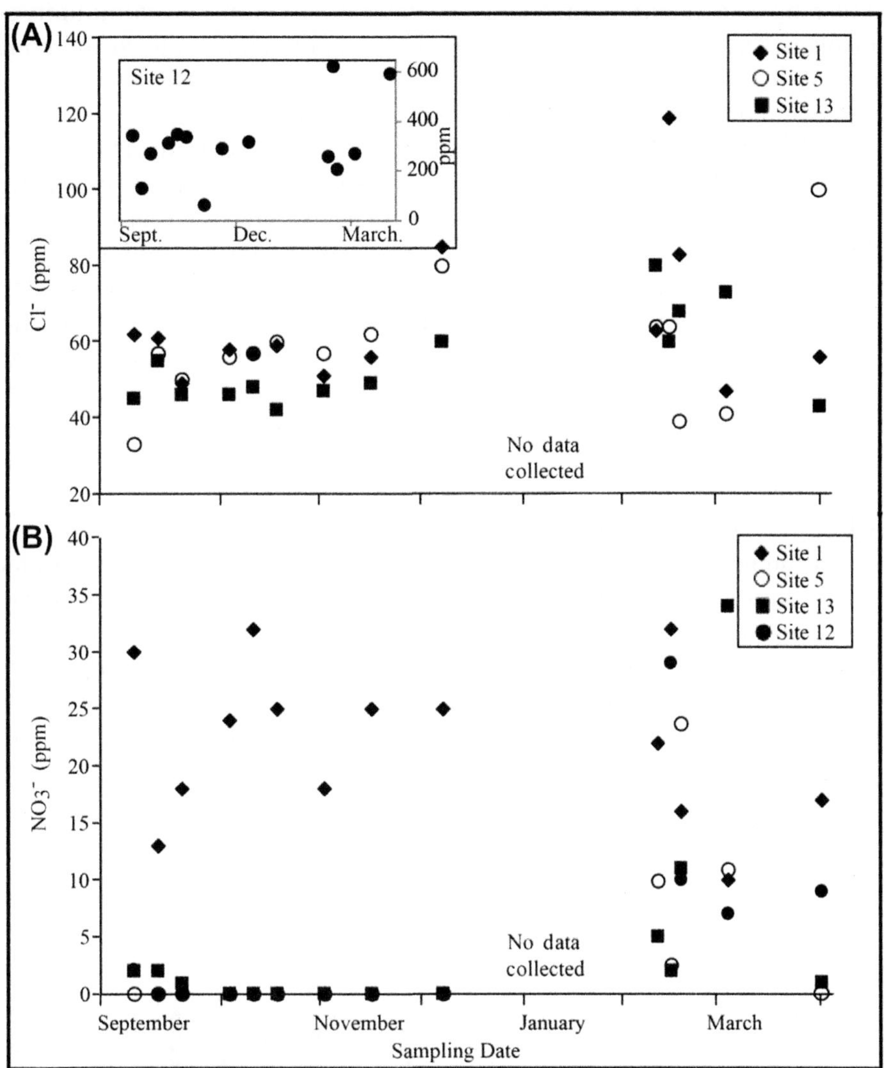

FIGURE 7-4 (A) Chloride ion concentrations, in ppm, for Sites 1, 5, and 13. The inset depicts the chloride ion concentration in ppm for Site 12. (B) Nitrate ion concentrations, in ppm, for sites 1, 5, 12, and 13.

freezing precipitation, i.e. the use of road salts in the winter months. The clearest seasonal trends were evident at Sites 5 and 13, which show the most direct rainfall-runoff response as determined through precipitation-discharge comparisons. A similar seasonal pattern is noted for the nitrate ion concentrations. The highest measured nitrate concentrations were associated with snowfall and snowmelt. Site 1 had measurable concentrations of nitrate for all the collected water samples, while the other sites recorded "low" or "not detected" for the more uneventful weather days (Figure 7-4). As noted in Table 7-1, Site 1 is surrounded by an agricultural area and also downstream from the community sewage treatment plant. One or both of these stressors probably contribute to the consistent nitrate load in this part of the watershed.

The average concentrations of chloride ions from the sampling sites between September 2010 and April 2011 varied from 41 ppm at Site 4 to 310 ppm at Site 12, a stream with a reputation for high conductivity values, according to the watershed-management data. Site 12 chloride concentrations consistently exceeded those of other sites by an order of magnitude (Figure 7-4). Analyzing specific ion concentrations can offer information on major nonpoint pollution stressors to the Salt Creek watershed. Chloride ions are considered conservative ions, dissolved substances with no true natural "sink" or means of elimination. Nonpoint sources of chloride in Indiana watersheds are water-softener discharges, septic systems, and road salts. Without natural processes to eliminate or utilize Cl^-, these dissolved ions accumulate. Interestingly, the chloride ion concentration and conductivity at Site 12 dropped to their lowest values of 208 ppm and 892 mmhos/cm respectively, during the large-discharge event associated with heavy snowmelt on February 18, 2011. This apparent dilution effect was reflected by low conductivity values at all sites, even though chloride ion concentrations were not directly variable. In fact, the conductivity values shown in Table 7-2 for Site 13 decreased after the snowmelt, contrary to the ion concentrations and especially the ion load data. This data highlights the complexity of the issues related to nonpoint-source pollution. High-discharge events can lead to deceptively low values in conductivity because of dilution effects, while the total load of chloride remains high, especially in areas that apply road salts (Figure 7-5).

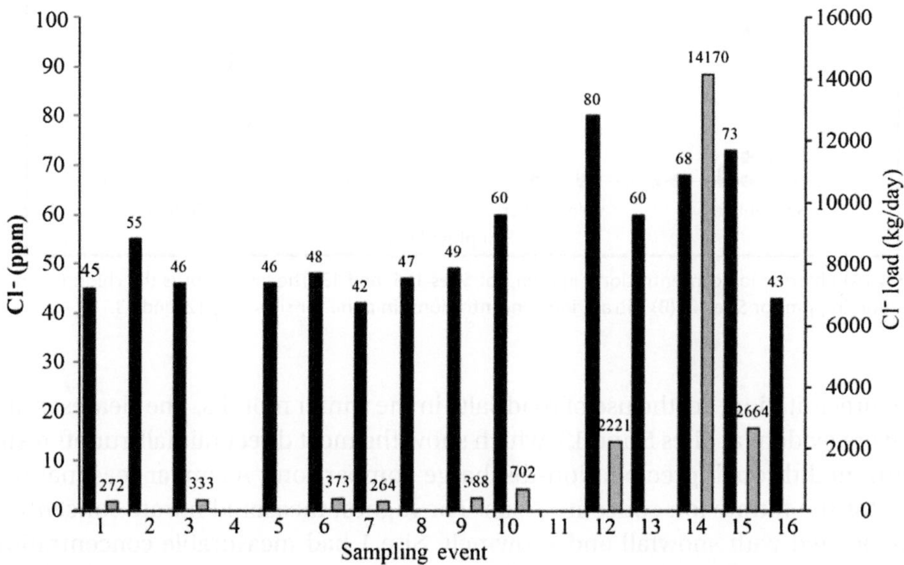

FIGURE 7-5 Chloride ion concentrations, in ppm, (black) and load, in kg/day, (gray) for Site 13 over the 16 sampling events beginning October 9, 2010.

Chloride ion concentrations in Lake Michigan have been rising steadily since the turn of the twentieth century [20–22]. This is confirmed by the results of our studies, where the measured chloride ion concentrations were consistently higher in the tributaries. The average concentration of chloride ions flowing through the Salt Creek watershed streams throughout our study period was 102 ppm. Salt Creek flows into the Little Calumet River, (which we tested and determined to be 105 ppm Cl^-), which then flows into Burns Ditch and to Lake Michigan. The average chloride ion concentration in Burns Ditch (drainage area = 331 square miles), sampled during fall 201, was 47 ppm. In addition to the nonpoint impacts, wastewater treatment plant effluents add significant amounts of chloride to the lake. Treated wastewater from one of the reclamation facilities which discharges into the Little Calumet River was measured at 157 ppm. In spring 2009, the average chloride ion concentration in the open waters of Lake Michigan was reported as 12 ppm [22], up from 8 ppm in the late 1970s, and significantly higher than the reported values of 1–3 ppm in the early 1900s [23]. For comparison, during the 2011 fall months, the chloride ion concentration averaged 16 ppm in the beach waters along the southern shoreline of Lake Michigan, which are impacted by several tributaries such as Burns Ditch.

Both nonpoint- and point sources of chloride are contributing to the rising concentration in Lake Michigan, according to the water samples tested in our studies. To more quantitatively assess the nonpoint-source contributions, load values were calculated using the discharge measurements at the sampling sites. Figure 7-5 shows the chloride ion concentrations and chloride ion loads (kg/day) for Site 13 which were calculated by multiplying ion concentrations by discharge rates. While the concentration for this site varied from 43 to 80 ppm Cl^-, the load, during and after the snowmelt, was significantly higher than the average loading. This seems to indicate that winter months with heavy precipitation can have a significant adverse impact on the watershed. The increase in chloride ions in surface water from the application of road salts was also reported in a recent study in Southern Indiana [24].

The specific chemical data offer insights into the main sources of nonpoint pollution entering the watershed. In contrast, specific dissolved substances (indicators) and seasonal patterns are not ascertained in the conductivity measurements collected as part of the watershed management plan (Table 7-2). The importance of monitoring the watershed throughout the year and with more specific measurements is clearly illustrated with the ion load data and further exemplified by the TSS loads.

TSS concentrations were determined for individual samples and converted to load (mg/s) using corresponding discharge data. Discharge values for Site 13 are shown in Table 7-2 and TSS concentrations and loads are shown in Figure 7-6. Transport of suspended sediment is considered a concern in local rivers as sediment loading contributes to stream embeddedness, degradation of aquatic habitat, and increased water temperatures. It also provides a mechanism for the transport of pollutants adsorbed to particle surfaces. While monitoring is not required in the SCWMP, there

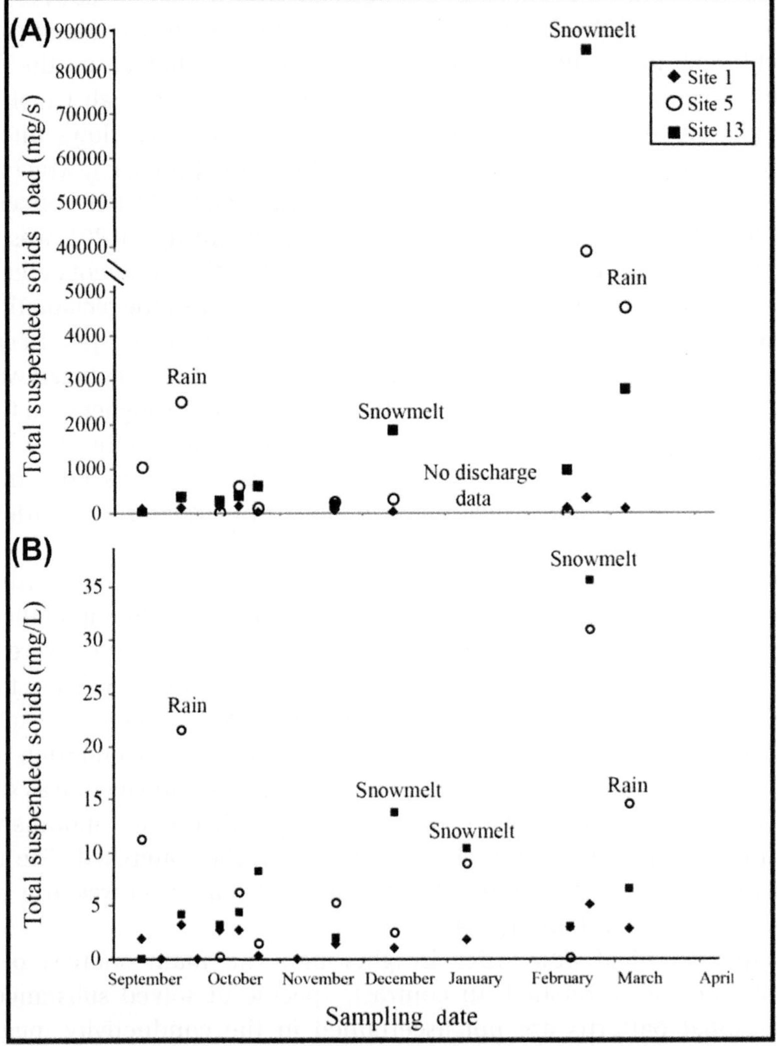

FIGURE 7-6 (A) TSS load in mg/s for Sites 1, 5, and 13. (B) TSS concentrations (ppm) for Sites 1, 5, and 13.

was great interest in understanding sediment dynamics in local rivers and streams in relation to the health of biotic communities. While total maximum daily loads have not been established for TSS in Salt Creek, the watershed management plan suggests that targets should be maintained below 80 mg/L, with levels ideally below 25 mg/L, based on assessed impacts on fish concentrations [25]. Data from Sites 4 and 12 were eliminated from the analysis since high levels of TSS were probably due to sediment disturbance during sampling of the shallow streams. Stream flow at Site 1 is affected by an upstream dam, which has historically had issues of sedimentation build-up behind the dam, probably affecting TSS and preventing a direct correlation between TSS and weather events.

Concentrations of TSS typically remained below 25 mg/L (the level recommended by the SCWMP), though a high degree of variability, reported in a previous study of Salt Creek by Morris and Simon, is also observable in this data [12]. However, the significance of studying TSS lies in documenting the variability in response to precipitation and discharge events. Sites 5 and 13, which are larger streams in diameter and discharge, consistently yield higher concentrations of TSS (Figure 7-6). Concentrations for TSS at Sites 5 and 13 responded to individual precipitation and snowmelt events with the most direct correlation observed at Site 13 (Figure 7-6B). The highest concentrations of TSS occurred at all sites in response to the meltwater events after the record snowfall of February 1, 2011. TSS values collected on February 18, 2011 exceeded 30 mg/L at Sites 5 and 13. When TSS concentration is combined with discharge data, the TSS load reflects the volume of sediment that was mobilized in Salt Creek in response to individual precipitation and melt events (Figure 7-6A).

The TSS data highlight the importance of hydrologic processes operating during the cold seasons when ground is frozen. Increased rates of runoff from the reduced infiltration capacity of the landscape produce flashier stream dynamics characterized by a rapid rise to peak flow and rapid fall to normal base flow levels. These conditions promote landscape and instream erosion that contribute to higher sediment loads. The results of this study and the prediction for increased winter rainfall and snowmelt under projected climate-change scenarios suggest that steam erosion and sediment loading will continue to increase and degrade water quality as well as the habitat in the Great Lakes region in the twenty-first century [26].

The outcomes of the first-year study integrating research science with the watershed management plan in monitoring the Lake Michigan watershed in northern Indiana demonstrate the substantial increase in scientific knowledge gained from such partnerships. While basic measurements of pH, dissolved oxygen and other parameters offer an idea of the waterway's health, especially within a long-term context or with land-use changes, the determination of nonpoint sources and impacts requires more specific, analytical, measurements and assessments. Concentrations and loads of specific ions point to certain probable nonpoint sources of pollution, and vary with weather conditions. This type of information was not seen in the conductivity measurements, which were affected by many factors and interferences. A major contribution of this study lies in the recognition that instream water chemistry data in combination with stream-discharge data at the subwatershed scale, can provide a more comprehensive understanding of the potential impacts of pollution downstream in Lake Michigan.

7.5 Conclusions

This 1-year pilot study merged the strengths of the Salt Creek Watershed Management Plan in Porter County Indiana with analytical science measurements and analyses, to more accurately assess and understand the nonpoint sources of pollution that potentially affect Lake Michigan's water quality. In combination with the watershed management plan's basic measurements of pH, dissolved oxygen, and conductivity, a design to better understand the waterway's health and nonpoint pollution challenges was initiated. The first-year study focused on measurements of dissolved anions, FWAs, total suspended solids, and discharge loads from September 2010 through April 2011, with assistance from people intricately involved in the Salt Creek watershed. The main conclusions were that the chloride, nitrate, and TSS concentrations and loads were associated with seasonal and weather stresses, and the chloride and nitrate concentrations were linked to specific nonpoint pollution sources. The absence of FWAs in all water samples prompted laboratory experiments, which indicated that the compounds are more inclined to adsorb to sediments in waterways with high sediment-to-water ratios.

Acknowledgments

Student support for this project was provided by the Indiana University Northwest Undergraduate Research Fund and the National Science Foundation's Louis Stokes Alliance for Minority Participation Grant. Special gratitude is extended to Jennifer Birchfield, the Water Program Director for Save the Dunes of Northwest Indiana for her expertise, advice and assistance on the Salt Creek Watershed.

References

[1] USEPA. A screening assessment of the potential impacts of climate change on combined sewer overflow (CSO): mitigation in the Great Lakes and New England regions. Washington, DC: United States Environmental Protection Agency; 2008.

[2] USEPA. Combined sewer overflows to the Lake Michigan Basin. Washington, DC: United States Environmental Protection Agency; 2007.

[3] Kolpin DW, et al. Pharmaceuticals, hormones, and other organic wastewater contaminants in US Streams, 1999–2000: a national reconnaissance. Environ Sci Technol 2002;36(6):1202–11.

[4] Lindqvist N, Tuhkanen T, Kronberg L. Occurrence of acidic pharmaceuticals in raw and treated sewages and in receiving waters. Wat Res 2005;39(11):2219–28.

[5] Kasprzyk-Hordern B, Dinsdale R, Guwy AJ. The occurrence of pharmaceuticals, personal care products, endocrine disruptors and illicit drugs in surface water in South Wales, UK. Water Res 2008;42(13):3498–518.

[6] USEPA. Treating contaminants of emerging concern. A literature review database. Washington, D.C: U.S. Environmental Protection Agency, Office of Water; 2010.

[7] USEPA. Areas of concerns (AOCs). Available from: http://www.epa.gov/glnpo/aoc/; [accessed 07.09.11].

[8] Labus P, Whitman RL, Nevers MB. Picking up the pieces: natural areas in the post-industrial landscape of the Calumet region. Nat Areas J 1999;19(2):180–7.

[9] Simon TP, Bright GR, Rudd J, Stahl J. Water quality characterization of the Grand Calumet River basin using the index of biotic integrity. Proc Indiana Acad Sci 1988;98:257–65.

[10] USEPA. Monitoring guidance for determining the effectiveness of nonpoint source controls. Washington, D.C: U.S. Environmental Protection Agency; 1997.

[11] USEPA. Watershed Assessment, Tracking & Environmental Results. Available from: http://iaspub.epa.gov/waters10/attains_state.report_control?p_state=IN&p_cycle=2010&p_report_type=A; 2010 [accessed 09.01.12].

[12] Morris CC, Simon TP. Evaluation of watershed stress in an urbanized landscape in southern Lake Michigan. In: Vaughn JC, editor. Watershed: management, restoration and environmental impact. Nova Science Publishers, Inc; 2009.

[13] Patz JA, Vavrus SJ, Uejio CK, McLellan SL. Climate change and waterborne disease risk in the Great Lakes region of the US. Am J Prev Med 2008;35(5):451–8.

[14] Parry M, Canziani O, Paultikof J, van der Linden P, Hanson C. Climate change 2007: impacts, adaptation and vulnerability. Intergovernmental Panel on Climate Change. In: Parry M, et al, editors. Contribution of working group II to the fourth assessment report of the intergovernmental panel on climate change. Cambridge, UK: Cambridge University Press; 2007.

[15] Boving TB, Meritt DL, Boothroyd JC. Fingerprinting sources of bacterial input into small residential watersheds: fate of fluorescent whitening agents. Environ Geol 2004;46(2):228–32.

[16] Hagedorn C, Weisberg SB. Chemical-based fecal source tracking methods. In: Hagedorn C, Blanch AR, Harwood VJ, editors. Microbial source tracking: methods, applications and case studies. New York: Springer; 2011. p. 189–206.

[17] Glassmeyer ST, et al. Transport of chemical and microbial compounds from known wastewater discharges: potential for use as indicators of human fecal contamination. Environ Sci Technol 2005;39:51–7.

[18] Fenner K, et al. Developing methods to predict chemical fate and effect endpoints for use within REACH. CHIMIA 2006;60(10):683–90.

[19] Iamazaki ET, et al. A morphological view of the sodium 4,4'-distyrylbiphenyl sulfonate fluorescent brightness distribution on regenerated cellulose fibers. J Appl Polym Sci 2010;118(4):2321–7.

[20] Sonzogni WC, Richardson W, Rodgers P, Monteith TJ. Chloride pollution of the Great Lakes. J Water Poll Control Fed 1983;55(5):513–21.

[21] Chapra SC, Dove A, Rockwell DC. Great Lakes chloride trends: long-term mass balance and loading analysis. J Great Lakes Res 2009;35:272–84.

[22] USEPA. Great Lakes Limnology Program. Available from: http://www.epa.gov/glnpo/monitoring/limnology/index.html; 2011.

[23] Rockwell DC, DeVault DS, Palmer MF, Marion CV, Bowden RJ. Lake Michigan intensive survey. Great Lakes National Program Office. Chicago: United States Environmental Protection Agency; 1978.

[24] Gardner KM, Royer TV. Effect of road salt application on seasonal chloride concentrations and toxicity in south-central Indiana streams. J Environ Qual 2010;39(3):1036–42.

[25] Salt Creek watershed management plan. Porter County, IN: IDEM; 2008.

[26] Angel JR, Kunkel KE. The response of Great Lakes water levels to future climate scenarios with an emphasis on Lake Michigan-Huron. J Great Lakes Res 2010;36(SI):51–8.

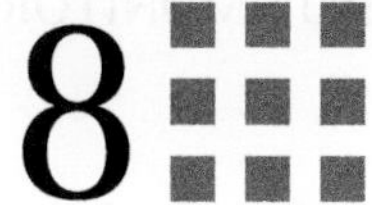

8 Real-Time and Near Real-Time Monitoring Options for Water Quality

Natalie Linklater, Banu Örmeci*

DEPARTMENT OF CIVIL AND ENVIRONMENTAL ENGINEERING, CARLETON UNIVERSITY, OTTAWA, ON, CANADA

*CORRESPONDING AUTHOR

CHAPTER OUTLINE

Monitoring Water Quality, http://dx.doi.org/10.1016/B978-0-444-59395-5.00008-X

8.1 Introduction

Current water quality issues are complex and involve a wide range of chemical and microbial contaminants of concern. Coupled with increasing urban populations and aging infrastructure, poor water quality can have a large impact on public health. Modern water management requires reliable and quick characterization of contaminants to allow for a timely response. Real-time monitoring enables a quick response to water quality concerns that arise from natural or intentional contamination and allows for the greatest protection of public health.

Ideally, all water-quality monitoring would be performed in real-time to allow for the most accurate and precise view of water quality. However, there are relatively few technologies that are able to provide true real-time measurements. Traditional water quality monitoring involves on-site sampling and transportation of samples to testing facilities to determine the physical, chemical, and biological characteristics. Laboratory methods are lengthy, expensive and samples may be compromised during transportation. With ever changing water characteristics, results obtained from these trials do not necessarily reflect current characteristics of water.

Recent advancements in both sensor technology and computer networks have brought about new technologies that can determine water characteristics in a dynamic way. Online real-time monitoring and screening methods can be used to provide up-to-date information about water systems and warn against contamination. This enables faster response times and quicker adjustments to treatment methods, which reduces the risk to public health. Real-time information can be used to assess changes to water quality and identify trends, determine the state of water quality and ecosystem health, identify emerging issues and contaminants, and achieve rapid screening of water for toxic substances and pathogens. Another important application is the monitoring and optimization of water- and wastewater-treatment processes and ensuring compliance to water-quality standards. In spite of the need, presently there are limited options for the implementation of real-time or near real-time (within 1–4 h) monitoring of water quality.

With limited funds and increasingly complex water-monitoring issues, municipalities and water utilities are looking to apply water-quality monitoring strategically. Typically, a tiered approach to water-quality monitoring is preferred. Such a system would monitor for general changes in water quality and screen for possible contaminants. Then, more specific testing would identify the type and extent of contamination [1]. A strategic approach to water monitoring may also include the selection of specific contaminants to be tested and frequency of testing. The number of testing methods selected may depend on financial restriction as well as regulatory requirements. Prioritization should be given to tests that are easy to perform (especially if they can be done on site), can detect substances that present high risks to public health, or can give a wide description of the quality of the water [2].

Real-time and near real-time detection methods are more limited compared to laboratory methods, but their variety and capabilities are continuously improving.

The majority of the current real-time monitoring applications are based on one or more of the following water quality parameters: turbidity, conductivity, temperature, dissolved oxygen (DO), pH, and chlorophyll-*a*. More recently systems that can measure ultraviolet absorbance at 254 nm (UVA) and total organic carbon (TOC) have also been used as indirect measurements of dissolved organic matter (DOM). These parameters are relatively easy to measure and provide useful information on the daily, weekly, monthly and seasonal changes in water quality. Sudden and uncharacteristic changes in these parameters also serve as an early warning system against intentional contamination of water with toxic chemicals or biological agents. However, none of these parameters can identify a specific chemical or a biological agent, and they merely serve as a screening method. Efforts continue to employ advanced spectroscopy and molecular methods in in-situ detectors, but the majority are still in the research phase and are not ready for field applications. Other methods include using organisms such as fish, clams, mussels, daphnia, and algae as biomonitors and the use of bioluminescence-based assays for toxicity testing.

The goal of this study is to provide an in-depth literature review of existing and emerging technologies that can be used for real-time or near real-time monitoring of water quality. Technologies that are well established or in the last stages of development are included in this review and those that are still in the early research phase have been omitted due to space limitations. Commercially available instruments are also included in the review with a description of their capabilities and limitations. It is important to note that the authors do not endorse or recommend the instruments mentioned in the review and do not have any relationship with the manufacturers. The review provides a summary of a wide range of physical, chemical, and microbial techniques that can be used for monitoring water quality in reservoirs, treatment plants, and distribution systems; and it aims to be a valuable resource for researchers, practitioners, and policy makers in the field.

8.2 Monitoring General Water Quality Parameters

Online sensors that monitor physical and chemical characteristics of water are widely available and are used by most modern water-treatment facilities and municipalities. They can measure single or multiple parameters at once. Available single-parameter sensors can measure chlorine residual, temperature, DO, oxidation-reduction potential (ORP), pH, conductivity, nitrogen and ammonia. Membrane, ion-selective, colorimetric, electrochemical or optical electrodes are typically used in these sensors. Table 8-1 provides a summary of technologies used for different types of single-parameter sensors. Multi-parameter sensors, on the other hand, are able to measure multiple parameters with a single instrument and are typically composed of a number of single-parameter sensors with a shared casing.

Two strategies have been suggested to use single- and multi-parameter sensors to screen for contaminants in water. First, a sensor could be used to monitor changes in the

Table 8-1 Technology used in Online Sensors to Monitor for Physical and Chemical Parameters

Parameter	Technology
Chlorine	Calorimetric and membrane electrode
	Ion-selective electrode (Cl^-)
	Electrochemical reduction of chlorine species at a microelectrode
Temperature	Thermistor
Dissolve oxygen	Membrane electrode
	Optical sensor
	Electrochemical voltammetric
Oxygen reduction potential	Potentiometric method
pH	Glass bulb electrode
	Potentiometric measurement with reference electrode
Turbidity	Mephelometric method
	Optical sensor
Conductivity	Conductivity cell method
Nitrogen compounds	Ion-selective electrode (NO_3)
Ammonia compounds	Ion-selective electrode (NH_4)

Source: Hasan et al., 2005 [1].

"state" of water, defined by the sum of physical and chemical parameters. An expected or typical state of water is described by baseline conditions determined for a specific site or water sample. A change of state, marked by a significant deviation from baseline conditions of any one or a combination of the parameters, would suggest that a contaminant has been introduced into the water. The second strategy attempts to establish contaminant signatures (e.g. UV and visible absorbance signatures) that could indirectly suggest a specific contaminant if introduced in the water [1].

The use of surrogate (or indicator) parameters has also been suggested as an initial step to warn of contamination. Parameters already being monitored online, such as pH, chlorine residuals, and conductivity, can change in the presence of a contaminant. If these changes can be identified, they can serve as a warning of contamination, trigger additional testing, and be used as a tool for early identification and possible solutions to the problem.

The United States Environmental Protection Agency (USEPA) performed a variety of studies to determine the long-term performance of multi-parameter sensors and also established baseline conditions for various waters. One of these studies [3] evaluated water-quality parameters for their ability to detect changes due to chemical, physical, and microbial contaminants. The study examined a small number of single- and multi-parameter commercial sensors to determine which water quality parameters were changed after the introduction of a variety of contaminants (Tables 8-2 and 8-3). The single-parameter sensors that were evaluated included ATI, Hach Model Cl-17, Hach 1720D, GLI Model phD, GLI Model 3422, and Hach Astro TOC UV Process Analyzer. The multi-parameter sensors that were evaluated included Dascore Six-Cense Sonde, YSI

Table 8-2 Single-Parameter Sensors Evaluated by Hall et al. [11]

Sensor Manufacture/Name	Parameter(s) Measured
ATI	Free chlorine
Hach Model A-15 Cl-17	Free/total chlorine
Hach 1720D	Turbidity
GLI Model phD	pH
GLI Model 3422	Specific conductance
Hach Astro TOC UV Process Analyzer	Total organic carbon

Table 8-3 Multi-Parameter Sensors Evaluated by Hall et al. [11]

Sensor Manufacture/Name	Parameters Measured
Dascore Six-Cense Sonde	Specific conductance, DO, ORP, pH, temperature and free chlorine
YSI 6600 Sonde	Specific conductance, DO, ORP, pH, temperature, ammonia–nitrogen, chloride and turbidity
Hydrolab Data Sonde 4a	Specific conductance, DO, ORP, pH, temperature, ammonia–nitrogen, chloride, nitrate–nitrogen and turbidity

6600 Sonde, and Hydrolab Data Sonde 4a. The contaminants investigated included secondary wastewater (non-chlorinated), potassium ferricyanide, malathion (pesticide), glyphosate (herbicide), arsenic trioxide, nicotine, aldicarb (pesticide), and *Escherichia coli* K-12. Contaminants were injected into an uncontaminated pipe loop where the sensor responses could be observed. The results showed that each contaminant induced a response from at least one parameter. As predicted, TOC increased with the addition of carbon containing contaminants such as wastewater, nicotine, herbicides and pesticides. In addition, the specific conductance was responsive to the introduction of ionic compounds such as wastewater and potassium ferricyanide. The ORP was responsive to oxidizing compounds or an increase in chlorine demand, which was the case with wastewater, potassium ferricyanide, pesticides and the herbicide. The most responsive parameters were free chlorine, TOC, specific conductance and ORP. Free chlorine was the only parameter that responded to all contaminants except potassium ferricyanide. Chlorine reacts readily with many compounds and thus a decrease of free chlorine is a good indicator of contamination. However, the single use of free chlorine as an indicator of contamination was not recommended as some contaminates may not readily react with free chlorine or provide false positives [3].

Hach Corporation (Colorado, USA) manufactures the GuardianBlue Early Warning System, which detects, alerts, and classifies a wide variety of contaminants including cyanide, anthrax, arsenic and pesticides. The system is also able to detect and classify events such as water-main breaks, caustic overfeeds to cross connections and aging infrastructure problems such as corrosion. The GuardianBlue Early Warning System is comprised of a water panel, a TOC analyzer, an automatic sampler, and an event monitor. The water panel is a multi-parameter instrument that is able to measure free or total chlorine, conductivity, pH, turbidity, temperature and pressure. Total organic carbon is also measured with a TOC analyzer. Every 60 s the water is analyzed and compared to baseline data via the event monitor. An alarm is triggered if a user-set threshold exceeds the baseline data. Events are then compared to "fingerprints" stored in the device's library and a second alert is given if a match is found. If no match is found, the event is stored in the library for future reference. A digital signal transmits data between the water panel and event monitor and finally the automatic sampler will collect a sample from water if an event is 'triggered' to allow for more detailed analysis.

Hach also manufactures a multi-parameter in-pipe probe, the PipeSonde, for water quality monitoring in distribution systems. Designed to be placed directly in a water-distribution pipe, this sensor measures pH, ORP, conductivity, turbidity, DO, line pressure and temperature. Data is also available remotely through a supervisory control and data acquisition system. To perform similar operations as the GuardianBlue system, a water distribution panel, a TOC analyzer and an event monitor trigger system can be added. The water distribution panel, coupled with the event-monitor trigger system provides real-time analysis of the state of the water and scans for contamination.

Horiba, Ltd. (Kyoto, Japan) has a series of portable single-parameter sensors that measure pH, DO, ORP, conductivity and ions such as sodium, potassium and nitrate. Horiba's multi-parameter sensor can simultaneously measure pH, ORP, DO, conductivity, salinity, total dissolved solids (TDS), turbidity, and water depth. Hanna Instruments (Rhode Island, USA) also provide single- and multi-parameter sensors. Single-parameter sensors are available to measure DO, ORP, pH, turbidity, conductivity, TDS, and various ions. Multi-parameter sensors can measure pH, TDS, ORP, conductivity, turbidity, ammonium, nitrate, chloride, resistivity, TDS, salinity, and seawater specific gravity. Rosemount Analytical Inc. (California, USA) manufactures smart wireless sensors for liquid analytical measurements including pH, ORP, conductivity, dissolved and gaseous oxygen, chlorine, ozone, and turbidity.

The Six-CENSE Sonde (Dascore Inc., Florida, USA) is a multi-parameter in-line sensor that is able to monitor free chlorine, monochloramine, DO, pH, conductivity, oxidation-reduction potential, and temperature. The instrument utilizes electrochemical technology placed on a small ceramic chip and data can be remotely downloaded. As part of increased security measures, the Six-CENSE Sonde was used for water quality monitoring in the distribution system by the Mohawk Valley Water Authority in New York, USA [4].

The YSI 6600 V2 monitoring instrument (YSI Incorporated, Ohio, USA) is equipped with four optical ports available to measure conductivity, temperature, pH and ORP, as well as DO, turbidity, chlorophyll and blue-green algae. Measurements are done in situ and the instrument is able to record data for up to 75 days in 15-min intervals (limited by the life of the battery). The instrument can also calculate total dissolved solids, resistivity, and specific conductance. Each optical sensor is equipped with automated wipers to clean the lenses to prevent biofouling.

Liquid monitoring systems have spectrometric, ion-selective, electrochemical, and optical probes that can measure biochemical oxygen demand (BOD); chemical oxygen demand (COD); total, dissolved, and assimilable organic carbon; nitrate; nitrite; ammonia; potassium; free chlorine; chloride; TSS; turbidity; pH; oxygen reduction potential (ORP); conductivity; benzene-toluene-xylene; oxygen; ozone; and hydrogen sulfide. Alarm software that reacts to a change in the UV-absorption range is also available.

Absorption of different wavelengths of UV and visible light in water can be correlated to different contaminants. For example, DOM was shown to have a good correlation with UV absorbance at 254 nm; and 5-day biochemical oxygen demand (BOD_5) was correlated to absorbance at 440 nm [5]. In addition to UV-visible (UV-vis) spectroscopy, fluorescence spectroscopy can also be used for water monitoring. Fluorescence spectroscopy is an optical technique that uses three-dimensional excitation-emission matrices that are recorded at two distinct wavelengths. However, fluorescence measurements are affected by environmental factors and may not always be suitable for in situ measurements [6]. Synchronous fluorescence spectroscopy is a two-dimensional fluorescence technique that requires a single scan where the excitation and emission wavelengths are measured synchronously. This method is more suited as a real-time monitoring technique. Hur et al. [7] used synchronous fluorescence spectroscopy and found an excellent relationship with BOD in river water samples.

Dissolved organic matter is a mixture of humic acids, fulvic acids, acids of low molecular weights, carbohydrates, and proteins that contain both light-emitting and light-absorbing compounds. Because of this, analysis of DOM includes the use of light-absorption and fluorescence techniques. High-performance, size-exclusion chromatography (HPSEC) is able to provide insight into the composition of DOM as retention time in the chromatography is directly related to the size of a molecule [8]. HPSEC is often coupled with UV–vis or UVA measurements [6]. Her et al. [9] modified a commercially available TOC analyzer to improve the detection of DOM as a function of molecular weights. The modified instrument was used in conjunction with an HPSEC-UVA system. Data could be displayed in real time [9].

The majority of single- and multi-parameter sensors summarized above utilize established technology; however, emerging sensor technologies are also being developed. Dybko et al. [10], designed a multi-parameter sensor that utilized optical fibers to measure pH, concentration of calcium ions, and heavy metals. This device immobilized a series of indicators using chemical bonding on polymeric beads. Indicators used included neutral red for pH values from 6.5 to 7.8 and bromothymol blue for pH values from 7.7 to 9.4. Chlorophosphonazo III was used to indicate calcium concentration, which was sensitive to concentrations between 10^{-5} and 10^{-3} M. The indicator for total heavy metal concentration was 4-(2-pirydylazo) resorcinol, which is sensitive at concentrations between 10^{-7} and 10^{-4} M. The beads were placed in a glass tube through which water was allowed to flow. Tubes were illuminated with optical fibers and independent LED lights were used to measure each parameter at specific wavelengths:

590 or 630 nm for pH (depending on the pH range), 680 nm for calcium and 510 nm for heavy metal ions.

Innovations in microfabrication and sensor technology have lead to the replacement of large and bulky electrochemical sensors with miniaturized instruments that rely on new technologies, such as thick-and-thin film technology, silicon-based techniques and photolithography [11]. Microsensors offer several advantages including small size, portability, successful use in field applications, stability, efficiency, low energy usage, and acceptance of small samples and reagent volumes. Recent advances in automation and communication have also allowed sensors to be completely automated and equipped with flow-based delivery systems and wireless communication [11]. These developments allow true real-time measurements and continuous monitoring of a wide range of biological and chemical contaminants in water and open the doors to exciting innovations.

8.3 Monitoring Microbiological Contaminants

A microbiological contaminant is any living organism that can negatively affect the health of the public or the receiving body of water. Testing for microbiological contaminants is traditionally done using culture-based methods, where indicator bacteria are cultivated on selective media and the resulting number of colonies is counted. Depending on the type of agar used and organism selected, a plate count typically takes 24 h or more and requires the use of laboratory equipment, such as an autoclave and an incubator. In addition, if colonies are grouped or difficult to distinguish, plate counts can be prone to human error.

An indicator organism is a non-pathogenic organism that is easy to isolate and is present in polluted waters but not in treated waters [12]. Indicator organisms are selected because they are easy to culture and provide insight into the microbial quality of the water. A common indicator organism, *E. coli,* is typically found in feces and its presence in water is a good indication of fecal contamination. However, new pathogens have emerged, causing acute problems in humans and raising levels of concern that indicator organisms are not sufficient to describe the microbial quality of water. Pathogens of concern include but are not limited to *Cryptosporidium, Legionella, Campylobacter, Yersinia pestis, Mycobacterium avium* complex (Mac), *Aeromonas hydrophila* and *Helicobacter pylori* [13]. There is also an increased awareness of the vulnerability of water systems as vehicles for bioterrorism. An act of bioterrorism is the use of a living organism to cause harm, and typically a human-disease-causing strain of a bacteria, virus or protozoa is employed [14]. A good candidate for a bioterrorism agent is contagious, has a small infectious dose, is stable and can be dispersed in water, can cause a severe disease that is difficult to cure, and is resistant to chlorine. Possible candidates include *Bacillus anthracis, Clostridium botulinum* toxin, *Y. pestis, Variola major, Francisella tularensis,* filoviruses, and arenaviruses [15].

New and innovative monitoring technologies are needed to detect a large variety of microorganisms quickly and accurately. These technologies should ideally (1) require no sophisticated laboratory equipment, (2) require minimal sample preparation, (3) yield results quickly, (4) have low detection limits, (5) have high specificity for the targeted pathogens, and (6) differentiate between viable and nonviable pathogens. Sensors for living organisms typically target genetic material, proteins, or the adenosine triphosphate (ATP) of cells. Most sensors based on biological interactions have biological components and detection is performed by various methods such as light production or mass change. Concentration of the target molecule is typically needed.

Based on advances in microbiology, molecular biology, chemistry, and computerized imaging, a wide range of methods are being studied and developed for the rapid detection, identification, and quantification of microorganisms. In general, these methods rely on growth-based, viability-based, nucleic acid-based, cellular-component or artifact-based technologies, automated methods, or a combination of these [16]. Available microbiological methods can be listed as follows: ATP bioluminescence assay, adenylate kinase, autofluorescence, biochemical assays and physiological reactions, biosensors and immunosensors, carbon dioxide detection, changes in headspace pressure, colorimetric detection of carbon-dioxide production, concentric arcs of photovoltaic detectors with laser scanning, direct epifluorescent filter technique (DEFT), DNA sequencing, endospore detection, enzyme-linked immunosorbent assay (ELISA), flow cytometry, fluorescent probe detection, fatty acid profiles, Fourier transform infrared (FTIR) spectroscopy, gram stains, impedance, immunological methods, lab-on-a-chip (LOC) arrays, microarrays, and microchips, limulus amebocyte lysate (LAL) endotoxin testing, mass spectrometry, microcalorimetry, micro-electro-mechanical systems (MEMS), nanotechnology, near infrared spectroscopy (NIRS), nucleic acid probes, optical particle detection, polymerase chain reaction (PCR), Raman spectroscopy, ribotyping/molecular typing, solid phase, laser scanning cytometry, southern blotting/restriction fragment-length polymorphism, spiral plating, and turbidimetry [16]. Well-developed methods suitable for rapid monitoring of water supplies are summarized below.

8.3.1 Immunoassays

Immunoassays are typically performed on grab samples and are based on antigen–antibody reactions, which have high specificity to one another. A positive immunoassay will produce a visible light or color. An immunoassay requires a capture and a target molecule. A capture or recognition molecule is the part of the sensor that interacts directly with the sample, and can be either an antigen or an antibody. The target molecule of the microorganism interacts with the capture molecule of the sensor. Immunoassays are easy to use and require little specialized knowledge. However, they are prone to false–positive responses because molecules other than the target molecules may react with the capture molecules [1].

A strip test or lateral flow assay is a common type of immunoassay, used to detect antigens in water. A well-known example is the home pregnancy test. For this assay, an adsorbent membrane is fixed on a solid piece of plastic called the strip and each membrane can have a number of different stripes. Each stripe contains high concentrations of a specific antibody labeled with either a colored dye or a fluorescent agent. A liquid sample is applied to one end of the strip and allowed to diffuse. A color change or fluorescence response in any of the stripes indicates the presence of an antigen in the sample. A control stripe is often placed at the end of the strip to indicate the completion of the test. These tests are rapid, with results typically obtained within 15 min; however, they provide no quantitative information [1].

One of the issues with immunoassays is that they have relatively high levels of detection. A study tested three commercially-available, lateral–flow, immunoassay kits that were designed to test for *B. anthracis* in water, and reported that all three assays were able to detect *B. anthracis* at 10^6 spores [17]. BADD (Osborne Scientific, Arizona, USA) and SMART II (New Horizons Diagnostics, Maryland, USA) were able to consistently detect 10^5 spores, whereas Anthrax BTA (Tetracore Inc., Maryland, USA) detected 10^5 spores once in eight attempts. Both BADD and Anthrax BTA had a specificity of 100%, while SMART II showed a positive reaction with *Bacillus thuringiensis* resulting in a specificity of 75% [17]. It may be possible to lower the measurement limits if an electronic reader is used. Overall, lateral flow assays for the detection of biological contaminants in water are not suitable for trace elements but more suitable for large-scale contamination events or for preliminary screening purposes. Dilution of the contaminant in water may also reduce the effectiveness of an assay. Following an immunoassay, confirmation of contamination should be obtained by more precise measures. Concentration of a water sample using ultrafiltration techniques or centrifugation, is likely to improve detection limits [18].

Several immunoassays are available on the market for the detection of the main bioterrorism agents of interest, including anthrax, botulinum toxin, brucella, plague, ricin, and tularemia. These tests typically require small sample volumes and results are produced within 5–15 min. Some of these tests include RAMP (Response Biomedical Corporation, British Columbia, Canada) assays for anthrax, botulinum toxin, ricin, and smallpox; BioThreat Alert (Tetracore Inc., Maryland, USA) assays for abrin, anthrax, botulinum toxin, brucella, orthopox, plague, ricin, SEB (Staphylococcal enterotoxin), and tularemia; SMART (New Horizons Diagnostic, Maryland, USA) assays for anthrax, cholera, ricin, SEB, plague, tularemia, and botulism toxin; and BADD (ADVNT Biotechnologies, Arizona, USA) assays for anthrax, botulinum, ricin, SEB, and plague.

NASA's Jet Propulsion Laboratory (Pasadena, California, USA) developed a quantitative lateral-flow test strip to test water for *E. coli* in space. Test stripes are chemically treated with different compounds that change color in the presence of specific antigens. Bacteria with the target antigen cause the stripe to change color, and different color intensities indicate the concentration of the bacteria present in the water [19]. This type of test is advantageous because in addition to the confirmation of the presence of a biological agent, quantitative results can be obtained within minutes.

Zhao et al. [20] used nanogold-labeled probes to develop an immunochromatographic (IC) strip for *E. coli* O157:H7 based on the biotin-streptavidin system. The detection of

E. coli O157:H7 could be completed in less than 10 min. The IC strip was tested in 265 water samples and was shown to have 99.2% specificity and 100% sensitivity compared with culture-based methods.

Rule and Vikesland [21] prepared Raman labels by conjugating gold nanoparticles with commercial antibodies and dye molecules. Using surface-enhanced resonance Raman spectroscopy (SEERS), which is a technique that enhances Raman scattering by molecules adsorbed on metal surfaces, they were able to measure and differentiate *Cryptosporidium parvum* oocysts and *Giardia lamblia* cysts in water samples. Immunogold signal intensities were optimized to provide a sensitive method for multipathogen monitoring.

Zhao et al. [22] developed a method that simultaneously detects *Salmonella typhimurium*, *Shigella flexneri*, and *E. coli* O157:H7, using antibody-conjugated semiconductor quantum dots as fluorescence markers and magnetic microparticles for enrichment. Quantum dots are nanoparticles that are 20 times brighter and 100 times more stable than traditional fluorescent dyes [99]. Antibody-conjugated quantum dots could attach to the surface of bacterial cells selectively and specifically. The detection limit in a food matrix was 10^3 cfu/ml, and the method could be completed in <2 h. The method provides the potential to detect a panel of pathogens in environmental and food samples and to be applied for water assessment applications [22,23].

8.3.2 Real-Time Polymerase Chain Reaction

PCR is able to detect microorganisms, including bacteria, protozoa, and viruses through amplification of a target sequence of microbial DNA. The choice of primer, a specific DNA sequence, allows for the selection of different strains, species, and genus of a microorganism. PCR amplifies the selected DNA sequence several orders of magnitude, generating thousands to millions copies. Multiple primer pairs can also be used to simultaneously amplify several sequences for the detection of single or multiple strains of pathogens in a single reaction, which is known as multiplex PCR.

PCR amplification can be initiated with as little as one strand of DNA. This allows for much lower detection limits compared to other detection methods. In addition, microorganisms that cannot be cultured can be detected with PCR. However, PCR is not without its disadvantages. Amongst them, PCR requires specialized knowledge and equipment. In addition, microbial DNA must be extracted using a cell lysing procedure prior to the PCR run. Postprocessing techniques such as gel electrophoresis and photodocumentation are also required. The PCR method qualitatively determines if a certain DNA target is present. Since DNA is able to persist after the death of a cell, PCR is not able to determine the viability of the microorganism identified [24,25].

Improving on PCR, real-time quantitative PCR (qPCR) allows for the quick determination and quantitation of DNA and RNA using fluorescent DNA-binding probes. Compared to conventional PCR techniques, real-time polymerase chain reaction (RT-PCR) is able to reduce the time needed to produce results from 3 h to <1 h. While a preprocessing step is required to extract DNA, postprocessing steps are eliminated. During the PCR reaction, fluorescence increases as the reaction progresses and more of the targeted DNA is amplified. Also, fluorescence is recorded during each amplification cycle. The cycle number at which amplification reaches a concentration threshold is called the C_t value which is inversely proportional to the quantity of DNA in solution. Using calibration curves formulated with known amounts of DNA, one can determine the amount of DNA present in a solution. The more of the template that is detected at the beginning of the RT-PCR reaction, the fewer the cycles needed to reach the detection levels [25,26].

Compared to conventional plate count methods, RT-PCR has higher sensitivity, precision, and reliability. It is independent of culture conditions, nutrient conditions, and bacterial metabolic states. It also has the ability to process more than one reaction at once and is able to quantify microbial DNA including viral DNA, as well as RNA and mRNA [24–27].

Real-time PCR has been widely applied to water monitoring especially for the detection of human pathogens. Vaitomaa et al. [28] used RT-PCR to identify the microcystin-producing genus of cyanobacteria, such as *Anabaena* and *Microcystis*. Donaldson et al. [29] developed a real-time reverse-transcriptase PCR method to detect the presence of enteroviruses in surface waters.

Sharma and Dean-Nystrom [30] used RT-PCR to distinguish between highly virulent forms of *E. coli* O157:H7 from less virulent serotypes. Aslan et al. [31] used RT-PCR to identify and quantify human adenoviruses and enteroviruses in recreation and bathing waters, and reported that adenovirus was strongly associated with point sources of human fecal pollution. Zhou et al. [25] used RT-PCR to determine the total bacterial DNA present in water. *Naegleria fowleri* was successfully detected with an efficiency of 99% using a duplex RT-PCR by Behets et al. [32]. Helmi et al. [33] monitored *G. lamblia* and *C. parvum* for 2 years in the largest drinking water reservoir of Luxembourg using microscopy and RT-PCR techniques. They reported that only 25% of the samples that were positive by microscopy were confirmed by RT-PCR. Clark et al. [34] used RT-PCR to detect contamination from bacterial pathogens *E. coli* O157:H7, *S. typhimurium, Campylobacter jejuni, Pseudomonas aeruginosa,* and *S. flexneri,* and reported that it was possible to detect pathogen loads as low as 10 cells/ml with high sensitivity and specificity.

Simplified RT-PCR technology has been incorporated into several commercially available tests. The Qualicon BAX System (DuPont, Delaware, USA) employs RT-PCR for rapid detection of pathogens in environmental and food samples. It is an automated system that is able to detect the presence or absence of a target microorganism within 4 h. The BAX System was used successfully by the Pittsburgh Water and Sewer Authority to measure *C. parvum, Salmonella* and *E. coli* O157:H7 [24]. The BAX System can also detect *Campylobacter, Enterobacter sakazakii, Listeria, Staphylococcus aureus, Vibrio,* yeast and mold. RAPID (Idaho Technology Inc., Utah, USA) is another portable RT-PCR device used to detect biological agents within 30 min. The system can identify anthrax, ricin, smallpox, plague, *Listeria, E. coli* O157:H7, *Salmonella, Campylobacter, C. parvum, Salmonella,* Tularemia, Ebola, Brucella, Avian Influenza, Botulism, and Marburg.

Biodetection Enabling Analyte Delivery System (BEADS) is an automated, flow-through RT-PCR, which includes automated cell-capture microcolumns. Developed by the Pacific Northwest National Library (Washington, USA), the system achieved automated sample collection, purification, amplification, labeling, and detection with a reported detection limit of 10 *E. coli,* O157:H7 cells in river water samples. The system was also able to detect, simultaneously, *E. coli Salmonella,* and *S. flexneri* with a limit of 100 cells [35]. BEADS allows for real-time monitoring of viruses as well. While false positives were not observed, false negatives are possible if the microbial concentration is below the detection limit of the system. In addition, the system is prone to overestimate the pathogen quantities. As with all RT-PCR techniques, the technology identifies a target DNA sequence but cannot discriminate among viable, nonviable or damaged sequences [36].

The hand-held advanced nucleic acid analyzer (HANNA) developed by the Lawrence Livermore National Laboratory (California, USA) uses RT-PCR to detect bioterrorism agents. It is small in size and battery operated, making it very suitable for field evaluations. HANNA was evaluated for use in water monitoring for the detection of *C. parvum* and *E. coli* [37]. In water that contained *E. coli* from a beef cattle farm, HANNA was able to detect an *E. coli* indicator gene, lacZ, in $<$13 min, and a virulence gene in 25 min. While the instrument allowed for quick PCR results, there was a large amount of variability in the real-time results [37].

8.3.3 Nucleic Acid Microarrays

Conventional pathogen detection methods allow for the detection of one gene and thus one pathogen per experiment. Microarray technology allows for the simultaneous detection of hundreds or even thousands of pathogens during one experiment [38]. A DNA microarray is a series of genes immobilized on a solid surface such as a glass slide or a silicone chip. Immobilized DNA molecules, also known as probes, are ordered in a pre-determined order. A complementary series of polynucleotides to a probe is called a target. The selected target will hybridize with its complementary probe. Fluorescent labeling of the target will allow for the portion of the array where hybridization has occurred to be identified. DNA microarrays are thus able to identify thousands of genes of a small sample simultaneously [26,38]. Microarray technology can be used to detect a specific strain within an entire population of microorganisms [39], to determine whether specific genes are turned on or off [40], and to categorize microorganisms based on their genotype [41]. Therefore, microarray technology provides a unique tool for pathogen detection, enumeration, and genome-wide analysis for water samples as well as for the rapid detection of bioterrorism threats [42].

Bacterial waterborne pathogens were detected using a microarray equipped with short oligonucleotide probes targeting 16S rRNA sequences. Detection sensitivity of the microarray for *Aeromonas hydrophila* was determined to be approximately 10^3 cells per sample [43]. The method was also successful in detecting multiple bacterial pathogens in wastewater during chlorine and ultraviolet disinfection processes [43]. A DNA microarray with 21 oligonucleotide probes was developed to detect common waterborne protozoan pathogens, and had a detection limit of approximately 50 *C. parvum* oocysts [44]. An integrated system for monitoring a wide range of pathogens is possible with microarray technology, and has great potential as a microbiological monitoring tool for source water, drinking water and wastewater.

The application of microarray technology to complex water samples still presents a challenge. DNA extraction from cells and sample processing can be difficult in the presence of substances that interfere with DNA isolation and amplification [45]. It should also be noted that, similar to other molecular methods that rely on the amplification of nucleic acid, DNA microarrays cannot distinguish between dead and live bacteria since DNA can persist for significantly long periods of time after a cell is dead [46].

Recent developments in nanobiotechnology have allowed the application of lab-on-chip technology to DNA microarrays. Lab-on-chip systems use miniature microchips that hold a microfluidic network of samples and reagents. Chips are connected to a detection instrument and also to a computer system for data analysis and transfer, and allow real-time collection, analysis, and communication of data. Because of low fluid volumes, they require very small volumes of samples and reagents, and enable faster analysis and response times because of the efficiency of the chips. Jenkins et al. [47] developed a lab-on-chip system for direct profiling of microbial populations in a water sample. High molecular weight rRNA were fragmented into approximately 400 basepaires of rRNA, and were hybridized on a DNA microarray with a recirculating system, which could all be integrated into a credit-card-sized platform. Rantala et al. [48] developed a DNA chip for the detection and identification of hepatotoxin-producing cyanobacteria. The specificity, sensitivity and ability of the DNA chip to simultaneously detect the main hepatotoxin producers suggested that the method could be used for high throughput analysis and monitoring of water samples.

The Biolog System (Biolog Inc., California, USA) is a biochemical identification kit used to identify unknown bacteria. The test is comprised of a 96 well microplate, where each well contains a different selection of carbon sources, dried nutrients, and biochemicals. An isolate of an unknown bacterium is inoculated on either a gram-positive or a gram-negative microplate. After an incubation period between 4–6 h and/or 12–24 h, bacterial respiration has utilized the carbon sources of selected wells to produce a purple color in those wells. The pattern created by the purple wells is then cross-referenced with a library through computer software, which is able to identify approximately 339 gram-positive and 526 gram-negative species of bacteria [24].

8.3.4 Flow Cytometry

Flow cytometry measures the scatter and absorption patterns of cells as they flow through a light beam. The extent and nature of the scattering and absorption are intrinsic properties of microorganisms, which can be correlated to the size, shape, and density of microorganisms present in the sample. Fluorescent tags that bind to specific cell components such as DNA, RNA, antigens, or other target molecules can be added to samples. The combination of scattered and fluorescent light is picked up by detectors, which allow for the determination of total bacteria population, of selected types of bacteria and viability of bacteria. Flow cytometers allow simultaneous multiparametric analyses of physical and chemical characteristics of cells. It is very rapid with results provided in a matter of minutes.

Flow cytometry was originally used to identify bacteria in hospital samples as well as in the food and beverage industry. In water monitoring, flow cytometry has been used to identify *C. parvum* and *Giardia* using fluorescent-labeled antibodies that bind specifically to these microorganisms [49]. This method was shown to be very precise and to produce results much faster than traditional plate counting [50].

The Micro Pro (Advanced Analytical Technologies, Iowa, USA) is a fully automated optical flow cytometer that can process up to 42 samples at once. Samples are loaded in the instrument and the system automatically adds reagent, mixes, and injects the sample. Both quantitative and qualitative results can be obtained. With direct enumeration, results are available within 4 min and screening for microbial contamination within 18–24 h. The reagents used allow for labeling and fluorescent markers to be attached to specific compounds. Cells are then passed through a red diode laser and two photomultiplier tubes collect emitted signals from fluorescence and scatter. The system can reportedly detect microbial cells as small as 0.1 μm and its dynamic range is from 10^1–10^6 cfu/ml. Test kits are available to provide total viable organisms, total biomass, or total dead cell counts. The Micro Pro system was evaluated by Schreppel [4] to identify *E. coli* O157:H7 and *Cryptosporidium*. In both cases, microorganisms were first separated from water samples using immunomagnetic separation and labeled with specific fluorescent antibodies. The entire process allowed for microorganism identification in <4 h. The Micro Pro flow cytometer was used by the Mohawk Valley Water Authority (New York, USA) as part of an online monitoring system [51]. Micro Pro was also used to differentiate methicillin-resistant *S. aureus* and methicillin-susceptible *S. aureus* isolates after 2 h of incubation in an oxacillin-containing liquid culture medium [98].

The FlowCAM (Fluid Imaging Technologies, Maine, USA) automatically counts and characterizes cells in a continuous water flow in real-time or in a discrete sample. A digital image of each cell is captured and associated data is presented in a spreadsheet. Each image can be stored with up to 26 different measurements such as length, width, area, and aspect ratio. The FlowCAM has been used to detect microorganisms, such as zebra mussel veligers and cyanobacteria, as well as to measure yeast growth during fermentation processes. Depending on the selected magnification, the FlowCAM can reportedly detect particles ranging between 3 μm and 3 mm. The FlowCAM has also been used to enumerate and determine the size structure of a plankton community in surface water [52].

The Microcyte Aqua and Microcyte Field instruments (BioDetect, Texas, USA) uses flow cytometry to quantify algae and other microorganisms. The instruments are able to differentiate between biological and nonbiological particles, require little sample preparation, and provide results in minutes. The Microcyte Field is a suitcase-sized portable flow cytometer where data can be presented as a two-color histogram or dot plots directly on the instrument or downloadable to a PC. The Microcyte Aqua is a stationary device suited for online and continuous water surveillance. Both instruments are equipped with a laser diode that has a wavelength of 635 nm and uses fluorescence and lightscattering to count microorganisms and particles in the water. The detection limit of the instruments depends on the background levels of the sample; however, a fluorescent particle, such as algae, may be detected to levels below 10 cells per ml. Instruments can reportedly detect particles from 0.4 to 15 μm. The Microcyte Field instrument was also used to monitor cell number, viability, and apoptosis in mammalian cell culture [53].

8.3.5 ATP Bioluminescent Assays

The basis of an ATP bioluminescent assay is a luminescent reaction between an ATP molecule and a mixture of luciferin and luciferase extracted from fireflies [54]. ATP is involved in the energy transfer of all living cells. If a cell dies, ATP concentration quickly declines. Light produced from the reaction between luciferin–luciferase and ATP is thus an indication of the amount of living cells present in the water. Unlike DNA assays, the ATP assay is able to determine the amount of viable cells in a water sample [55]. This is the main advantage of ATP assays compared with other assays used for water monitoring.

Many ATP assay kits are available and require only a small sample volume to produce results within minutes. Most tests can be performed on-site and require no sample preparation. Kits are available that can distinguish between free (extracellular) ATP, microbial (intracellular) ATP, and total ATP. Separation of intracellular ATP from extracellular ATP typically involves a filtration step where extracellular ATP is filtered out and collected cells are lysed to release intracellular ATP. An ATP assay is a quick way to determine the viable microbial activity of a water sample.

The concentration of ATP in microorganisms is variable and depends on the species, strain, and environmental and metabolic factors. Thus, ATP assays can only be used as an approximation of the amount of cells in a solution. To identify microbial contamination, it is advised that baseline conditions be established and monitored regularly. Any fluctuations from the baseline may indicate a change of microbial concentration and contamination [1].

There are several ATP assay kits on the market, some of which specifically designed for water and wastewater monitoring. Quench-Gone is a series of ATP assays produced by LuminUltra Technologies (New Brunswick, Canada) to measure the total cell population. Each test is specifically designed for water, wastewater, industrial, pharmaceutical, and food and beverage industries. A complete test, including sample preparation time, can be completed within 5 min. To complete a test, a water sample is first pushed through a filter attached to a syringe to capture bacterial cells on the filter paper. Then, a lysing agent is added to release ATP from microorganisms. Finally, the solution is diluted, a luminescent agent is added, and the luminosity of the sample is measured by a luminometer. ATP assays that use microplate readers are also available from various companies including Perkin Elmer, Promega, Merck, Molecular Probes, Roche, and BioAssay Systems.

The Profile-1 (New Horizons Diagnostic, Maryland, USA) is an ATP Bioluminescence technology that makes use of a "Filtravette" to determine the presence of bacterial ATP in a water sample. The Filtravette is a combination of a filter and a cuvette. It first removes free ATP from a solution and all bacterial ATP remains on a filter paper. The Filtravette is then placed in a micro-luminometer and a lysing agent is added, releasing the bacterial ATP. The released bacterial ATP is then mixed with luciferin–luciferase and the light emission is recorded. It is estimated that the Profile-1 system is able to detect quantities of ATP equivalent to 200 cells (approximately) [51].

Lee and Deininger [56] used immunomagnetic separation of *E. coli* from beach water samples and the ATP bioluminescent assay to detect *E. coli*. Superparamagnetic polystyrene beads coated with *E. coli* antibodies were used to bind to *E. coli* in water. The bead–*E. coli* suspension was then easily separated using magnets. Prefiltration using a 20 μm nylon filter was also used to initially remove any plant material. The entire method produced results within an hour and had a detection limit of 20 cfu/100 ml [56].

Delayahe et al. (2003) used the ATP assay to monitor water quality in the drinking-water distribution network in Paris. A linear relationship between the quantity of ATP in water samples and heterotrophic plate counts were observed. The ATP assay did not appear to be affected by changes in the temperature and chlorine concentration. Based on the simplicity of the test and the rapid generation of the results, it may be possible to use the test as an alarm system for microbial contamination [97].

8.3.6 Light Scattering

Particles scatter light when passed through a beam of light. Photo diodes collect scattered light and translate it into a usable signal. The quantity of scatter created by particles in water is called turbidity and is caused by microscopic organisms as well as organic and inorganic suspended matter. Highly turbid waters can encourage bacterial growth and limit disinfection. Turbidity meters detect particles within a certain size range but give no information on the nature of particle. Thus, a microorganism and a suspended inorganic particle of similar size are considered equal. For this reason, using turbidity as a parameter for monitoring microorganisms has low levels of sensitivity and may result in false positives. On the other hand, the simplicity and rapidness of the method are important advantages.

Particle counting is a more precise and informative method of monitoring particles in water compared with turbidity meters. Particle counters provide detailed information on the number, size, shape, and size-distribution of particles in water samples, and with recent advancements in equipment technology, they are very sensitive and accurate.

For most particle counters, the lower detection limit for particle size is <1 μm and the new generation particle counters that employ specialized optics, lasers, and detectors have pushed the limits of detection down to 50 nm. Particle counters are commonly used at water treatment plants to monitor and evaluate the performance of sand filters, and they can potentially be used as surrogates for monitoring waterborne pathogens such as protozoa, bacteria, and possibly viruses.

The DPA4100 (Brightwell Technologies Inc., Ottawa, Canada) particle counter uses Micro-Flow Imaging technology for rapid measurement of particle size, shape, count, and concentration. Water samples are automatically drawn through a flow cell and imaged. Particles are then detected and measured using advanced image-analysis software. Douglas et al. [57] reported that the instrument was fast, sensitive, and able to accurately detect high particle concentrations when used in a water treatment plant.

New light-scattering technologies are reducing the quantity of false positives by using multiple collection angles and algorithms to distinguish between light-scattering patterns. This allows for the distinction between types of particles such as a grain of sand and a micro-organims of similar size. Water samples can be scanned to determine the size and number of a particular light-scattering pattern.

The BioSentry™ (JMAR Technologies Inc., California, USA) is designed to be used as an early warning system for the detection of pathogens in water. The system uses bio-optical signatures created by multi-angle light scattering (MALS) to distinguish between different organisms such as *Giardia, E. coli* and *Cryptosporidium*. This instrument is also able to identify different states of the *Cryptosporidium* oocysts including the type of treatment that was performed (ozone, heat, etc.) [58]. The BioSentry is fully automated, can be remote-controlled, and allows for continuous, real-time monitoring of water distribution systems. Data is also available online. The detection limit of the system depends on water quality and user-defined sampling intervals. In addition, water of poor quality will have a greater number of false positives due to interfering particles. When tested with filtered water, the system was able to detect less than one bacterium per 100 ml in 4 h and 22 bacteria in 100 ml within 5 min [58].

8.4 Monitoring Chemical Contaminants

Effects of microbiological contaminants are often acute and, as a result, prescribed regulations require frequent monitoring to assure public safety. On the other hand, effects of chemical contaminants are typically chronic in nature or are altogether unknown. In addition, changes in water chemistry tend to be long-term unless a specific contamination event has occurred. Regulatory testing requirements are thus less frequent or altogether nonexistent [2].

The urban water matrix is a complex mixture of low concentrations of chemical compounds. These compounds appear in water because they are continuously used in industry, agriculture, and by urban populations, and have limited removal rates during treatment. Persistent compounds can accumulate to concentrations that can affect human health. In addition, recent technological developments and associated research activities have allowed for the detection of contaminants that would previously go undetected. Increased awareness about the effect of persistent molecules will increase monitoring requirements [2]. The main contaminants of concern in water include heavy metals, pesticides, herbicides, hormones, pharmaceutical compounds, personal care products, fluorinated and brominated compounds, and toxins produced by algae.

A number of difficulties arise when developing monitoring techniques for chemical contaminants. To begin with, extraction and concentration procedures are typically required for low-concentration contaminants. These procedures are complex, require expert personnel and are inherently difficult to automate. Real-time chemical monitoring requires probes to be in constant contact with water. This is problematic because instruments are prone to fouling, which results in losses in sensitivity and reproducibility. It can also be difficult to determine the exact chemical associated with a specific symptom or series of symptoms. However, with the development of new methods of detection and research efforts, new and emerging chemical contaminants are raising awareness of the importance of testing for low-concentration chemical contaminants. There is a future need for real-time monitoring of chemical contaminants to assure long-term public health [59].

Currently, there is a large amount of research activity being performed to develop new detection methods and techniques for a wide range of chemical contaminants of concern. Some of the main contaminants of concern and the available detection technologies are described below.

8.4.1 Electrochemical Methods

Electrochemical sensors are based on the measurement of electrical quantities, such as current, potential, or charge, and their relationship to chemical properties. The majority of electrochemical methods use amperometry and voltammetry, which rely on the oxidation or reduction of electro-active species or potentiometry, which utilizes ion-selective electrodes and conductivity to measure the resistance of a solution [11]. Amperometry and voltammetry are commonly used for measuring oxygen concentrations, and potentiometry for measuring pH and ions of interest in water monitoring.

Electrochemical sensors offer several advantages. They use well-established technologies and provide rapid, reliable, sensitive, selective, and inexpensive measurements for a wide range of organic and inorganic contaminants in water. They are also one of the few sensor types that can provide truly real-time data with wireless communication devices. Electrochemical sensors are widely used for monitoring water-quality parameters in source waters as well as in distribution systems. The previous section on monitoring water-quality parameters provides information on some of the widely used electrochemical sensors (e.g. DO, ORP, pH, conductivity, chlorine, nitrogen, ammonia). In addition, electrochemical sensors have also been developed for emerging chemical contaminants including carcinogens [60], pesticides [61], phenols [62], explosives [63], and nerve agents [64].

Monitoring heavy metals is one of the main application areas of electrochemical sensors. Anodic stripping voltammetry (ASV) measures the current induced in an electrochemical cell. This method is typically used to determine the amount of metal ions, such as mercury and arsenic in water. The working principle of ASV is simple. A potential applied to a solution will reduce a metal ion to its metal form, which then accumulates on the surface of the electrode. The potential is increased at a constant rate to oxidize the metal, and the current produced by the oxidation is measured and plotted as a function of the potential applied. To determine the original concentration of the metal ion in solution, a calibration curve is compared to the area under the curve of the tested solution. The oxidizing potential is specific to the metal concentration of concern [1].

TraceDetect (Washington, USA) designs a number of trace metal analyzers with a focus on close to real-time procedures. The ArsenicGuard is an automated online arsenic analyzer able to monitor up to four sample streams with one system. The system will automatically grab a sample and perform the analysis. Results are obtained within 30 min and the system has a reported detection limit of 1 μg/L. The SafeGuard II, another model, is designed to automatically analyze grab samples, and the system can be configured for arsenic, copper, cadmium, lead, zinc, or mercury. The reported detection limit for all analytes is 1 μg/L and results are obtained within 30 min.

The PDV6000plus (Cogent Environmental Ltd., Cambridge, UK) is a portable device that uses ASV technology to measure the concentration of over 20 metals in water and soil samples. Before measurements are taken, a 30-min calibration procedure is required followed by 5–7 min needed to complete the measurements. The OVA 5000 model, on the other hand, can achieve continuous real-time heavy metal monitoring for a range of different metals (As, Cd, Cr, Cu, Hg, Ni, Pb, Se, etc.) to reportedly single figure ppb levels. The online OVA 5000 can be used for drinking water, wastewater, river water and process water.

8.4.2 Colorimetric Methods

Colorimetric methods are portable color-reaction kits typically used for rapid on-site analysis. To complete a test, a water sample is mixed with a series of reagents. A color will appear if a certain contaminant is present. Colorimetric methods are available for a wide variety of contaminants, are easy to use, and easy to transport. The concentration of the contaminant is determined by using a simple spectrophotometer or by comparing the gradient of the color produced to that of standard gradients.

Colorimetric methods have several advantages. They are simple, low–cost, and accurate and they can typically be completed within 5–10 min using a portable spectrophotometer. There are hundreds of well-established colorimetric methods developed for the detection of a wide range of organic and inorganic contaminants, and these test kits are available commercially from several manufacturers. For example, Hach

(Colorado, USA) test kits are well-known for water quality analysis, and can measure over 60 different contaminants and water-quality parameters, many of which are USEPA compliant.

To determine the quantity of arsenic in water, a colorimetric method called Gutzeit method is often applied. This method reduces soluble forms of arsenic to arsine gas with the addition of zinc in acidic conditions. The arsine gas (AsH_3) is then trapped in a silver diethyldithiocarbamate solution or with mercury bromide. Comparing the resulting color to a standardized color chart allows for the quantification of arsenic in water. It should be noted that the Gutzeit method has difficulties quantifying small quantities of arsine and the production of hydrogen sulfide may interfere with the results [65].

Arsenic Quick (Industrial Test Systems, South Carolina, USA) test kits utilize a series of three reactants and an arsenic test strip. Reactants are added to a water sample in a specific order and allowed appropriate amounts of time to react. A test strip is then inserted into the bottle and the color produced on the strip is compared to a color chart to determine the concentration. During the reactions, inorganic arsenic (As^{+3} and As^{+5}) is converted to arsine (AsH_3) using ferrous and nickel salts as catalysts. When arsine is applied to the mercuric bromide test strips, a color change from white to yellow to brown is observed. The test has a reported detection limit that ranges between 0 and 0.5 mg/L and concentrations up to 2.5 mg/L can be detected. Results are obtained within 15 min. The AS 75 (Peters Engineering, Austria) is another test that can be used in the field to determine the arsenic concentration of a sample. Indicator tablets are placed into a water sample. The intensity of the color produced is compared visually to a color gradient chart. A battery-operated tester can also be used.

Recently, a low-cost, biomimetic robotic fish, known as WANDA (wireless aquatic navigator for detection and analysis) was developed [66]. As WANDA moves through a number of static sensing stations, which respond colorimetrically to pH changes, its on-board camera is able to detect differences in color that are then correlated to chemical concentrations. The presence of a pH plume was successfully identified by WANDA [66].

8.4.3 Molecular and Immunological Methods

In addition to the detection of microorganisms, molecular and immunological methods can also be used for detection of chemicals in water. For example, immunoassays are widely used to screen for a wide group of contaminants in water. Immunoassays have high sensitivity and specificity, and they are rapid and cost-effective. They are also easy to use, require small-volume samples, and no pretreatment. In addition, simultaneous measurement of a number of samples is possible and tests are available to screen for a number of different contaminants. Two popular immunoassays are the ELISA and immunoaffinity chromatography (IAC) [2]. IAC is particularly suitable for polar substances.

There are over 60 different types of congeners of microcystin toxins produced by cyanobacteria, and the ELISA is useful to determine the total amount of microcystins. The ELISA uses an antibody to detect a unique amino acid called ADDA found in most microcystin toxins. Mountfort et al. [67] have used an ELISA in conjunction with a phosphatase inhibition assay called PP-2A to measure toxicity in mixtures. The authors developed an index called a response ratio which is created by dividing the amount determined by PP-2A test by the total amount determined by the ELISA. A high ratio indicates a highly toxic mixture. While the PP-2A is just as rapid as an ELISA, there are no field portable versions and the test must be performed in the lab [67].

Research is ongoing to develop immunoassays for new substances. The River Analyser (RIANA) uses total internal reflection fluorescence (TIRF) to detect very low levels of contaminants. Fluorescently-tagged antibodies are given time to react with a water sample before being streamed through a flow cell. A laser is directed at the flow cell and a signal created by the antibody–antigen complexes, is sent through polymer fibers to photo diodes that are connected to a computer. The entire process requires about 12 min to complete and has a limit of detection as low as 0.20 ng/L for estrone. The system can potentially detect pharmaceuticals, antibiotics, hormones, endocrine-disrupting chemicals, and pesticides in water [68]. Supported by funding from the European Commission, RIANA has been used in the development of the Automated Water Analyzer Computer Supported System (AWACSS), which is a multisensory system to monitor water pollution, where the immunosensors are linked by a communication network.

DNA microarray technology can also be used for the detection of chemicals in water [69]. A DNA microarray is a sequence of genes immobilized on a solid surface in a particular order. The hybridization between immobilized gene probes and a target gene produces a signal that can be detected [26,38]. Significant concentrations of a pharmaceutical can have an effect on gene molecules of different species. Using the transformed gene, a probe can be created and immobilized on a surface. The detection of the transformed gene would indicate that the chemical necessary to create the change is present in the water. The bioassay would thus assess the change induced by the chemical and not the chemical directly. The challenge is to find a gene transformation that is specific to a particular chemical [70]. In addition, the use of DNA microarrays requires DNA to first be extracted. This preprocessing step is lengthy, time consuming, and limits the real-time use of the DNA microarray. Though successful proof-of-concept studies are currently being performed, there are a number of obstacles that must be overcome before the wide use of DNA microarrays [70].

8.4.4 Biosensors

A biosensor is an analytical device that uses a biological or biologically-derived sensing element (e.g. enzymes, antibodies, microorganisms, or DNA) to measure the increased or decreased response in the presence of a contaminant [71]. A biosensor can use a physical (e.g. temperature), a chemical (e.g. pH) or a combination of physical and chemical signals, which can be interpreted by a transducer. In turn, a transducer can use an optical, electrochemical, or piezoelectric signal [71,72]. Biosensors have the advantage of having a low cost, are portable, are easy to use, can provide a continuous signal, and most importantly, can operate in situ [71]. There are typically two categories of biosensors, those that (1) obtain a signal from a catalyst such as an enzyme or a microorganism, and those that (2) obtain a signal from some form of irreversible binding of a target molecule such as affinity sensors on antibodies or nucleic acids. The first type of sensor is most often seen. One of the main limitations of the biosensor is that it is not compatible with chlorinated waters, and therefore not suitable for use in distribution systems [1].

Biosensors are widely used in wastewater to determine the BOD of a sample. The typical structure of a microbial BOD sensor includes an oxygen probe and immobilized microorganism [73]. The BOD measurement is based on the consumption of DO by the immobilized bacteria. Lehmann et al. [74] developed a biosensor using a salt-tolerant yeast, *Arxula adeninivorans,* to accurately measure BOD from wastewater in coastal and other high salt wastewaters. Pang et al. [75] employed the luminescent *Stenotrophomonas maltophilia* bacteria. The bacteria were loaded on a modified silica oxygen sensing film and luminescence measurements were done with a microplate reader. Results were obtained within 20 min. Similarly, Sakaguchi et al. [76] immobilized the *Photobacterium phosphoreum* bacteria on a small acrylic chip and the luminescence was converted to BOD using a commercial digital camera and a laptop computer. Results were obtained within 20 min and the use of a chip allowed for a high throughput and on-site detection of BOD. The system had a lower detection limit of 16 ppm.

It should be noted that immobilization of bacteria on membranes is not without disadvantages; as the immobilization membranes age, measurements begin to drift. Furthermore, DO probes have short lifetimes and require careful maintenance [77]. Membrane fouling of a biosensor is a problem and frequent maintenance may be required [78]. To eliminate membranes, Vaiopoulou et al. [77] immobilized bacteria on glass and measured the respiration activity (CO_2 concentration) of both immobilized and fluidized bacteria. This method produced continuous BOD measurements in real time.

Microbial fuel cells (MFC) use bacteria to degrade organic matter and to produce electricity. The electron transfer from the interior to the exterior of the cell can be facilitated by chemical mediators, or transferred directly to a metal electrode [78,79]. An MFC was used as a BOD sensor in wastewater. A linear relationship was found up to BOD values of 100 mg/L, and higher BOD values could be measured using a model-fitting method or a lower feeding rate.

Non-immobilized bacteria can also be used as biosensors where a luminescence response is measured by a portable luminometer. Bhattacharyya et al. [80] tested five bacterial responses to pollution as a complement to an existing chemical analysis at a contaminated groundwater remediation site. The luminescent bacteria used included the metabolic bacteria *Vibrio fischeri, Pseudomonas fluorescens* 10,568, and *E. coli* HB101, as well as catabolic bacteria *E. coli* DH5α and *Pseudomonas putida* TVA8. The catabolic biosensors detected toluene and middle-chain alkanes and estimated their bioavailability. The metabolic biosensors were used to appraise the bioremediation potential.

As previously mentioned, biosensors convert a biological response into a detectable signal. As such, biosensors have also been applied to water applications to detect phenolic compounds, bisphenol A, heavy metals, organic phosphates, polycyclic aromatic hydrocarbons, pesticides, and herbicides [81]. Rodriguez et al. [82] used a biosensor derived from the algae *Chlorella vulgaris* to detect herbicides in water. Fluorescence patterns of the chloroplasts of algae were shown to change when exposed to herbicides [82]. The fluorescence induction curves produced by *C. vulgaris* were used to detect chemicals such as methyl parathion, cyanide, and the herbicide paraquat dichloride in surface waters. The algae, immobilized on a sensor, was able provide continuous in situ measurements in a simulated river flow [82].

8.4.5 Gas and Liquid Chromatography

Gas chromatography (GC) is a laboratory procedure used for the identification of organic compounds that can be vaporized without decomposition. GC is often combined with mass spectrometer (MS), thermal conductivity detector (TCD), surface acoustic wave (SAW), electrolytic conductivity detector (ECD), electron capture detector, flame ionization detector (FID), and photo ionization detector (PID) [2].

GC analysis can be performed on grab samples a number of times per day or performed automatically at regular intervals. However, the use of GC devices requires skilled operators and regular maintenance. The use of GC may also be too expensive to be run on a continuous basis. GC is typically laborious, time consuming, and involves extensive pretreatment procedures [1]. Portable GC-MS devices are now available for the use in field measurements. Compared to laboratory testing, field measurements offer quicker results and reduced risk of contamination. This allows for rapid determination of contaminants, such as chemical weapons and other toxic pollutants [18].

The Scentograph CMS500 (by Inficon, New York, USA) is an automated gas chromatograph used for the online monitoring of volatile organic compounds (VOCs). The system requires no sample pretreatment or filtration steps, and it can reportedly measure concentrations from ppm to ppt levels. Scentograph CM500 was shown to work well for monitoring treated water for trihalomethane (THM) disinfection by-products.

The HAPSITE (by Inficon, New York, USA) device is a portable GC-MS system used to measure hydrocarbons in air, soil and water samples. The gas chromatographer is able to detect VOCs with molecular weights ranging between 45 and 300 g/mole. The MS uses the electron multiplier detector and has a library of 170,000 organic compounds. The HAPSITE system is also equipped with an in situ probe that automatically collects and concentrates samples, and provides results within minutes. The reported detection limits for the compounds are in the ppm to ppt range. In addition to VOCs, HAPSITE was also used for the detection of toxic industrial chemicals, chemical warfare agents, and select, semi-VOCs.

In recent years, liquid chromatography (LC) coupled to mass spectrometry (MS) has become the preferred method for the detection of emerging contaminants such as hormones, pharmaceuticals, and personal care products. Compared to GC, LC-MS requires reduced sample pretreatment and enhanced selectivity. However, pre-concentration of the sample is still required. LC-MS also suffers from matrix effects such as ion suppression and it can be difficult to separate highly polar substances. Recent research has focused on the development of new methods to remedy this problem [83,84].

A variety of LC techniques have been applied for the detection of cyanobacterial toxins. While LC is typically coupled with MS including tandem mass spectrometric techniques (MS/MS), methods such as UV absorption at fixed frequencies and selected ion monitoring (SIM) can also be used [85]. A well-optimized method would be selective and sensitive with low limits of detection and quantification to produce results in a reasonable amount of time.

Liquid chromatography-electrospray ionization mass spectrometry (LC-ESI-MS) was used to screen for toxic forms of cyanobacterial toxins such as major types of microcystins and nodularins. This method provided a high throughput of samples with a total analysis time of 2.8 min. A detection limit of 50–100 pg/injection was achieved. However, this analysis time did not include sample pretreatments that would be required if dealing with a field sample [86].

In another study, cyanobacterial toxins were separated in <30 min using reverse-phased liquid chromatography columns and a time of flight mass spectroscopy (LC/TOF-MS). The TOF-MS allows for the acquisition of full-spectrum data and real time, exact mass-to-charge ratio measurements. In this case, toxins were eluted from the water using extraction disks, containing very fine C-18 silica particles in a Teflon matrix [85].

A separate study used liquid chromatography and electro-spray ionization triple quadrupole mass spectrometry (LC-ESI-MS/MS) to detect cyanobacterial toxins, microcystins and nodularin in both lake water and drinking water. The method was able to optimize the mobile-composition of MS. This method used a reversed-phase guard cartridge system as the analytical column, thus eliminating the need for sample pre-concentration procedures and allowing for a short detection time of 10 min. The achieved limit of detection was 0.27 μg/L and the limit of quantification was 0.9 μg/L [87].

Microchip technology can miniaturize a number of different analytical techniques. For example Yuan and Carmichael [88] developed a microchip to identify the cyanobacterial toxins microcystin and nodularin. The microchip used surface-enhanced laser desorption ionization (SELDI) mass spectrometry to complete the capture, purification, analysis, and processing of a complex biological mixture. The chip was used to identify microcystins and nodularins in flood and liver tissue sample with a detection limit of 2.5 pg of microcystin-LR detected in 2 μg of water (or 1.2 μg/L) [88].

Solid-phase extraction (SPE) is a process that extracts compounds from a liquid mixture based on their physical and chemical properties. SPE is used to isolate and concentrate compounds for subsequent analysis with gas or liquid chromatography. Compared to other sample extraction and concentrating procedures, SPE is advantageous because it can be automated and depending on the choice of stationary phase, can be very selective. Online SPE units allow for the reduction of sample preparation time, an increase sample throughput, decrease risk of contamination, and a lack of degradation or loss of the analyte. Online SPE units perform analysis on the entire sample, thus also leading to lower limits of detection. Application of an online SPE-LC-MS has been used in the water industry for the detection of pesticides, phenols, polycyclic aromatic hydrocarbons, caffeine, disinfection by-products, and endocrine-disrupting chemicals [89].

The PROSPEKT (Spark Holland, Netherlands) is a commonly used programmable online SPE unit. The Prospekt-2 decreases the time needed to produce results by working in parallel with two separate samples. The GX-271 ASPEC and the GX-274 ASPEC (Gilson, Wisconsin, USA) automate the entire SPE and liquid-handling processes from initial conditioning steps to the final transfer. With these instruments, the SPE process can be completed in approximately 10 min and the instruments can be operated in both batch and sequential modes. The GX-247 is able to process up to four samples in parallel.

8.5 Monitoring Overall Toxicity

Considering the large number and variety of contaminants, it is costly and impractical to test for each contaminant separately. Toxicity testing assesses the general safety of the water quality. An ideal toxicity assay should require little time, skill, and money to complete, and detect the presence of toxic substances at concentrations that would cause a risk to the aquatic ecosystem or to public health. A toxicity test should also be quantitative and reproducible. Toxic substances to be considered include heavy metals, industrial chemicals, biotoxins, trace organic compounds, and chemical warfare agents. A positive result indicates the need for more testing but a negative result may not always assure the safety of the water [18].

LuminoTox (Lab_Bell Inc., Quebec, Canada) is a portable biosensor that indicates the presence of toxic chemicals in water. The assay uses an extract of enzymes derived from plants and algae. These extracts consist of photosynthetic enzymatic complexes that emit fluorescence when exposed to light. An analytical instrument measures the inhibition fluorescence in the presence of toxic substances, such as herbicides, organic solvents, ammonia–nitrogen, and organic amines. The sensitivity of the test is increased with increased incubation. An online version of LuminoTox test called Robot LuminoTox is also available. This instrument is self-cleaning, is able to provide measurements every 30 min and stores data in an Excel file.

The Eclox Rapid Toxicity System (Severn Trent Services, England) was developed by the UK Armed Forces to evaluate water quality in the field. Eclox is a chemiluminescent test that provides a quantitative assessment of water quality. The test relies on the reaction between luminol and an oxidant in the presence of horseradish peroxidase, which is a catalyst. This reaction produces a flash of light that is measured with a luminometer. If pollutants are in the water, then the reaction is inhibited and the light produced is less. Free radicals, antioxidants as well as phenols, amines, and heavy metals all interfere with the reaction. To complete the test, 1 ml of water is added to the reagents and placed in the illuminometer for 4 min. Results are compared to a control of deionized water. The luminometer can hold up to 60 measurements that can be downloaded to a computer.

The Rapid Field Enzyme Test (Severn Trent Services, England) uses inhibition of the cholinesterase enzyme to detect nerve agents and pesticides. A membrane disk saturated with cholinesterase is dipped into a water sample for 1 min and then pressed against another disk containing an ester for 3 min. If no pollutants are present, the enzyme is not inhibited and the disk turns blue. If pollutants are present, then the disk remains white. Results are obtained within 5 min and reported detection ranges for the carbamate pesticides, thiophosphates, and organophosphates are 0.1–5, 0.5–5 and 1–5 mg/L respectively.

Toxicity testing of a water supply system can also be achieved with living organisms such as bacteria, water fleas, mussels, and fish where physiological or behavioral changes are most often monitored. Also referred to as biosensors, living organisms provide a rapid response to a wide range of contaminants. Typically, the response of a living organism is monitored and compared to a baseline response. If a threshold is exceeded, then an alarm is triggered [90]. Responses can also be monitored in real time to provide continuous toxicity screening.

Biosensors are typically unable to determine the concentrations of individual contaminants. To maintain the health of an organism, an appropriate amount of care must also be given. Pathogens are species specific and thus biological organisms are typically not able to detect human pathogens [90,91].

An organism suitable for use as a biosensor must be sensitive to chemicals at low enough concentrations that will prove useful, have a known response to toxicity, and have a response that lasts long enough to be detected. The use of cells is increasingly popular because they can be modified to contain a specific response, such as bioluminescence, and can be miniaturized. This makes them suitable for use in portable instruments [1,90].

Bacteria can rapidly detect toxins in water. Disruption of cell metabolism or respiration rate is typically monitored by a reduction in bioluminescence. Cell growth patterns by means of turbidity or cell density is also a useful indication of toxicity [91]. Kits of freeze-dried bacteria are widely available and convenient to use and test results can be obtained within 30 min. Bacteria are able to detect pesticides, herbicides, chlorinated hydrocarbons, and heavy metals. The problems encountered when using bacteria biomonitors include clogging, continuous growth of the bacteria, and the need to add nutrients and media [1,91]. In addition to respiration rate and bioluminescence, more recently-developed biosensors are based on such measuring principles as the transformation of carbon, sulfur, nitrogen, enzyme activity, and glucose uptake [2].

Microtox and DeltaTox (Strategic Diagnostics Inc., Delaware, USA) are acute toxicity bioassays based on the bioluminescence of the bacteria *V. fischeri.* Chemical energy produced by the metabolic activity of the bacteria is converted into visible light. A change in cell respiration, results in a measurable change in luminescence. The change in luminescence is compared with a control sample and software is able to identify the toxic substance. Results can be obtained in <1 h and multiple tests can be run simultaneously. Test bacteria are freeze-dried and do not require any special care. The Microtox test is for use in the laboratory while the DeltaTox is the portable field version. Equipped with a self-calibrating photometer, a photomultiplier tube, and data collection software, results are promptly available within 5–15 min. Both instruments are sensitive to chlorine and thus not suitable for use in the distribution system.

The ToxScreen bioassay developed by CheckLight Ltd. (Israel) uses the luminescent bacterium *Photobacterium leiognathi* to detect μg/L concentrations of toxic organic and inorganic pollutants. This bioassay has been shown to be easy to use and is able to detect heavy metals, pesticides, polycyclic aromatic hydrocarbons (PAHs), and chlorinated hydrocarbons within 20–45 min, using a luminometer. Light emitted from the bacteria changes in response to chemophysical and biological toxicant that affect cell respiration and the rate of protein or lipid synthesis. Each toxin is distinguished by differences in the bacteria's response. Replacement of a buffered solution and fresh freeze-dried bacteria are required every 14 days.

Estrogenic compounds can have an effect on the endocrine and reproductive systems of humans and other animal species. The LUMI-CELL ER (Xinobiotic Detection Systems, Inc., North Carolina, USA) bioassay uses a human ovarian carcinoma cell that contains an estrogen-responsive luciferase reporter gene (BG1Luc4E2) that emits light in response to estrogenic compounds in a time, dose-dependent, and chemical-specific manner. This test is only available off-line because replacement of the organisms is needed after each measurement. The assay has a reported detection limit of approximately 0.04 pg/g. The Chemical Activated Luciferase Expression Assay (CALUX) (Xinobiotic Detection Systems, Inc., North Carolina, USA) uses a genetically modified mouse cell line inserted with the gene for firefly luciferase to detect dioxins, PCB, or furans. The assay can quickly determine the total concentrations of all dioxins and dioxin-like compounds.

Algae are typically used to detect herbicides in water through measurements of cell growth, oxygen production, chlorophyll fluorescence, or the mobility of the flagellate [91]. Measurements taken from algae have been performed on a semi-continuous basis with measurements taken at intervals of 30 min. Algal suspensions are typically used though standardization procedures are difficult [1,2,91].

The algae toximeter (by bbe Moldaenke, Germany) places algae in a regulated fermenter where the activity of the algae is monitored. Water samples are injected into the instrument and changes in the algae's levels of fluorescence are monitored. If the algae's activity is constant, then no toxic substances are present in the water. This test is not suitable for use in chlorinated water.

The behavioral patterns of bivalves, mainly shell closure, can also be monitored. Two commercial monitoring systems that employ bivalves include the Dreissena-Monitor (Germany) and Musselmonitor (Delta Consult, The Netherlands). The Musselmonitor correlates the valve position of mussels, *Dreissena polymorpha*, to temperature, concentration of suspended material (turbidity and chlorophyll concentration) and general toxicity of the water. The monitoring system was developed by studying the normal behavior of mussels considering seasonal variations [92]. The behavior of eight mussels is continuously monitored for up to 2 months and a continuous supply of nutrients is supplied. Software enables the graphical representation of the individual behavior of each mussel. The Musselmonitor was used by Waterworks of Budapest Hungary [1].

The first toxicity test to employ a living organism recorded the dynamic movements of the water flea or *Daphnia magna* as it passes through an infrared light beam. With the advancement of image recognition equipment, video-based technology is being developed to monitor the swimming behavior of individual water fleas to provide real time results. To maintain a healthy population of water fleas, some level of skill is required. The water fleas require a continuous algae feed [2,91].

The IQ-Tox (Aqua Survey Inc., New Jersey, USA) toxicity test utilizes *D. magna* and is able to detect chemicals and biological toxins. In the presence of a toxin, the metabolism of the *D. magna* is reduced, which can be detected through fluorescent tagging. Results are available within 75 min and these organisms are sensitive to chlorine residual. In 2003, the USEPA Environmental Technology Verification program found that amongst other toxicity tests evaluated; the IQ-Tox test identified the highest number of chemicals (aldicarb, colchicine, cyanide, dicrotophos, thalium, Botulinum toxin, ricin, and soman) below human lethal-dose concentrations.

The Daphnia Toximeter (by bbe Moldaenke, Germany) allows for the swimming speed, altitude, number of turns, circling movements, growth rates and number of live daphnia to be used as a toxicity screening method. Data is available online and measurements are performed at 30-min intervals.

The Multispecies Freshwater Biomonitor (LimCo International, Germany) monitors the behavioral response of up to 96 different microvertebrates and is able to collect readings every 10 min [93]. Each specimen is placed in a chamber where the instrument records movements and ventilation patterns. Software analyzes the collected signal which is then correlated to toxicity of the water. The MFB can be used for both sediment and water analysis [94].

When a fish is employed to monitor the toxicity of water, the rheotaxis of the fish is visually monitored [91]. The rheotaxis is the tendency of the fish to face the current of the water. If downstream movement is observed, then fish are trying to move to a more favorable water condition and toxicity can be inferred [91]. Other methods record the ventilation rate, cough rate, heart rate, and other swimming behaviors [91]. Fishes are sensitive and can provide real-time response if standardization problems can be overcome. Fish cells are a possible alternative to the use of whole organisms; however, cells are typically not as sensitive [2].

In particular, the Bluegill fish (*Lepomis machrochirus*) has been used and a total of four parameters can be measured: (1) ventilatory rate, (2) cough rate, (3) ventilatory depth and (4) movement of each fish in 15-s intervals [90,95]. Basic water quality parameters such as temperature, DO, conductivity, and pH are also monitored. A toxicity index is created by combining baseline behavioral data with basic water characteristics. Should the index exceed a certain threshold for more than 70% of the fish, then an alarm is triggered. The Bluegills are especially sensitive to chemicals that cause direct gill damage such as zinc or residual chlorine [90,95]. Bluegill fish have been used successfully to detect toxic contaminant in Fort Detrick (USA) and by the New York City Department of Environmental Protection (USA) [91].

A research group at the University of Wisconsin's Great Lakes Water Institute have created a biosensor for use in the distribution system using transgenic Zebrafish [96]. After fertilization, Zebrafish embryos are injected with pollution-responsive reporter genes designed to detect 18 chemical contaminants. The presence of one of these contaminants in the water supply will trigger a response which will produce luciferase in the fish. The light emitted by the luciferase can then be detected [96].

The Fish-Bio-Sensor (Biological Monitoring Inc., Virginia, USA) monitors the electric field of 8–12 fish. Any deviation from normal condition is indicative of the presence of a toxin. When this occurs, an alarm is sounded and a water sample is collected to allow for further analysis. The system is equipped with an automatic feeder and a dechlorination module can be added. The Fish-Bio-Sensor has been used in Singapore, Australia and South Africa.

8.6 Conclusions

Modern water management requires reliable, rapid, and accurate characterization of contaminants to allow for a timely response. Currently, there are limited options for the implementation of real-time or near real-time monitoring of water quality, and there is an urgent need for developing rapid methods for the detection of microbiological and chemical contaminants in water. This is important not only for the protection of public health, but also for the protection of water resources and the environment. The majority of currently available real-time or near real-time methods are based on monitoring simple water quality parameters such as temperature, pH, conductivity, DO, and turbidity. These parameters are useful as screening methods but they are unable to identify specific chemical or biological agents.

Continuous monitoring for microbiological agents is a challenge because of the large number of microorganisms of interest and the low densities of these microorganisms in water. In addition, water characteristics and composition may interfere with detection methods. Molecular methods offer great potential for the rapid detection and identification of microorganisms; however, these methods also have some limitations. They typically require skilled personnel, are expensive, and may take hours to complete. Methods that rely on nucleic-acid amplification, such as PCR, qPCR, and real-time polymerase chain reaction (RT_PCR) have the ability to detect very low numbers of microorganisms but cannot distinguish between DNA from dead and live cells. The lateral flow immunoassays are very easy to use and can be completed within a few minutes. There are several commercially available immunoassays that can detect a large number of pathogens and toxins quickly. However, these tests are not sensitive enough to detect low densities of microorganisms and toxins, and are prone to false-positive responses. Contrary to DNA-based methods, ATP assays are able to determine cell viability and this is their main advantage. The assay can be completed within 5 min and gives a good measure of the overall viable activity in water. However, the ATP assay cannot provide specific information of target micro-organisms and it is not sensitive enough to detect low levels of microbial activity.

There are a series of difficulties faced when monitoring for chemical contaminants as well. Chemical contaminants are typically present at very low concentrations and water quality and composition may interfere with detection methods. Well-established technologies monitor simple water-quality parameters, such as pH, temperature, conductivity, and DO; but there is a lack of quick, reliable, and established methods for the detection of emerging contaminants. Electrochemical sensors are the most widely-used sensors for the detection of chemical contaminants. They are rapid, sensitive, and inexpensive. They can provide true real-time data with wireless communication devices. Gas and liquid chromatography are powerful methods to detect organic compounds, particularly when coupled to mass spectrometry. Portable GC-MS and LC-MS are now available for field use, which helps to offer quicker results, decrease sample-preparation time, increase sample throughput and reduce the risk of contamination. However, pretreatment and preconcentration of samples are still needed, and these are complex methods that require highly skilled personnel. Detection and identification of chemical contaminants are also possible using molecular and immunological methods, which are finding an increasing application area due to the rapid advances that have been achieved in these fields.

Considering the large number and variety of chemical and biological contaminants, it is costly and impractical to test for individual contaminants; instead, testing for overall toxicity of water may be preferable. Toxicity assays require little time, skill, or money to complete and provide information on whether toxic substances are present in water that may pose a risk to the health of the public or aquatic organisms. Toxicity assays utilize a wide range of organisms such as bacteria, fish, algae, mussels, and water fleas and they are valuable as an initial screening method.

Microbiological and chemical methods used for water monitoring are not yet perfect and require further development to improve accuracy, precision, completion times, and effectiveness in different water matrices. The majority of methods do not have true real-time capabilities, but considering the rapid advances in sensor and communication technologies, this is certainly within reach. Standardization of methods is also necessary for their validation, routine use, and acceptance by national and international water regulations and agencies.

Acknowledgments

Prof. Banu Örmeci would like to thank the Ontario Centers of Excellence for providing funding for this study under the Interact Program.

References

[1] Hasan J, Goldbloom-Helzner D, Ichida A, Rouse T, Gibson M. Technologies and techniques for early warning system to monitor and evaluate drinking water quality: a state-of-the-art review. accession number ADA452625. Washington, DC: US Environmental Protection Agency; 2005.

[2] Gheewala SH, Babel MS, Gupta AD, Babael S. Rapid assessment techniques for chemicals in raw water sources. J Water Supply Res Technol 2003;52(7):521–7.

[3] Hall J, Zaffiro AD, Marx RB, Kefauver PC, Krishnan ER, Haught RC, et al. On-line water quality parameters as indicators of distribution system contamination. J Am Water Works Assoc 2007;99(1):66–77.

[4] Schreppel C. Setting the alarm for an early warning. Opflow 2003;29(6):1–8.

[5] Ahmad SR, Reynolds DM. Monitoring of water quality using fluorescence techniques: prospects of on-line process control. Water Res 1998;33(9):2069–74.

[6] Nollet LML. Handbook of water analysis. 2nd ed. Boca Raton, FL: CRC Press; 2007.

[7] Hur J, Hwang S-J, Shin J-Ki. Using synchronous fluorescence technique as a water quality monitoring tool for an urban river. Water Air Soil Pollut 2008;191:231–43.

[8] Her N, Amy G, McKnight D, Sohn J, Yoon Y. Characterization of DOM as a function of MW by fluorescence EEm and HPLC-SEC using UVA, DOC and fluorescence detection. Water Res 2003;37:4295–303.

[9] Her N, Amy GA, Foss D, Cho J, Yoon Y, Kosenka P. Optimization of method for detecting and characterizing NOM by HPLC-size exclusion chromatography with UV and on-line DOC detection. Environ Sci Technol 2002;36:1069–76.

[10] Dybko A, Wroblewski W, Rozniecka E, Pozniakb K, Maciejewski J, Romaniuk R, et al. Assessment of water quality based on multiparameter fiber optic probe. Sensors and Actuators, B 1998;51:208–13.

[11] Hanrahan G, Patil DG, Wang J. Electrochemical sensors for environmental monitoring: design, development and applications. J Environ Monit 2004;6:657–64.

[12] Dufour A. Assessing microbial safety of drinking water: improving approaches and methods. London, UK: IWA Publishing; 2003.

[13] Health Canada. Guidelines for Canadian drinking water quality – technical documents. Available online at: http://www.hc-sc.gc.ca/ewh-semt/pubs/water-eau/index-eng.php#guide; 2011.

[14] Paul DH. Bioterrorism and water treatment – part 1. Water Conditioning & Purification 2003. October 2003.

[15] Centers for Disease Control and Prevention (CDC). Bioterrorism agents/diseases. Available online at: http://www.bt.cdc.gov/agent/agentlist-category.asp; 2011.

[16] Moldenhauer J. Overview of rapid microbiological methods. In: Zourob M, et al, editors. Principles of bacterial detection: biosensors, recognition receptors and microsystems. Springer Science + Business Media, LLC; 2008.

[17] King D, Luna V, Cannons A, Cattani J. Performance assessment of three commercial assays for direct detection of *Bacillus anthracis* spores. J Clin Microbiol 2003;41(7):3454–5.

[18] Casson LW, States SJ, Wichterman J, Zimmerman A. Thwarting terrorists. Water Environment Laboratory Solutions 2006;13(1):1–8.

[19] NASA, National Aeronautics and Space Administration. Securing the home planet. Space Res 2003;2(3):6–11.

[20] Zhao X, He XW, Li WM, Liu Y, Yang LS, Wang JH. Development and evaluation of colloidal gold immunochromatographic strip for detection of *Escherichia coli* O157. Afr J Microbiol Res 2010;4(9):663–70.

[21] Rule KL, Vikesland PJ. Surface-enhanced resonance Raman spectroscopy for the rapid detection of *Cryptosporidium parvum* and *Giardia lamblia*. Environ Sci Technol 2009;43(4):1147–52.

[22] Zhao Y, Ye M, Chao Q, Jia N, Ge Y, Shen H. Simultaneous detection of multifood-borne pathogenic bacteria based on functionalized quantum dots coupled with immunomagnetic separation in food samples. J Agric Food Chem 2009;57(2):517–24.

[23] Chao QG, Jia NQ, Ge Y, Shen HB. Simultaneous detection of multifood-borne pathogenic bacteria based on functionalized quantum dots coupled with immunomagnetic separation in food samples. J Agric Food Chem 2009;57(2):517–24.

[24] States S, Scheuring M, Kuchta J, Newberry J, Casson L. Microbial screening methods for ensuring security of public water supplies. American Water Works Association – Water Quality Technology Conference; 2002.

[25] Zhou W, Kageyama K, Li F, Yuasa A. Monitoring of microbial water quality by real-time PCR. Environ Technol 2007;28:545–53.

[26] Dorigo U, Volatier L, Humbert J-F. Molecular approaches to the assessment of biodiversity in aquatic microbial communities. Water Res 2005;39:2207–18.

[27] Schmitten TD. Real-time quantitative PCR. Methods 2001;25:383–5.

[28] Vaitomaa J, Rantala A, Halinen K, Rouhiainen L, Tallberg P, Mokelke L, et al. Quantitative real-time PCR for determination of microcystin synthetase E copy numbers for *Microcystis* and *Anabaena* in lakes. Appl Environ Microbiol 2003;69(12):7289–97.

[29] Donaldson KA, Griffin DW, Paul JH. Detection, quantitation and identification of enteroviruses from surface waters and sponge tissue from the Florida Keys using real-time RT-PCR. Water Res 2002;36:2505–14.

[30] Sharma VK, Dean-Nystrom EA. Detection of enterohemorrhagic *Escherichia coli* O157:H7 by using a multiplex real-time PCR assay for genes encoding intimin and Shiga toxins. Vet Microbiol 2003;93(3):247–60.

[31] Aslan A, Xagoraraki I, Simmons FJ, Rose JB, Dorevitch S. Occurrence of adenovirus and other enteric viruses in limited-contact freshwater recreational areas and bathing waters. J Appl Microbiol 2011;111(5):1250–61.

[32] Behets J, Declerck P, Delaedt Y, Verelst L, Ollevier F. A duplex real-time PCR assay for the quantitative detection of *Naegleria fowleri* in water samples. Water Res 2007;41:118–26.

[33] Helmi K, Skraber S, Burnet JB, Leblanc L, Hoffmann L, Cauchie HM. Two-year monitoring of *Cryptosporidium parvum* and *Giardia lamblia* occurrence in a recreational and drinking water reservoir using standard microscopic and molecular biology techniques. Environ Monit Assess 2011;179(1–4):163–75.

[34] Clark ST, Gilbride KA, Mehrvar M, Laursen AE, Bostan V, Pushchak R, et al. Evaluation of low-copy genetic targets for waterborne bacterial pathogen detection via qPCR. Water Res 2011;45(11):3378–88.

[35] Straub TM, Dockendorff BP, Quiñonex-Díaz MD, Valdex CO, Shutthanandan JI, Tarasevich BJ, et al. Automated methods for multiplexed pathogen detection. J Microbiol Methods 2005;62:303–16.

[36] Jian S. Molecular alternatives to indicators and pathogen detection: real-time PCR. WERF Report ES-01-HHE-2a. Washington, DC, USA: Water Environment Research Foundation; 2006.

[37] Jenkins M, Feyer R, Trout J, Palmer R, Short K, Cosio M, et al. Hand-held advanced nucleic acid analyzer (HANNAA) for waterborne pathogen detection. WERF Report 99-HHE-4-ET. Washington, DC, USA: Water Environment Research Foundation; 2001.

[38] Brousseau R. Application of DNA microarray technology for wastewater analysis, WERF report O1HHE1. Washington DC, USA: Water Environment Research Foundation; 2005.

[39] Call DR, Borucki MK, Loge FJ. Detection of bacterial pathogens in environmental samples using DNA microarrays. J Microbiol Method 2003;53:235–43.

[40] Weiner III J, Zimmerman C-U, Gohlmann HWH, Herman R. Transcription profiles of the bacterium *Mycoplasma pneumoniae* at different temperatures. Nucleic Acids Res 2003;31:6306–20.

[41] van Leeuwen, Jay C, Snijders S, Durin N, Lacroix B, Verbrugh HA, et al. Multilocus sequence typing of *Staphylococcus aureus* with DNA array technology. J Clin Microbiol 2003;41:3323–6.

[42] Dennis P, Edwards EA, Liss SN, Fulthrope R. Monitoring gene expression in mixed microbial communities by using DNA microarrays. Appl Environ Microbiol 2003;69(2):769–7778.

[43] Lee DY, Lauder H, Cruwys H, Falletta P, Beaudette LA. Development and application of an oligonucleotide microarray and real-time quantitative PCR for detection of wastewater bacterial pathogens. Sci Total Environ 2008;398(1–3):203–11.

[44] Lee DY, Seto P, Korczak R. DNA microarray-based detection and identification of waterborne protozoan pathogens. J Microbiol Methods 2010;80(2):129–33.

[45] Gilbride KA, Lee DY, Beaudette LA. Molecular techniques in wastewater: understanding microbial communities, detecting pathogens, and real-time process control. J Microbiol Methods 2006;66(1):1–20.

[46] Masters CI, Shallcross JA, Mackey BM. Effect of stress treatments on the detection of listeria-monocytogenes and enterotoxigenic *E. coli* by the polymerase chain reaction. J Appl Bacteriol 1994;77(1):73–9.

[47] Lee HY, Yager P. Microfluidic lab-on-a-chip for microbial identification on a DNA microarray. Biotechnol Bioprocess Eng 2007;12(6):634–9.

[48] Rantala A, Rizzi E, Castiglioni B, de Bellis G, Sivonen K. Identification of hepatotoxin-producing cyanobacteria by DNA-chip. Environ Microbiol 2008;10(3):653–64.

[49] Ferrari BC, Vesey G, Weir C, Williams KL, Veal DA. Comparison of *Cryptosporidium*-specific and giardia-specific monoclonal antibodies for monitoring water samples. Water Res 1999;33(77): 1611–7.

[50] Chesnot T, Marly X, Chevalier S, Estévenon O, Bues M, Schwartzbrod J. Optimized immunofluorescence procedure for enumeration of *Cryptosporidium parvum* oocysts suspensions. Water Res 2002;36(13):3283–8.

[51] Schreppel CK, Tangorra PA, Eaton DD, Donovan SP. On-line real-time monitoring –peace of mind? American Water Works Association – Water Technology Conference; 2002.

[52] Jakobsen HH, Carstensen J. Flow Cam: sizing cells and understanding the impact of size distributions on biovolume of planktonic community structure. Aquat Microb Ecol 2011;65(1):75–87.

[53] Harding CL, Lloyd DR, McFarlane CM, Al-Rubeai M. Using the microcyte flow cytometer to monitor cell number, viability, and apoptosis in mammalian cell culture. Biotechnol Prog 2000;16(5):800–2.

[54] Tanaka H, Shinji T, Sawada K, Monji Y, Seto S, Yajima M, et al. Development and application of a bioluminescence ATP assay method for rapid detection of coliform bacteria. Water Res 1997;31(8):1913–8.

[55] Shimomura O. Bioluminescence: chemical principles and methods. New Jersey, USA: World Scientific; 2006.

[56] Lee JY, Deininger RA. Detection of *E. coli* in beach water within 1 hour using immunomagnetic separation and ATP bioluminescence. Luminescence 2004;19:31–6.

[57] Douglas I, Thomas D, Guthmann J, Russel S, Springthorpe S. A new technology for optimizing particle removal in a water treatment plant. American Water Works Association, 2004 Water Quality Technology Conference Proceedings; 2004.

[58] Adams JA, McCarty D. Real-time on-line monitoring of drinking water for waterborne pathogen contamination warning. Int J High Speed Electr Syst 2007;17(4):643–59.

[59] Bonnastre A, Ors R, Capella JV, Fabra MJ, Peris M. In-line chemical analysis of wastewater: present and future trends. Trends Anal Chem 2005;24(2):128–37.

[60] Barek J, Cvačka J, Muck A, Quaiserová V, Zima J. Electrochemical methods for monitoring of environmental carcinogens. Anal Bioanal Chem 2001;369(7–8):556–62.

[61] Budnikov GK, Evtyugin GA, Rizaeva EP, Ivanov AN, Latypova VZ. Comparative assessment of electrochemical biosensors for determining inhibitors-environmental pollutants. J Anal Chem 1999;54:864–71.

[62] Wang J, Chen W. Remote electrochemical biosensor for field monitoring of phenolic compounds. Anal Chim Acta 1995;312:39–44.

[63] Wang J, Bhada R, Lu J, MacDonald D. Remote electrochemical sensor for monitoring TNT in natural waters. Anal Chim Acta 1998;361:85–92.

[64] Wang J, Chen L, Mulchandani A, Chen W. Remote biosensor for in situ monitoring of organophosphate compounds. Electroanalysis 1999;11:866–90.

[65] Dhar RK, Zheng Y, Rubenstone J, van Green A. A rapid colorimetric method for measuring arsenic concentrations in groundwater. Anal Chim Acta 2004;526:203–9.

[66] Fay C, Lau K-T, Beirne S, Conaire CO, McGuinness K, Corcoran B, et al. Wireless aquatic navigator for detection and analysis (WANDA). Sensor Actuat B-Chem 2010;150(1):425–35.

[67] Mountfort DO, Holland P, Sprosen J. Method for detecting classes of microcystins by combination of protein phosphatase inhibition assay and ELISA: comparison with LC-MC. Toxicon 2005;45(2):199–206.

[68] Tschmelak J, Proll G, Gauglitz G. Optical biosensor for pharmaceuticals, antibiotics, hormones, endocrine-disrupting chemicals and pesticides in water: assay optimization process for estrone as example. Talanta 2005;65:313–23.

[69] Poynton HC, Vulpe CD. Ecotoxicogenomics: emerging technologies for emerging contaminants. J Am Water Res Assoc 2009;45(1):83–96.

[70] Kullman SW, Hilton DE, Linden KG. Moving towards and innovative DNA array technology for detection of pharmaceuticals in reclaimed water, WERF Report 703-684-2470. Washington, DC, USA: Water Environment Research Foundation; 2009.

[71] Dennison MJ, Turner APF. Biosensors for environmental monitoring. Biotechnol Adv 1995;13: 1–12.

[72] Hart BT, McKelvie ID, Benson RL. Real-time instrumentation for monitoring water quality: an Australian perspective. Trends Anal Chem 1993;12(10):403–11.

[73] Bourgeois W, Burgess JE, Stuetz RM. On-line monitoring of wastewater quality: a review. J Chem Technol Biotechnol 2001;76:337–48.

[74] Lehmann M, Chan C, Lo A, Lung M, Tag K, Kunze G, et al. Measurement of biodegradable substances using the salt-tolerant yeast *Arxula adeninivorans* for a microbial sensor immobilized with poly(carbamoyl)sulfonate (PCS) Part II: application of the novel biosensor to real samples from coastal and island regions. Biosens Bioelectron 1999;14:295–302.

[75] Pang H-L, Kwok N-Y, Chan P-H, Yeung C-H, Lo W, Wong K-Y. High-throughput determination of biochemical oxygen demand (BOD) by a microplate-based biosensor. Environ Sci Technol 2007;41:4038–44.

[76] Sakaguchi T, Morioka Y, Yamasaki M, Iwanaga J, Beppu K, Maeda H, et al. Rapid and onsite BOD sensing system using luminous bacterial cells-immobilized chip. Biosens Bioelectron 2007;22: 1345–50.

[77] Vaiopoulou E, Melidis P, Kampragou E, Aivasidis A. On-line load monitoring of wastewaters with a respirographic microbial sensor. Biosens Bioelectron 2005;21:356–71.

[78] Chang IS, Jang JK, Gil GC, Kim M, Kim HJ, Cho BW, et al. Continuous determination of biochemical oxygen demand using microbial fuel cell type biosensor. Biosens Bioelectron 2004;19:607–13.

[79] Chang S, Hyunsoo M, Jang JK, Kim BH. Improvement of a microbial fuel cell performance as a BOD sensor using respiratory inhibitors. Biosens Bioelectron 2005;20:1856–9.

[80] Bhattacharyya J, Read D, Amos S, Dooley S, Killham K, Paton GI. Biosensor-based diagnostic of contaminated groundwater: assessment and remediation strategy. Environ Pollut 2005;134:485–92.

[81] Farré M, Kantiani L, Perez S, Barcelo D. Sensors and biosensors in support of EU directives. Trends Anal Chem 2009;28(2):170–85.

[82] Rodriguez Jr M, Sanders CA, Greenbaum E. Biosensors for rapid monitoring of primary-source drinking water using naturally occurring photosynthesis. Biosens Bioelectron 2002;17:843–9.

[83] Richardson SD, Ternes TA. Water analysis: emerging contaminants and current issues. Anal Chem 2005;77:3807–38.

[84] Rodriguez-Mozaz S, Lopez de Alda MJ, Barceló D. Advantages and limitation of on-line solid phase extraction coupled to liquid chromatography-mass spectrometry technologies versus biosensors for monitoring of emerging contaminants in water. J Chromatogr A 2007;1152:97–115.

[85] Maizels M, Budde WL. A LC/MS method for the determination of cyanobacteria toxins in water. Anal Chem 2004;76(5):1342–51.

[86] Meriluoto J, Karlsson K, Spoof L. High-throughput screening of ten microcystins and nodularins, cyanobacterial peptide hepatotoxins, by reverse-phase liquid chromatography-electrospray ionisation mass spectrometry. Chromatographia 2004;59(5/6):291–8.

[87] Allis O, Dauphard J, Hamilton B, Shuilleabhain A Ni, Lehane M, James KJ, et al. Liquid chromatography – tandem mass spectrometry application, for the determination of extracellular hepatotoxins in Irish lake and drinking waters. Anal Chem 2007;79(9):3436–47.

[88] Yuan M, Carmichael WW. Detection and analysis of the cyanobacterial peptide hepatotoxins microcystins and nodularin using SELDI-TOF mass spectrometry. Toxicon 2004;44:561–70.

[89] Richardson SD. Water analysis: emerging contaminants and current issues. Anal Chem 2009;81: 4645–77.

[90] Mikol YB, Richardson WR, van der Schalie WH, Shedd TR. An online real-time biomonitor for contaminated surveillance in water supplies. Am Water Works Assoc J 2007;99(2):107–14.

[91] Gerhardt A, de Bisthoven LJ, Schmidt S. Automated recording of vertical negative phototactic behavior in *Daphnia magna* straus (Crustacea). Hydrobiologia 2006;559:433–41.

[92] Sluyts H, Van Hoof F, Cornet A, Paulussen J. A dynamic new alarm system for use in biological early warning systems. Environ Toxicol Chem 1996;15(8):1317–23.

[93] Gerhardt A, Schmidt S. The multi-species freshwater biomonitor: a potential new tool for sediment biotests and biomonitoring. J Soils Sediments 2002;2(2):67–70.

[94] De Bisthoven LJ, Gerhardt A, Soares AMVM. Effects of acid mine drainage of larval *chironomus* (diptera: chironomidae) measured with the multispecies freshwater biomonitor. Environ Toxicol Chem 2003;23(5):1123–8.

[95] van der Scharlie WH, Shedd TR, Widder MW, Brennan LM. Response characteristics of an aquatic biomonitor used for rapid toxicity detection. J Appl Toxicol 2004;24:387–94.

[96] Kusik BW, Carvan III MJ, Udvadia AJ. Detection of mercury in aquatic environments using EPRE reporter Zebrafish. Mar Biotechnol 2008;10(6):750–7.

[97] Delahaye E, Welte B, Levi Y, Leblon G, Montiel A. An ATP-based method for monitoring the microbiological drinking water quality in a distribution network. Water Res 2003;37(15):3689–96.

[98] Shrestha NK, Scalera NM, Wilson DA, Procop GW. Rapid differentiation of methicillin-resistant and methicillin-susceptible *Staphylococcus aureus* by flow cytometry after brief antibiotic exposure. J Clin Microbiol 2011;49(6):2116–20.

[99] Walling MA, Novak JA, Shepard JRE. Quantum dots for live cell and in vivo imaging. Int J Mol Sci 2009;10(2):441–91.

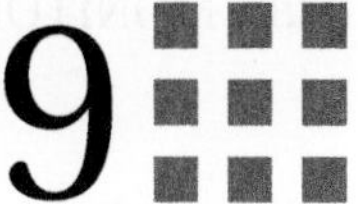

Advanced Oxidation and Reduction Process Radical Generation in the Laboratory and on a Large Scale: An Overview

Stephen P. Mezyk*, Kimberly A. Rickman, Charlotte M. Hirsch, Michelle K. Dail, Jeremy Scheeler, Trent Foust

DEPARTMENT OF CHEMISTRY AND BIOCHEMISTRY, CALIFORNIA STATE UNIVERSITY, LONG BEACH, CA, USA

**CORRESPONDING AUTHOR*

CHAPTER OUTLINE

Monitoring Water Quality, http://dx.doi.org/10.1016/B978-0-444-59395-5.00009-1

9.1 Introduction

The adverse ecological impacts of anthropogenically generated chemicals in our waters, especially endocrine-disrupting compounds, antibiotics, and pesticides/herbicides [1–6], are of major concern to regulatory groups and the public. Traditional water treatment relies primarily on adsorptive, chemical–physical, and microbial-based processes to transform or remove these unwanted organic contaminants. However, standard large-scale water treatments are not always sufficient [6], with quantitative removal of small (nanograms per liter) levels of dissolved chemicals being complicated by the presence of much higher levels of water constituents such as carbonates and dissolved organic matter (DOM).

The quantitative removal of trace levels of these chemicals requires additional process treatments, and therefore, the use of in situ generated radical species has been proposed. These approaches are generally referred to as advanced oxidation/reduction processes (AO/RPs) [7–11]. Radicals can be generated in water by using a variety of techniques (see Table 9-1 for a summary of some of the major AO/RP processes and the radicals they generate), and this requires either external energy deposition into the water or directly into a deliberately added chemical such as persulfate. Other radical-producing processes, or combinations of these processes, have also been used for contaminant treatment. Although direct destruction of chemical contaminants by ultraviolet (UV) light itself, or pulsed UV, is also widely used, this approach does not usually generate oxidizing or reducing radicals in water, and, hence, is not considered further here.

The most widespread AO/RPs are based on the use of the oxidizing hydroxyl radical ($^{\cdot}OH$), which is a powerful oxidant ($E^0 = 2.8$ V [12]) that reacts with both organic and

Table 9-1 Summary of Major AO/RPs and Their Reactive Species Generated

AO/RP	$^{\cdot}OH$	e^-_{aq}	$H^{\cdot}$	$SO_4^{-\cdot}$
H_2O_2/UV	X			
O_3/UV-C	X			
O_3/H_2O_2	X			
Fentons (Fe(II)/H_2O_2) or Photo-Fentons (Fe(II)/H_2O_2/UV)	X			
$TiO_2/h\nu$	X	X*		
$ZnO/h\nu$	X	X*		
Sonolysis (ultrasound)	X		X	
Supercritical Water	X		X	
Electrohydraulic Cavitation Plasma Discharge	X		X	
Electron-beam Irradiation γ-radiolysis	X	X	X	
Persulfate and UV/heat/transition metal ion				X

*Conduction band electron initially generated.

inorganic chemicals. For large-scale water treatment, the hydroxyl radical is usually generated by using some combination of O_3/H_2O_2, O_3/UV-C, or H_2O_2/UV-C. Additional techniques such as ionizing radiation (electron beam and gamma irradiation), Fenton's reaction, UV irradiation of titanium dioxide, and sonolysis also generate this radical and facilitate the study of AO/RP radical chemistry in the laboratory. As noted in Table 9-1 some AO/RPs generate other radicals, such as reducing hydrated electrons (e^-_{aq}) and hydrogen atoms ($H^\cdot$). Although these reducing radicals can also react to destroy chemical contaminants in water, their use in real-world waters is problematic because of the presence of air, where the relatively high level of dissolved oxygen ($[O_2] \sim 2.5 \times 10^{-4}$ M) preferentially scavenges these radicals to create the inert superoxide radical, $O_2^{\cdot-}$ [12]. Another approach has been to preferentially convert these two reducing radicals to another oxidizing radical, specifically the sulfate radical ($SO_4^{\cdot-}$), through the deliberate addition of persulfate [13]. The sulfate radical is also strongly oxidizing ($E^0 = 2.3$ V [14]), and thus, it can also react with most organic contaminants.

The optimal, quantitative removal of chemical contaminants from waters by using these processes requires a thorough understanding of the redox chemistry that occurs between free radicals and the chemicals of concern. This can be accomplished if absolute kinetic rate constants and efficiencies are determined for all the reactions that occur in the AO/RP system. These kinetic and mechanistic data allow for the quantitative computer modeling of AO/RP systems to establish the efficiency and large-scale applicability of using radicals for specific contaminant removal [15]. One important aspect of this understanding is how these systems generate their radicals. The generation and reactivity of AO/RP radicals are the focus of this overview chapter.

9.2 Experimental

The radical generation chemistry for all these AO/RPs has been established in many studies; the important reactions involved are briefly summarized here.

9.2.1 H_2O_2/UV-C

The deep UV photolysis (usually 254 nm) of added hydrogen peroxide results in the breaking of the O–O bond in this molecule,

$$\text{H–O–O–H} + h\nu \rightarrow {}^{\cdot}\text{OH} \tag{9.1}$$

with the quantum yield (production efficiency) for the hydroxyl radical being $\Phi_1 = 1.0$ [16]. The subsequent reaction of this radical with the added peroxide is slow [12]:

$$\cdot\text{OH} + H_2O_2 \rightarrow H_2O + HO_2^{\cdot} \quad k_2 = 2.7 \times 10^7\ \text{M}^{-1}\,\text{s}^{-1} \tag{9.2}$$

so even at the relatively high amounts of H_2O_2 added (usually millimolar levels are required for sufficient UV absorbance to take place), hydroxyl radical oxidation of chemical contaminants can still occur. The $HO_2^{\cdot}$ species is a relatively inert radical, existing in the equilibrium [17]:

$$HO_2^{\cdot} \rightleftharpoons O_2^{-\cdot} + H^+ \quad pK_{a_3} = 4.8 \text{ at } 25\,°C \tag{9.3}$$

9.2.2 O_3/UV-C

Similar photolysis chemistry occurs for UV light irradiation of added ozone in water, where the initial breaking of an O–O bond in this molecule ($\Phi_4 \sim 0.5$ for 254 nm light [18]):

$$O_3 + h\nu \rightarrow O^{\cdot} + O_2 \tag{9.4}$$

is followed by the fast reaction of the primary $O^{\cdot}$ radical with water [12]:

$$O^{\cdot} + H_2O \rightarrow H_2O_2 \quad k_5 = 1.8 \times 10^6\ M^{-1}\ s^{-1} \tag{9.5}$$

to generate hydrogen peroxide. Subsequent photolysis of H_2O_2, and/or ozone reaction with any $O_2^{-\cdot}$ present [19,20]

$$O_2^{-\cdot} + O_3 \rightarrow O_3^{-\cdot} + O_2 \quad k_6 = 1.6 \times 10^9\ M^{-1}\ s^{-1} \tag{9.6}$$

$$O_3^{-\cdot} + H^+ \rightarrow HO_3^{\cdot} \quad k_7 \sim 5 \times 10^{10}\ M^{-1}\ s^{-1} \tag{9.7}$$

$$HO_3^{\cdot} \rightarrow {}^{\cdot}OH + O_2 \quad k_8 = 1.4 \times 10^5\ s^{-1} \tag{9.8}$$

generates the hydroxyl radical.

9.2.3 O_3/H_2O_2

The peroxone treatment depends on the prior ionization of the peroxide [12]

$$H_2O_2 \rightleftharpoons HO_2^- + H^+ \quad pK_{a_9} = 11.8 \tag{9.9}$$

with subsequent reaction of the HO_2^- anion with ozone to generate $O_3^{-\cdot}$ [21],

$$HO_2^- + O_3 \rightarrow O_3^{-\cdot} + HO_2^{\cdot} \quad k_{10} = 2.8 \times 10^6\ M^{-1}\ s^{-1} \tag{9.10}$$

which results in hydroxyl radical production. Analogously, the decomposition of ozone through its reaction with hydroxide

$$OH^- + O_3 \rightarrow O_2^{-\cdot} + HO_2^{\cdot} \quad k_{11} = 70\ M^{-1}\ s^{-1} \tag{9.11}$$

can also occur, and this results in hydroxyl radical production through reactions (9.3), (9.6) and (9.8).

9.2.4 Fenton's Chemistry

The generation of hydroxyl radicals using a combination of Fe(II) and hydrogen peroxide in acidic media results in the following chemical reactions [22]:

$$Fe^{2+} + H_2O_2 \rightarrow Fe^{3+} + OH^- + {}^{\cdot}OH \quad k_{12} = 76\ M^{-1}\ s^{-1} \tag{9.12}$$

$$Fe^{3+} + H_2O_2 \rightarrow Fe^{2+} + HO_2^{\cdot} + H^+ \quad k_{13} \tag{9.13}$$

The latter reaction is slow and catalytically complex. Some loss of hydroxyl radicals may also occur from the side reaction of this species with the ferrous ion present [12]:

$$Fe^{2+} + {}^{\cdot}OH \rightarrow Fe^{3+} + OH^- \quad k_{14} = 3 \times 10^8\ M^{-1}\ s^{-1} \tag{9.14}$$

9.2.5 $TiO_2/h\nu$ and $ZnO/h\nu$

These semiconductor photocatalysts require UV light for photoexcitation (TiO_2, <380 nm [23], ZnO, <375 nm [24]). On photon absorption in these solids, a single electron is promoted to the conduction band, generating what is known as an electron–hole pair. The conduction band electron (e^-_{cond}) is available for reduction, and the valence band hole ($h_{vb}{}^+$) is available for oxidation. Most of the charge-separated pairs simply recombine to regenerate the photon (or heat), but in the presence of an electron scavenger such as dissolved oxygen, the formation of adsorbed superoxide

$$e^-_{cond} + O_2(ads) \rightarrow O_2{}^{-\cdot}(ads) \tag{9.15}$$

can prolong the lifetime of the hole and can allow it to react. In condensed aerated solution, the surfaces of these semiconductors are hydroxylated so that the hole reaction results in adsorbed hydroxyl radicals being formed:

$$h_{vb}{}^+ + HO^-(ads) \rightarrow {}^{\cdot}OH(ads) \tag{9.16}$$

Subsequent diffusion of an organic contaminant to this adsorbed hydroxyl radical (active site) or diffusion of the adsorbed hydroxyl radical across the catalyst surface or into the solution to react with a substrate may also occur. For chemical contaminants that strongly absorb onto the oxide surface, direct oxidation by the hole may also compete with the formation of the hydroxyl radical.

9.2.6 Sonolysis

Ultrasonic irradiation of aqueous solutions results in the formation and subsequent collapse of gas bubbles (cavitation), which produces extremely high temperatures (up to 7000 K) and pressures (up to 1000 atm). This leads to homolytic water O–H bond breaking [25]:

$$H_2O\text{--})))))\rightarrow H^{\cdot} + {}^{\cdot}OH \tag{9.17}$$

The hydrogen atom produced will generally recombine or react with the dissolved oxygen present in the solution. The hydroxyl radical will partition between the three regions present during cavitation: the interior bubble gas phase, bulk liquid solution, and gas–liquid interface. The highest hydroxyl radical reactivity is in the gas phase. Significantly higher temperature and pressure conditions in the interface can accelerate the decomposition of contaminant chemicals through hydrolysis, low-temperature pyrolysis, and supercritical water oxidation (SCWO) processes.

9.2.7 Supercritical Water Oxidation

SCWO [26], also referred to as hydrothermal oxidation, is a thermal process that treats water at high temperatures and pressures above the critical point of water ($P_c = 220.6$ bar, $T_c = 374$ °C [27]). Under these extreme conditions, gas-phase, neutral radical species reactions mainly occur. The direct formation of a variety of intermediary species,

$$H_2O \xrightarrow{SCWO} H^{\cdot}, O^{\cdot}, {}^{\cdot}OH, HO_2^{\cdot} \tag{9.18}$$

results in a complex suite of reactions that quantitatively degrades any chemical contaminants that may be present. This ultimately results in completely oxidized products such as carbon dioxide, nitrogen, water, HCl, H_2SO_4, and solid metal oxides.

9.2.8 Electrohydraulic Cavitation/Plasma Discharge

The chemistry behind this AO/RP is similar to that of sonolysis, where cavitation of the solution occurs to pyrolytically destroy hazardous chemicals in water [28]. Pulsed-power discharge is an electrohydraulic phenomenon where periodic rapid release of electrical energy across a submerged electrode gap occurs [29]. This results in a highly ionized and pressurized plasma that dissociates, excites, and ionizes water to give a mixture of radical ($H^{\cdot}$, ${}^{\cdot}OH$) and molecular (H_2, H_2O_2) species. Additional reaction pathways are also generated from the direct reactions of the rapidly expanding plasma gases with the chemical contaminants, as well as from indirect production of ${}^{\cdot}OH$ radicals because of the release of soft X-rays and high-energy UV radiation from the energized plasma.

9.2.9 Electron-Beam Irradiation/γ-Radiolysis

The low linear energy transfer irradiation of aqueous solutions containing small amounts (<0.1 M) of chemical contaminants produces a mixture of radicals and molecular products of water ionization and bond breaking. In the pH range of 3–10, a homogeneous distribution of these products occurs at approximately 100 ns after irradiation occurs, according to the stoichiometry [12]:

$$H_2O \xrightarrow{radiation} [0.28]\,{}^{\cdot}OH + [0.27]e_{aq}^{-} + [0.06]H^{\cdot} + [0.07]H_2O_2 + [0.05]H_2 + [0.27]H_3O^{+} \tag{9.19}$$

The numbers in brackets indicate the yields (*G*-values) of each species in units of micromoles per joule of energy deposited and are the same for both electron and gamma irradiation. By adding suitable chemicals, each of the radical species, hydroxyl radicals (${}^{\cdot}OH$), hydrated electrons (e_{aq}^{-}), and hydrogen atoms (${}^{\cdot}H$), can be isolated for independent study.

9.2.10 Persulfate and UV/Heat/Transition Metal Ions

The sulfate radical ($SO_4^{-\cdot}$) is another powerful oxidant ($E^0 = 2.3$ V, [14]). This radical can be generated in water in several ways, notably through the reaction of added persulfate ($S_2O_8^{2-}$) with reducing e_{aq}^- or $H^{\cdot}$ atoms [12]:

$$S_2O_8^{2-} + e_{aq}^- \rightarrow SO_4^{2-} + SO_4^{-\cdot} \quad k_{20} = 1.2 \times 10^{10}\ M^{-1}\ s^{-1} \tag{9.20}$$

$$S_2O_8^{2-} + H^{\cdot} \rightarrow H^+ + SO_4^{2-} + SO_4^{-\cdot} \quad k_{21} = 1.4 \times 10^{7}\ M^{-1}\ s^{-1} \tag{9.21}$$

by photolytically induced cleavage of the S–S bond [13]:

$$S_2O_8^{2-} + h\nu \rightarrow SO_4^{-\cdot} \quad \Phi_{22} = 1.4 \tag{9.22}$$

or by reaction with a transition metal (M^{2+}) ion (e.g. Fe^{2+}, Ag^+, Ni^{2+}, Sn^{2+}):

$$S_2O_8^{2-} + M^{2+} \rightarrow M^{3+} + SO_4^{2-} + SO_4^{-\cdot} \tag{9.23}$$

9.2.11 Radical Reaction Studies

For the typical AO/RP radical reaction examples provided in the next section, the chemicals used in this study were purchased from Sigma–Aldrich Chemical Company at the highest purity available (steroids, >98%; KSCN, 99%; $K_2S_2O_8$, 99%; β-lactam antibiotics >98%; nitrosamines >98%). Parabens were purchased from AccuStandard Chemical Reference Materials at >98% purity. All chemicals were used as received. These commercially available chemicals were dissolved in high quality, Millipore Milli-Q, charcoal-filtered (total organic carbon < 13 ppb), deionized (>18.0 MΩ) water along with other chemicals required to isolate the specific radical of interest.

All rate constant data were collected using the pulsed electron linear accelerator facility at the Radiation Laboratory, University of Notre Dame. This irradiation and transient absorption detection system has been described in detail previously [30]. Absolute radical concentrations (dosimetry) were based on the transient absorption of $(SCN)_2^{\cdot-}$ at 475 nm, using 10^{-2} M thiocyanate (KSCN) in N_2O-saturated solution at natural pH with $G^*\varepsilon = 5.2 \times 10^{-4}\ m^2\ J^{-1}$ [31] performed daily.

All these experiments were conducted at ambient room temperature (20 ± 2 °C) with temperature variation in any given experiment being <±0.3 °C. Rate constant error limits reported here are the combination of experimental precision and compound purities.

9.3 Results and Discussion

The AO/RP radicals generated will react with contaminant chemicals in water through various mechanisms and with a range of rate constants. In this section, four specific examples are given for the major AO/RP radicals with different classes of chemical contaminants to illustrate this. However, it is important to note that the chemistry of all these AO/RP free radicals in water is *independent* of their generation method. Therefore, kinetic data obtained by one AO/RP are applicable to another. Laboratory studies on chemical contaminant removal have been reported using all the aforementioned systems, whereas on a large scale, the AO/RP-type treatments have been based mainly on the hydroxyl radicals generated by ozone, H_2O_2, and UV-C light combinations.

To establish the feasibility of using AO/RP radicals as $^{\bullet}OH$, $SO_4^{-\bullet}$, e_{aq}^-, and the $H^{\bullet}$ atom for large-scale water remediation treatment, the absolute kinetic rate constants and efficiencies for their reactions with individual contaminant classes have to be determined. Typically, we have used the well-established electron-pulse-radiolysis process methodology to perform these investigations. Electron-pulse radiolysis uses a short burst of electrons generated by a linear accelerator. In moderately dilute solutions, the energy will be absorbed predominately by the water and will result in the occurrence of the reactions as shown in Figure 9-1 [32,33]. The energized water molecules and ions in the solution can further react to form radicals. From 10^{-12} to 10^{-7} s after the energy deposition event, there are many side reactions that occur, which ultimately results in a mixture of radical and molecular products (H_2 and H_2O_2).

There are three radicals produced in the pulsed electron radiolysis of water: the hydroxyl radical ($^{\bullet}OH$), the hydrogen atom ($H^{\bullet}$), and the hydrated electron (e_{aq}^-). The other ($SO_4^{-\bullet}$) radical can be readily produced by the reaction of the hydrated electron and/or hydrogen atom with added persulfate, by using this AO/RP [13].

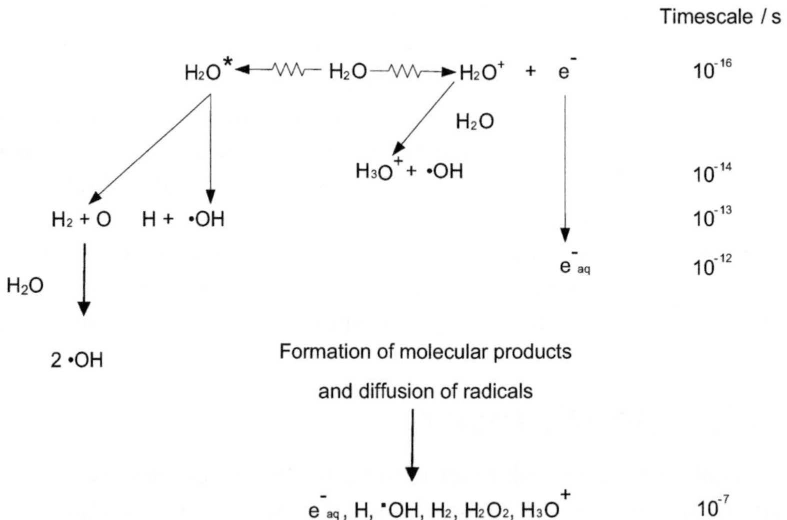

FIGURE 9-1 Summary of the reactions occurring in water after the absorption of energy from 10^{-16} to 10^{-7} s [32,33].

9.3.1 Hydroxyl Radical Reactions: β-Lactam Antibiotics

To isolate the hydroxyl radical in the electron-irradiated solution, they were presaturated with N_2O (g), which removes all the dissolved oxygen and then allows the following radical conversion reactions to occur [12]:

$$e_{aq}^{-} + N_2O(+ H_2O) \rightarrow N_2 + OH^{-} + {}^{\cdot}OH \quad k_{24} = 9.1 \times 10^9 \ M^{-1}\ s^{-1} \tag{9.24}$$

$$H^{\cdot} + N_2O \rightarrow N_2 + {}^{\cdot}OH \quad k_{25} = 2.1 \times 10^6 \ M^{-1}\ s^{-1} \tag{9.25}$$

Because of the high solubility of N_2O in water (26 mmol dm^{-3} at 25 °C [34]) and its fast reaction rate constant, all generated hydrated electrons are quantitatively converted to hydroxyl radicals. The hydrogen atom reacts with N_2O more slowly and therefore contributes less to the overall yield of hydroxyl radicals in the solution.

The use of hydroxyl radical reactions to remove chemical contaminants from water is demonstrated by their reactions with β-lactam antibiotics [35] (Figure 9-2). These particular antibiotics have been extensively used for many years [36,37], both for human consumption and in agriculture to prevent infection from occurring in plants and livestock [4,5,38]. The reaction of hydroxyl radicals with this class of antibiotics typically gave transient absorption spectra with peaks in the region of 300–380 nm (e.g. penicillin-G in Figure 9-3, inset). From the rate of change of the first-order growths of this transient absorption with antibiotic concentration (Figure 9-3), reaction rate constants for the oxidative hydroxyl radical reaction with each compound were determined (Figure 9-4). This approach was used for a library of β-lactam antibiotics; the obtained kinetic data are summarized in Table 9-2.

For the 5-member-ring β-lactam antibiotics, the average rate constant for hydroxyl radical reaction is $k_{av} = (7.9 \pm 0.8) \times 10^9\ M^{-1}\ s^{-1}$. The comparison of this value to literature data for model aromatic chemicals such as toluene ($k = 3.0 \times 10^9\ M^{-1}\ s^{-1}$ [39]) and thiophene ($k = 3.3 \times 10^9\ M^{-1}\ s^{-1}$ [40]), as well as the previously established rate constant of $(2.40 \pm 0.05) \times 10^9\ M^{-1}\ s^{-1}$ for (+)-6-aminopenicillanic acid core moiety [41], suggests that the initial hydroxyl radical reaction for all the antibiotics in this study would be partitioned between the peripheral aromatic ring on the lactam substituent (~70%) and the central double-ring system (~30%).

For the four 6-member ring antibiotics, the overall rate constant for hydroxyl radical oxidation measured was slightly slower, $k_{av} = (6.6 \pm 1.2) \times 10^9\ M^{-1}\ s^{-1}$. This again

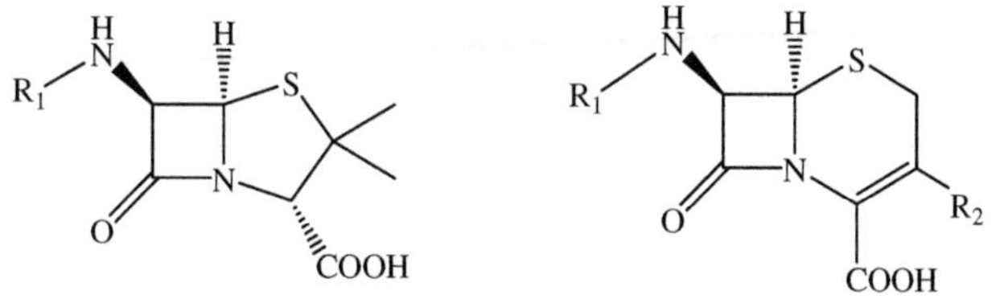

FIGURE 9-2 β-Lactam antibiotic structures: 5-member-ring β-lactams have a single R_1 substituent; 6-member-ring β-lactams have two substituents R_1, R_2.

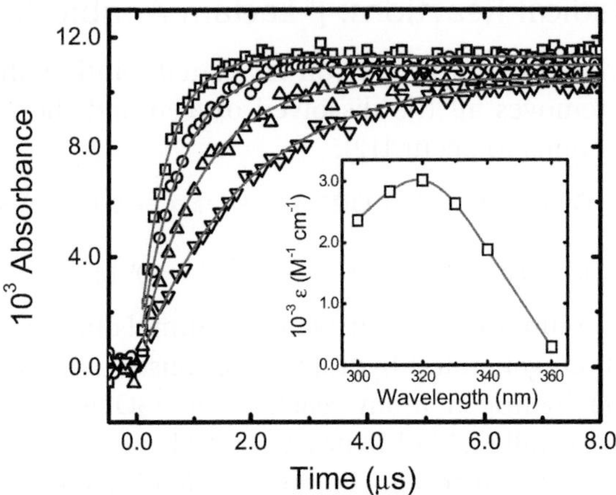

FIGURE 9-3 Measured first-order growth kinetics for hydroxyl radical reaction with penicillin-G at pH 7.40 and 23.0 °C. Inset: maximum transient absorption spectrum of oxidized 250.0 μM penicillin-G under the same conditions. (For color version of this figure, the reader is referred to the online version of this book.)

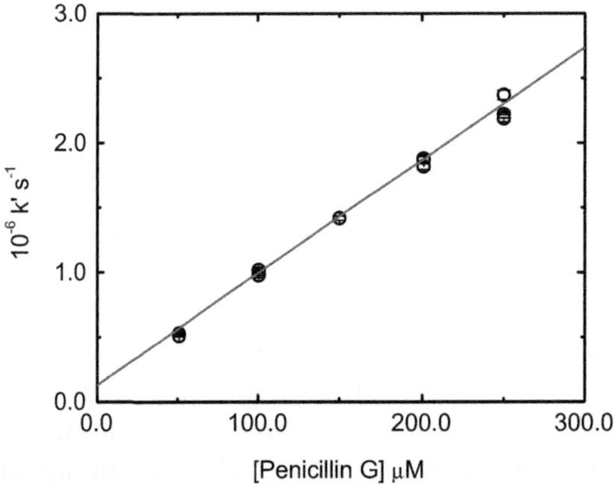

FIGURE 9-4 Second-order plot for hydroxyl radical reaction with penicillin-G at 23.0 °C, based on fitted exponential growth kinetics of Figure 9-3. The slope corresponds to the reaction rate constant of $k = (8.70 \pm 0.32) \times 10^9\ M^{-1}\ s^{-1}$ ($R^2 = 0.99$). (For color version of this figure, the reader is referred to the online version of this book.)

suggests that significant partitioning of the hydroxyl radical between the central rings and a peripheral moiety occurs for these antibiotics.

As the majority of the hydroxyl radical oxidation occurs at peripheral moieties such as aromatic rings in these antibiotics, ultimately, it would be expected that phenolic compounds would be formed [41]. However, these products could still function as antibiotics (because of the intact β-lactam ring), which implies that hydroxyl-generating AO/RP systems could be useful for the quantitative removal of antibiotics in real-world waters only if multiple radical-induced oxidations occurred. This reduces the feasibility of

Table 9-2 Summary of Hydroxyl Radical Reaction Rate Constants for 5- and 6-Member-Ring β-Lactam Antibiotics

5-Member-Ring β-Lactams			6-Member-Ring β-Lactams		
Species	**10^{-9} $k_{\cdot OH}$ M^{-1} s^{-1}**	**Temperature °C**	**Species**	**10^{-9} $k_{\cdot OH}$ M^{-1} s^{-1}**	**Temperature °C**
Ampicillin	8.21 ± 0.29	22.7	Cefaclor	6.00 ± 0.13	22.7
Carbenicillin	7.31 ± 0.11	18.0	Cefazolin	6.48 ± 0.48	22.4
Cloxacillin	6.27 ± 0.15	23.0	Cephalothin	5.51 ± 0.29	19.2
Penicillin-G	8.70 ± 0.32	23.0	Cefatoxime	8.22 ± 0.14	21.2
Penicillin-V	8.54 ± 0.27	21.3			
Piperacillin	7.84 ± 0.49	22.5			
Tircarcillin	8.18 ± 0.99	21.8			

using large-scale remediation of β-lactam antibiotic-contaminated waters by an AO/RP treatment.

9.3.2 Sulfate Radical Reactions: Parabens

The sulfate radical typically reacts by electron abstraction from electron-rich centers in molecules, or by aliphatic hydrogen atom abstraction [13]:

$$SO_4^{-\cdot} + R\text{–}H \rightarrow H^+ + SO_4^{2-} + R^{\cdot} \quad (9.26)$$

$$SO_4^{-\cdot} + R\text{–}H \rightarrow SO_4^{2-} + (R - H)^{+\cdot} \quad (9.27)$$

in contrast to hydroxyl radical-based oxidations that occur mainly by hydrogen atom abstraction and/or addition to aromatic constituents [12]. The sulfate radical exhibits strong absorbance at 450 nm [42], which allows it to be directly observed on using UV/visible absorption spectroscopy in real time.

Typical reactions for the sulfate radical in water are demonstrated here for the three lowest-molecular-weight parabens. Parabens are preservatives found in many commonly used cosmetics, shampoos, and toothpaste. Several ill effects of these contaminants have been reported, and these include their ability to act as estrogen mimics [43]. The three parabens investigated in this study, methyl-, ethyl-, and propylparaben, are shown in Figure 9-5.

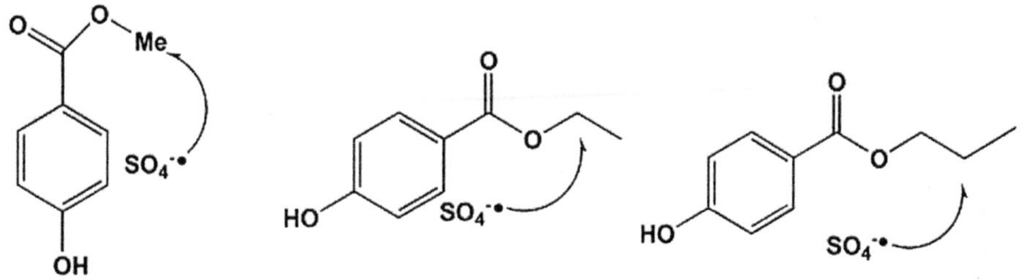

FIGURE 9-5 Methyl-, ethyl-, and propylparaben (left to right) and our proposed mechanism for reaction of sulfate radicals.

To generate the sulfate radical through reaction [20] by using electron-pulse radiolysis, typical conditions used included presaturating the solution with nitrogen gas to remove dissolved oxygen gas, plus adding 2.0 M *tert*-butanol to help dissolve the paraben and to scavenge the hydroxyl radicals and hydrogen atoms produced [12]:

$$(CH_3)_3COH + {}^{\cdot}OH \rightarrow {}^{\cdot}CH_2(CH_3)_2COH + H_2O \quad k_{28} = 6.0 \times 10^8\ M^{-1}\ s^{-1} \tag{9.28}$$

$$(CH_3)_3COH + H^{\cdot} \rightarrow {}^{\cdot}CH_2(CH_3)_2COH + H_2 \quad k_{29} = 1.7 \times 10^5\ M^{-1}\ s^{-1} \tag{9.29}$$

The resulting *tert*-butanol radical is relatively inert and does not react with the parabens. The isolated hydrated electron (which does not react with *tert*-butanol) will then react with added persulfate (typically 5–10 mM) to generate the sulfate radical (reaction (9.19)) on the sub-micro-second time scale.

Typical sulfate radical reaction kinetics are shown for ethylparaben in Figure 9-6. The decay of this radical occurs on the microsecond time scale and becomes faster with more added paraben solute (Figure 9-6a). Analysis of these data using pseudo-first-order kinetics gives the second-order plot shown in Figure 9-6b, whose slope is the absolute

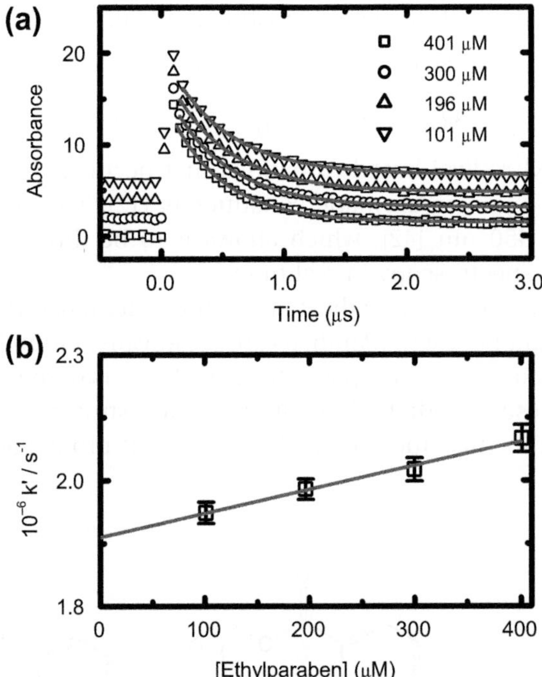

FIGURE 9-6 (a) Kinetic decay traces obtained for $SO_4^{\cdot-}$ radical decay in N_2-saturated aqueous solution containing 2.0 M *tert*-butanol at 450 nm in the presence of different amounts of ethylparaben. Solid lines are fitted exponential decay kinetics. (b) Second-order kinetic plot for the reaction of sulfate radical with ethylparaben using pseudo-first-order fits of (a). The slope corresponds to the reaction rate constant of $k = (1.30 \pm 0.14) \times 10^9\ M^{-1}\ s^{-1}$ ($R^2 = 0.992$). (For color version of this figure, the reader is referred to the online version of this book.)

Table 9-3 Summary of Rate Constants for Sulfate Radical ($SO_4^{-\bullet}$) Reaction with Parabens in Aqueous Solution

Compound	$k_{SO_4^{-\bullet}} \times 10^9$ M^{-1} s^{-1}
Methylparaben	1.55 ± 0.10
Ethylparaben	1.30 ± 0.14
Propylparaben	3.19 ± 0.19

reaction rate constant. This approach was used for all three parabens, with the kinetic results as summarized in Table 9-3.

The three parabens exhibited slightly different rate constants for the sulfate radical reaction. The values for methylparaben and ethylparaben were the same within error, whereas that of propylparaben was slightly faster (Table 9-3). For these sulfate radicals, it seems that most of the reaction takes place at the alkyl side chain, as shown in Figure 9-5, as these measured rate constants are greater than were both the previously measured rate constant for the reaction of benzene [44], 8.0×10^8 M^{-1} s^{-1}, as well as that of propane [45], 4.0×10^7 M^{-1} s^{-1}. Overall, these relatively fast oxidative rate constants for these species suggest that the sulfate-radical-based remediation of paraben-contaminated water may be feasible.

9.3.3 Hydrated Electron Reactions: Estrogenic Steroids

Previous work has shown that radical-based oxidation approaches can efficiently destroy estrogenic steroids in water [9–11,46,47], and this has encouraged further interest in the large-scale use of AO/RPs for this purpose. Although some hydroxyl radical and sulfate radical rate constants have been determined for several estrogenic steroids, the corresponding reduction of estrogenic steroids by direct radical reactions has been far less studied [48]. As stated previously, the hydrated electron is typically not a useful radical for remediating chemically contaminated waters because of the preferential reaction of this species with dissolved oxygen. However, if the chemical contaminant undergoes reduction extremely quickly, or for some AO/RPs with high radical production rates (e.g. electron beam) that can effectively convert all the oxygen to superoxide, then reduction of contaminant chemicals by this radical could occur even in aerated waters.

Two literature rate constants for estriol and estradiol reaction with the hydrated electron have been reported, with very high values of 2.6×10^{10} and 2.7×10^{10} M^{-1} s^{-1} [49], respectively, calculated from competition kinetics measurements using $Co(en)_3^{3+}$ as a standard. This rate constant is effectively diffusion controlled. However, there were some concerns regarding these measurements, particularly that this study was conducted at pH 13, which is not easily transferable to real-world water conditions.

To confirm these fast rate constants and to investigate the feasibility of using highly reducing radicals for the quantitative removal of other steroids in contaminated waters,

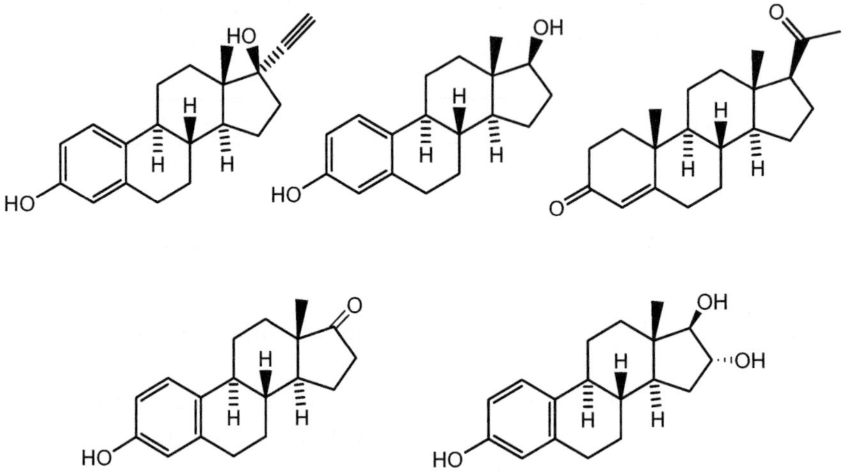

FIGURE 9-7 Top structures (left to right): ethynylestradiol (EE2), estradiol, and progesterone; bottom: estrone and estriol.

we determined absolute reaction rate constants for this radical based on a small library of estrogenic steroids (Figure 9-7) in water, using our electron-pulse-radiolysis methodology. These solutions were at natural water pH (measured as 6.8–7.2). To isolate the hydrated electron chemistry, the radiolysis experiments were conducted by using a constant high concentration (1.0–2.0 M) of *tert*-butanol as a cosolvent [12]. The presence of this alcohol immediately scavenges the radiolytically produced hydroxyl radicals and hydrogen atoms (see reactions (9.28) and (9.29)), and under these mixed water/alcohol conditions, the steroid solubility was sufficiently high (50–200 μM) to maintain pseudo-first-order kinetics.

The hydrated electron has a broad, intense, UV–visible absorption spectrum, and it peaks near 720 nm at room temperature [42]. Typical decay traces measured for different concentrations of estradiol are shown in Figure 9-8a. Under these conditions, excellent first-order kinetics were seen as evidenced by the exponential decays fitted to these curves. By plotting these first-order values against the estradiol concentration, the linear graph shown in Figure 9-8b was obtained, with the fitted weighted linear slope corresponding to the absolute rate constant for the reaction:

$$e_{aq}^{-} + \text{estradiol} \rightarrow \text{products} \quad k_{30} = (2.52 \pm 0.11) \times 10^{8}\ \text{M}^{-1}\,\text{s}^{-1} \tag{9.30}$$

This rate constant is >100-fold slower than the previously determined value of 2.7×10^{10} $\text{M}^{-1}\,\text{s}^{-1}$ [49]. Support for our value was obtained by the analogous hydrated electron rate constant measurement for ethynylestradiol (see Table 9-4 for all measured data for this radical), for which we obtained a slightly lower rate constant of $k = (1.46 \pm 0.16) \times 10^{8}\ \text{M}^{-1}\,\text{s}^{-1}$. Literature values for the hydrated electron reaction with the free common phenol moiety ($k = 3.0 \times 10^{7}\ \text{M}^{-1}\,\text{s}^{-1}$ [12]) as well as that for acetylene ($k = 2.0 \times 10^{7}\ \text{M}^{-1}\,\text{s}^{-1}$ [12]) in water were in agreement with our value.

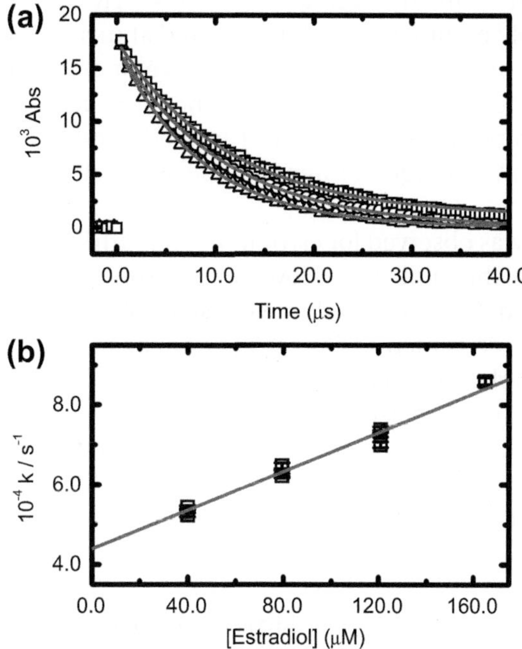

FIGURE 9-8 (a) Decay of hydrated electron absorption at 700 nm in N_2-saturated aqueous solution of 2.00 M *tert*-butanol at natural pH and 21.7 °C with 40.0 μM (□), 121 μM (○), and 165 μM (△) estradiol present. Solid lines are fitted exponential decays, corresponding to pseudo-first-order rate constants of $(5.43 \pm 0.04) \times 10^4$, $(7.01 \pm 0.08) \times 10^4$ and $(8.60 \pm 0.09) \times 10^4\ s^{-1}$, respectively. (b) Transformed second-order plot for hydrated electron reaction with estradiol. The solid line corresponds to weighted linear fit, with a slope of $k_{30} = (2.52 \pm 0.07) \times 10^8\ M^{-1}\ s^{-1}$, $R^2 = 0.99$. (For color version of this figure, the reader is referred to the online version of this book.)

Table 9-4 Summary of Measured Rate Constants for Hydrated Electron (e^-_{aq}) Reduction of Estrogenic Steroids in Aqueous Solution

Estrogen	$k_{e^-_{aq}}$ $M^{-1}\ s^{-1}$
Ethynylestradiol	$(1.46 \pm 0.16) \times 10^8$
Estradiol	$(2.52 \pm 0.11) \times 10^8$
Progesterone	$(8.65 \pm 0.15) \times 10^9$
Estrone	$(2.07 \pm 0.11) \times 10^{10}$
Estriol	$(2.37 \pm 0.03) \times 10^{10}$

In contrast, the measured rate constant for progesterone was much faster, with a value of $k = (8.65 \pm 0.15) \times 10^9\ M^{-1}\ s^{-1}$. By comparing these three steroid structures, it is probable that most of the progesterone reduction would occur at the ketone moiety in this steroid to give a ketyl radical intermediate. This faster rate constant is in good agreement with the value obtained for the model compound acetone in water, $k = (6–8) \times 10^9\ M^{-1}\ s^{-1}$

[12], and faster than that for methyl vinyl ketone ($H_2C{=}CHCOCH_3$, $k = 2.5 \times 10^9$ M^{-1} s^{-1} [12]), indicating that only minimal reduction occurred at this steroid constituent. Similar fast hydrated electron reactivity was seen for the ketone-containing steroid estrone. Unfortunately, the ketone moiety reduction would not be expected to result in any steroid destruction in real waters, as the ketyl radical formed would subsequently transfer its excess electron to dissolved oxygen or another species such as DOM to regenerate the original steroid.

The fastest reaction was observed for estriol, which is attributed to the reactivity of the adjacent diol moiety in this molecule. However, no suitable analog was found in the literature for comparison. Our measured rate constant of $(2.37 \pm 0.03) \times 10^{10}$ M^{-1} s^{-1} is actually in reasonable agreement with the previous steady-state determination of 2.6×10^{10} M^{-1} s^{-1}, in stark contrast to the values for estradiol [49]. However, the products of these reactions have yet to be established.

Ultimately, the low reaction rate constant for estradiol and the additional chemistry of the ketone-containing steroids suggest that reductive AO/RP technologies would not be feasible for large-scale remediation of the majority of estrogenic steroid-contaminated waters.

9.3.4 Hydrogen Atom Reactions: Nitrosamines

The reactivity of the hydrogen atom, like that for the hydrated electron, is limited in treatment of chemically contaminated waters because of its fast reaction with dissolved oxygen [12]:

$$H^{\cdot} + O_2 \rightarrow HO_2^{\cdot} \leftrightharpoons H^+ + O_2^{-\cdot} \quad k_{31} = 1.2 \times 10^{10}\ M^{-1}\ s^{-1} \tag{9.31}$$

However, under acidic water conditions, where hydrated electron reaction to this radical occurs,

$$e_{aq}^- + H^+ \rightarrow H^{\cdot} \quad k_{32} = 2.3 \times 10^{10}\ M^{-1}\ s^{-1} \tag{9.32}$$

or under sonolysis conditions in which deaeration of solutions can occur through cavitation, some hydrogen atom reaction with contaminant chemicals may also occur. As such, the kinetics and mechanisms of hydrogen atom reaction may also need to be established under certain treatment conditions.

Although considerable effort has been reported for AO/RP radical reactions with the carcinogenic nitrosamines [50–54], only one hydrogen atom reaction measurement has been performed [50]. For *N*-nitrosodimethylamine (NDMA), a direct measurement of this absolute value using the electron-pulse-radiolysis-based electron paramagnetic resonance (EPR) free-induction-decay (FID) methodology [55–57]

$$H^{\cdot} + (CH_3)_2NNO \rightarrow \text{products} \tag{9.33}$$

gave a rate constant of $k_{33} = (2.01 \pm 0.03) \times 10^8$ M^{-1} s^{-1} for this reaction [50]. A repeat measurement of this reaction in this study used electron-pulse radiolysis and UV/visible absorption spectroscopy, by competition kinetics with *p*-chlorobenzoic acid (pCBA) as a standard [12]:

$$H^{\cdot} + Cl–C_6H_4–COOH \rightarrow pCBA–H \text{ adduct} \quad k_{34} = 8.5 \times 10^8\ M^{-1}\ s^{-1} \tag{9.34}$$

This approach is based on the principle that neither the hydrogen atom nor the product species of reaction (9.33) shows any significant absorbance in the UV–visible wavelength range (250–800 nm). The H-atom adduct of pCBA has a strong absorbance at 355 nm (Figure 9-9a), which decreases when increasing concentrations of NDMA are added.

Under electron-pulse radiolysis under N_2-saturated acidic (pH = 1.0) solution conditions, the initially formed hydrated electrons react with the acid to produce more hydrogen atoms (reaction (9.32)). By adding a small amount of alcohol, 0.20 M *tert*-butanol, quantitative scavenging of the produced hydroxyl radicals occurs. Some hydrogen atom reaction with this alcohol also occurs (reaction (9.29)), but this reaction is sufficiently slow that most of the hydrogen atoms react with the added solute.

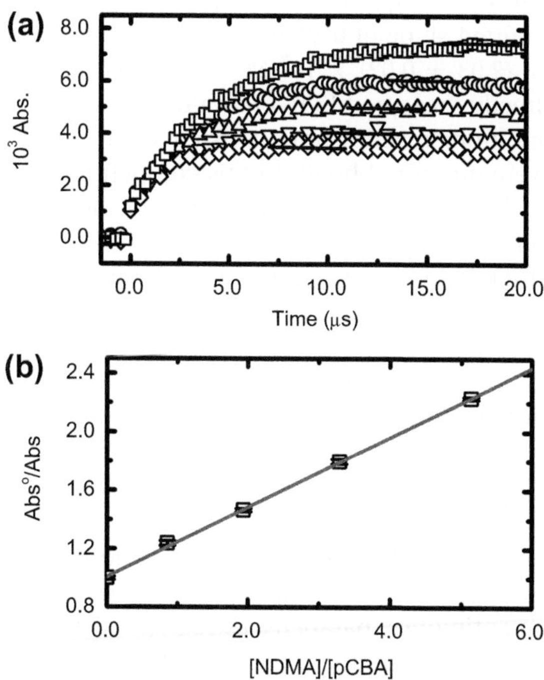

FIGURE 9-9 (a) Kinetic traces obtained for N_2-saturated solution of 225 μM pCBA at 355 nm in the presence of zero (□), 192 (○), 436 (△), 740 (▽), and 1154 (◇) μM NDMA. (b) Transformed competition kinetics plot, using the peak absorbances of (a) (see the horizontal lines). Error bars correspond to an average of four individual measurements. The solid line corresponds to a weighted linear fit of slope 0.2386 ± 0.0028 and intercept 1.007 ± 0.008 ($R^2 = 1.00$), which corresponds to a hydrogen atom reaction rate constant for NDMA of (2.03 ± 0.02) × $10^8\ M^{-1}\ s^{-1}$. (For color version of this figure, the reader is referred to the online version of this book.)

Table 9-5 Summary of Hydrogen Atom ($H^{\cdot}$) Reaction Rate Constants with Nitrosamines Measured Using pCBA Competition Kinetics in Aqueous Solutions at Room Temperature

Species	# CH_2 Groups	$10^{-8}\ k_H\ M^{-1}\ s^{-1}$
$(CH_3)_2NNO$	0	2.03 ± 0.02
$(C_2H_5)_2NNO$	2	2.39 ± 0.06
$(C_4H_9)_2NNO$	4	2.23 ± 0.08
c-$C_5H_{10}NNO$	5	2.29 ± 0.04

Based on the hydrogen atom competition of reactions (9.33) and (9.34), one can obtain an analytical solution of the form [12]

$$\frac{[\text{pCBA-H}]^0}{[\text{pCBA-H}]} = \frac{\text{Abs}^0}{\text{Abs}} = 1 + \frac{k_{33}[\text{NDMA}]}{k_{34}[\text{pCBA}]} \tag{9.35}$$

where Abs^0 and Abs are the absorbances of the pCBA-H adduct in the absence and the presence of NDMA, respectively. From a plot of Abs^0/Abs vs [NDMA]/[pCBA] (Figure 9-9b), one obtains a straight line of unity intercept and a slope of k_{33}/k_{34}. This transformed plot has a fitted slope of 0.2386 ± 0.0028, and if $k_{34} = 8.5 \times 10^8\ M^{-1}\ s^{-1}$ is known, one gets $k_{33} = (2.03 \pm 0.02) \times 10^8\ M^{-1}\ s^{-1}$. This rate constant is in excellent agreement with the directly measured EPR–FID value.

Additional hydrogen atom reaction rate constants with a series of low-molecular-weight nitrosamines were measured by using this competition kinetics approach; these values are summarized in Table 9-5. The consistency of these values suggests that the hydrogen atom reaction is with the common nitroso (>N–NO) moiety in these molecules. By analogy to the hydrated electron reactions with these carcinogens [52], which also occur at this moiety independent of chain substituents, we would not expect the hydrogen-atom-induced nitrosamine reduction to result in any permanent destruction under real-world AO/RP treatment conditions. However, the equivalent chemistry under sonolysis conditions has not yet been established.

9.4 Conclusions

A brief overview of the aqueous kinetic chemistry for the reaction of several different AO/RP radicals, the hydroxyl radical, sulfate radical, hydrated electron, and hydrogen atom, with beta-lactam antibiotics, parabens, estrogenic steroids, and aliphatic nitrosamines, respectively, has been presented. By comparing absolute second-order reaction rate constants for different species in each contaminant class, an insight into the efficacy of this radical-based approach for large-scale treatment of waters can be obtained.

From these four cases, it is clear that fast radical reaction rate constants are a necessary but not sufficient condition for consideration of AO/RP treatment of aqueous contaminant chemicals. For example, though the hydroxyl radical reaction with the β-lactam antibiotics is consistently fast, it was shown that most of the oxidation occurs at peripheral moieties, particularly at aromatic rings, which might not destroy antibiotic activity. As such, AO/RP treatment of β-lactam-antibiotic-contaminated waters might only be feasible if multiple hydroxyl radical reactions occur. For sulfate radical reaction with parabens in water, the similarity of the measured rate constants suggests that the predominant mode of reaction is hydrogen atom abstraction from a side chain, and it is not known as to whether this peripheral oxidation will destroy these chemicals. As such, AO/RP treatment of paraben-contaminated waters using sulfate radicals may be feasible, but multiple radical oxidations may again be necessary.

Similarly, the hydrogen atom reduction of nitrosamines gave consistent rate constants, suggesting a common reaction mechanism at the nitroso moiety, but they were too slow to compete with the efficient reaction of this radical with dissolved oxygen under real-world, aerated water, treatment conditions. In contrast, the hydrated electron reaction with estrogenic steroids occurs with a range of rate constants. However, the slow chemical reductions would again not be able to compete with the side reaction of dissolved oxygen, although the fast values for the ketone-containing chemicals would be expected to transfer their excess electron to the dissolved oxygen to regenerate the parent molecule.

Acknowledgments

The pulse-radiolysis measurements described herein were performed at the Radiation Research Laboratory, University of Notre Dame, which is supported in part by the Office of Basic Energy Sciences of the US Department of Energy. Partial funding for this work was from Research Corporation, under Cottrell College Science Awards CC6469 and CC7291.

References

[1] Ternes TA. Occurrence of drugs in German sewage treatment plants and rivers. Water Res 1998;32:3245–60.

[2] Synder SA, Westerhoff P, Yoon Y, Sedlak DL. Pharmaceuticals, personal care products, and endocrine disruptors in water: implications for the water industry. Environ Eng Sci 2003;20:449–69.

[3] Westerhoff P, Yoon Y, Snyder SA, Wert E. Fate of endocrine-disruptor, pharmaceutical, and personal care product chemicals during simulated drinking water treatment processes. Environ Sci Technol 2005;39:6649–63.

[4] Kummerer K. Antibiotics in the aquatic environment: a review Part I. Chemosphere 2009;75:417–34.

[5] Kummerer K. Antibiotics in the aquatic environment: a review Part II. Chemosphere 2009;75:435–41.

[6] Kolpin DW, Furlong ET, Meyer MT, Thurman EM, Zaugg SD, Barber LB, et al. Pharmaceuticals, hormones, and other organic wastewater contaminants in U.S. streams, 1999–2000: a national reconnaissance. Environ Sci Technol 2002;36:1202–11.

[7] Ikehata K, Nagashkar NJ, El-Din MC. Degradation of aqueous pharmaceuticals by ozonation and advanced oxidation processes: a review. Ozone Sci Eng 2006;28:353–414.

[8] Burbano AA, Dionysiou DD, Suidan MT. Effect of oxidant-to-substrate ratios on the degradation of MTBE with Fenton reagent. Water Res 2008;42:3225–39.

[9] Ning B, Graham N, Zhang YP, Nakonechny M, El-Din MG. Degradation of endocrine disrupting chemicals by ozone/AOPs. Ozone Sci Eng 2007;29:153–76.

[10] Lee Y, Escher BI, von Gunten U. Efficient removal of estrogenic activity during oxidative treatment of waters containing steroid estrogens. Environ Sci Technol 2008;42:6333–9.

[11] Huber MM, Ternes TA, von Gunten U. Removal of estrogenic activity and formation of oxidation products during ozonation of 17 alpha-ethynylestradiol. Environ Sci Technol 2004;38:5177–86.

[12] Buxton GV, Greenstock CL, Helman WP, Ross AB. Critical review of rate constants for reactions of hydrated electrons, hydrogen atom and hydroxyl radicals ($^{\cdot}OH/^{\cdot}O^-$) in aqueous solutions. J Phys Chem Ref Data 1988;17:513–886.

[13] Anipsitakis GP, Dionysiou DD. Transition metal/UV-base advanced oxidation technologies for water decontamination. Appl Catal B 2004;54:155–63.

[14] Huie RE, Clifton CL, Neta P. Electron transfer rates and equilibria of the carbonate and sulfate radical anions. Int J Radiat Phys Chem 1991;5:477–81.

[15] Crittenden JC, Hu S, Hand DW, Green SA. A kinetic model for H_2O_2/UV process in a completely mixed batch reactor. Water Res 1999;33:2315–28.

[16] Baxendale JH, Wilson JA. The photolysis of hydrogen peroxide at high light intensities. Trans Faraday Soc 1957;53:344–56.

[17] Bielski BHJ, Cabelli DE. Superoxide and hydroxyl radical chemistry in aqueous solution (Active Oxygen in Chemistry). Struct Energ React Chem Ser 1995;2:66–104.

[18] Gurol MD, Akata A. Kinetics of ozone photolysis in aqueous solution. AIChE J 1996;42:3283–92.

[19] Buhler RE, Staehelin J, Hoigne J. Ozone decomposition in water studied by pulse radiolysis. 1. HO_2/O_2^- and HO_3/O_3^- as intermediates. J Phys Chem 1984;88:2560–4.

[20] von Gunten U. Ozonation of drinking water: I. Oxidation kinetics and product formation. Water Res 2003;37:1443–67.

[21] Staehelin J, Hoigne J. Decomposition of ozone in water: rate of initiation by hydroxide ions and hydrogen peroxide. Environ Sci Technol 1982;16:676–81.

[22] Walling Cheves. Fenton's agent revisited. Acc Chem Res 1975;8:125–31.

[23] O'Shea KE, Beightol S, Garcia L, Aguilar M, Kalen DV, et al. Photocatalytic decomposition of organophosphonates in irradiated TiO_2 suspensions. J Photochem Photobiol 1997;107:221–6.

[24] Srikant V, Clarke DR. On the optical band gap of zinc oxide. J Appl Phys 1998;83:5447–51.

[25] Mason TJ. In: Crum LA, Mason TJ, Reisse JL, Suslick KS, editors. Sonochemistry and sonoluminescence. Dordrecht: Kluwer Academic; 1999. p. 363–70.

[26] Kritzer P, Dinjus E. An assessment of supercritical water oxidation (SCWO): existing problems, possible solutions and new reactor concepts. Chem Eng J 2001;83:207–14.

[27] Lide DR, editor. Handbook of chemistry and physics; 1991.

[28] Hoffman MR. Chemical applications of electrohydraulic cavitation for hazardous waste control. Proceedings from the 14th National Industrial Energy Technology Conference, Houston, TX, April 22–23, 1992.

[29] Namihira T, Wang D, Akiyama H. Pulsed power technology for pollution control. Acta Phys Polon A 2009;115:953–5.

[30] Whitman K, Lyons S, Miller R, Nett D, Treas P, Zante A, et al. Linear accelerator for radiation chemistry research at Notre Dame 1995. Proc. '95 Particle Accelerator Conference & International Conference of High Energy Accelerators, Dallas, TX, 1996.

[31] Buxton GV, Stuart CR. Re-evaluation of the thiocyanate dosimeter for pulse radiolysis. J Chem Soc Faraday Trans 1995;91:279–82.

[32] Sworski TJ. Kinetic evidence for H_3O as the precursor of molecular hydrogen in the radiolysis of water. J Am Chem Soc 1964;86(22):5034–5.

[33] Wishart JF, Nocera DG. Photochemistry and radiation chemistry; complementary methods for the study of electron transfer. Washington, DC: American Chemical Society; 1998.

[34] Ikehata K, Naghashkar NJ, Gamal El-Din M. Degradation of aqueous pharmaceuticals by ozonation and advanced oxidation processes: a review. Ozone Sci Eng 2006;28:353–414.

[35] Dail MK, Mezyk SP. Hydroxyl-radical-induced degradation oxidation of β-lactam antibiotics in water: absolute rate constant measurements. J Phys Chem A 2010;114:8391–5.

[36] Hirsch R, Ternes TA, Haberer K, Kratz K. Occurrence of antibiotics in the aquatic environment. Sci Total Environ. 1999;225:109–18.

[37] Bailón-Pérez MI, García-Campaña AM, Cruces-Blanco C, del Olmo Iruela M. Trace determination of beta-lactam antibiotics in environmental aqueous samples using off-line and on-line preconcentration in capillary electrophoresis. J Chromatogr A 2008;1185:273–80.

[38] Al-Ahmad A, Daschner FD, Kümmerer K. Biodegradability of cefotiam, ciprofloxacin, meropenem, penicillin G and sulfamethoxazole and inhibition of waste water bacteria. Arch Environ Contam Toxicol 1999;37:158–63.

[39] Dorfman LM, Taub IA, Harter DA. Rate constants for the reaction of the hydroxyl radical with aromatic molecules. J Chem Phys 1964;41:2954–5.

[40] Lilie J. Pulsradiolytische Untersuchung der oxydativen Ringoeffnung von Furan, Thiophen und Pyrrol. Z Naturforsch Teil B 1971;26:197–202.

[41] Song W, Chen W, Cooper WJ, Greaves J, Miller GE. Free radical destruction of beta-lactam antibiotics in aqueous solution. J Phys Chem A 2008;112:7411.

[42] Hug GL. Optical spectra of nonmetallic inorganic transient species in aqueous solution. Washington DC: US Department of Commerce, National Bureau of Standards, NSRDS-NBS 69.

[43] Harvey PW, Everett DJ. Significance of the detection of esters of p-hydroxybenzoic acid (parabens) in human breast tumours. J Appl Toxicol 2004;24(1):1–4.

[44] Roebke W, Renz M, Henglein A. Pulseradiolyse der anionen $S_2O_8{}^{2-}$ und $HSO_5{}^{-}$ in waessriger loesung. Int J Radiat Phys Chem 1969;1:39–44.

[45] Huie RE, Clifton CL. Rate constants for hydrogen abstraction reactions of the sulfate radical, $SO_4^{\cdot-}$. Alkanes and ethers. Int J Chem Kinet 1989;21:611–9.

[46] Huber MM, Canonica S, Park G-Y, von Gunten U. Oxidation of pharmaceuticals during ozonation and advanced oxidation processes. Environ Sci Technol 2003;37:1016–24.

[47] Snyder SA, Wert EC, Rexing DJ, Zegers RE, Drury DD. Ozone oxidation of endocrine disruptors and pharmaceuticals in surface water and wastewater. Ozone Sci Eng 2006;28:445–60.

[48] Rickman KA, Mezyk SP. Sulfate radical remediation of chemically contaminated waters: destruction of β-lactam antibiotic activity. Chemosphere 2010;81:359–65.

[49] Tsoni K, Mantaka-Marketou AE. γ-radiolysis of some estrogen hormones in alkaline solutions. Chem Chron 1985;14:237–42.

[50] Mezyk SP, Cooper WJ, Madden KP, Bartels DM. Free radical destruction of N-nitrosodimethylamine in water. Environ Sci Technol 2004;38:3161–7.

[51] Mezyk SP, Ewing D, Kiddle JJ, Cooper WJ, Madden KP. Kinetics and mechanisms of the reactions of hydroxyl radicals and hydrated electrons with nitrosamines and nitramines in water. J Phys Chem A 2006;110:4732–7.

[52] Landsman NA, Swancutt KL, Bradford CN, Cox CR, Kiddle JJ, Mezyk SP. Free radical chemistry of advanced oxidation process removal of nitrosamines in water. Environ Sci Technol 2007;41: 5818–23.

[53] Mezyk SP, Landsman NA, Swancutt KL, Bradford CN, Cox CR, Kiddle JJ, et al. Free radical chemistry of advanced oxidation process removal of nitrosamines in water. In: Karanfil T, Krasner SW, Westerhoff P, Xie Y, editors. Disinfection by-products in drinking water: occurrence, formation, Health effects, and control. American Chemical Society Symposium Series, vol. 995; 2008. p. 319–33.

[54] Mezyk SP, Rickman KA, McKay G, Hirsch CM, He X, Dionysiou DD. Remediation of chemically-contaminated waters using sulfate radical reactions: kinetics and product studies. In: American Chemical Society Symposium on Aquatic Redox Chemistry, 239th American Chemical Society (ACS) National Meeting, 21–25 March 2010, San Francisco, CA, 247–263.

[55] Bartels DM, Craw MT, Han P, Trifunac AD. Hydrogen/deuterium isotope effects in water radiolysis. 1. The mechanism of chemically induced dynamic electron polarization generation in spurs. J Phys Chem 1989;93:2412–21.

[56] Han P, Bartels DM. Hydrogen atom reaction rates in solution measured by free induction decay attenuation. Chem Phys Lett 1989;159:538–42.

[57] Bartels DM, Mezyk SP. EPR measurement of the reaction of atomic hydrogen with bromide and iodide in aqueous solution. J Phys Chem 1993;97:4101–5.

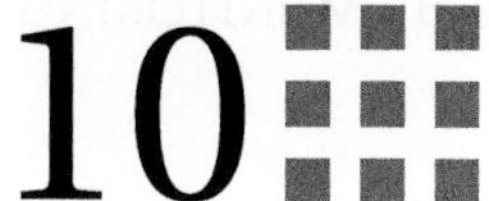

Cactus Mucilage as an Emergency Response Biomaterial to Provide Clean Drinking Water

Daniela Stebbins, Audrey L. Buttice, Dawn Fox, Deni Maire Smith, Norma A. Alcantar*

DEPARTMENT OF CHEMICAL AND BIOMEDICAL ENGINEERING, UNIVERSITY OF SOUTH FLORIDA, FL, USA

**CORRESPONDING AUTHOR*

CHAPTER OUTLINE

10.1 Introduction

The source of contaminants in drinking water can be associated with a range of chemical, biological, and geological materials released from natural and anthropogenic processes. Natural disasters can cause major damage to vulnerable water sources by releasing hazardous elements from local anthropogenic activities, toxic storage units, and other sources into the water bodies [1]. Further more, after an earthquake occurs, sediments and geological layers can release contaminant-bearing minerals into surface and ground waters and can expose humans to elevated concentrations of hazardous elements [1–6].

Monitoring Water Quality, http://dx.doi.org/10.1016/B978-0-444-59395-5.00010-8

Many studies have addressed the environmental impacts of natural disasters and have attempted to predict and modulate the effects of these events on contamination and subsequent bioavailability of potentially toxic chemicals [2,3,5,7–10]. Some of those reports are discussed later to illustrate the possible contamination that could be expected after a natural disaster occurs.

Contamination with trace metals (chromium (C), copper (Cu), nickel (Ni), cobalt (Co), lead (Pb), and zinc (Zn)) and major elements from dissolved salts (sodium, potassium, calcium, magnesium, and chlorine) occurred after a mega tsunami and subsequent huge tsunami waves in Sumatra, Indonesia [5].

Floodwaters contaminated with heavy metals, organic chemicals, and fecal coliform bacteria were pumped into neighboring Lake Pontchartrain (located in Louisiana) during dewatering after hurricane Katrina, with arsenic (As) levels from 20 to 26 μg/L and Pb levels from 0.05 to 44 μg/L [11]. Pardue et al. [1] also reported elevated levels of metals (Pb, As, Cu, Cr, Zn, Ni, and cadmium (Cd)) in bulk water samples from Lake Pontchartrain, where Cu and Zn were found to exceed acute and chronic fresh and saltwater aquatic life criteria. These authors compared their measured metal concentrations in the hurricane floodwaters to pre-hurricane storm water runoff. Metals were measured in three matrices, including New Orleans street mud, Lake Pontchartrain sediments, and suspended sediments. All the samples had elevated levels of Zn and Pb. The highest Zn levels were measured in suspended sediments and superficial sediments near the canal outfalls to Lake Pontchartrain. Pb levels were similarly elevated in all the three matrices. Johnson et al. [8] also reported high levels of Cu, Zn, Ni, Pb, and Cd in oyster tissue from two sites affected by hurricane Katrina. They concluded that metal concentrations were higher after the hurricane had occurred and speculated that the sources of these elevated metals were water and suspended sediments from the receding tidal surge and floodwater from the hurricane.

Provision of safe drinking water is a priority after the occurrence of a major natural disaster. Alternative methods of water treatment that could be easily implemented at the household level would significantly help to preserve the health and life of survivors of a disaster. Therefore, renewable, low-cost, and effective materials that can be used in robust and sustainable systems to provide safe drinking water are desired and are currently being studied.

This work investigates cactus-mucilage-based separation as a viable alternative treatment for rapid water purification in post-earthquake scenarios and focuses on the interaction of mucilage with metals, metalloids, and nonmetals dissolved in water. Mucilage was extracted from *Opuntia ficus-indica* (*Ofi*), commonly known as nopal or prickly pear cactus [12]. This readily available and inexpensive natural extract has been shown to remove turbidity, bacteria, and As from contaminated water [13–17].

The earthquakes on January 12, 2010, in Port-au-Prince, Haiti, offered an opportunity to test the effectiveness of mucilage and develop treatment parameters for earthquake-affected waters. The use of mucilage provides a potentially viable, readily transferable technology that could be adopted in future earthquake-affected regions of developing communities. Hence, the study investigates the mucilage of the cactus, *O. ficus-indica,* to purify water after a natural disaster had occurred in a vulnerable community.

10.2 Materials and Methods

10.2.1 Sample Collection

This study is focused on the city of Port-Au-Price, Haiti, and on surrounding areas (Figure 10-1), where the earthquake caused unprecedented loss of life and major damage to property [18]. The study involves one of the most important and sensitive areas in Haiti because of the high population density and considerable structural damage caused to the majority of landmark buildings. A research team visited areas affected by the earthquake in the first week of July 2010, 6 months after its occurrence. Water samples were collected at 10 different locations from representative samples of distribution trucks (A1, A2, and D), well water (sites B and C), tap water (E, F, G, and H), and commercialized drinking water (A3). More details of the sampling location and conditions are given in Figure 10-1. On return to our laboratory, the samples were stored at 4 °C until they were tested.

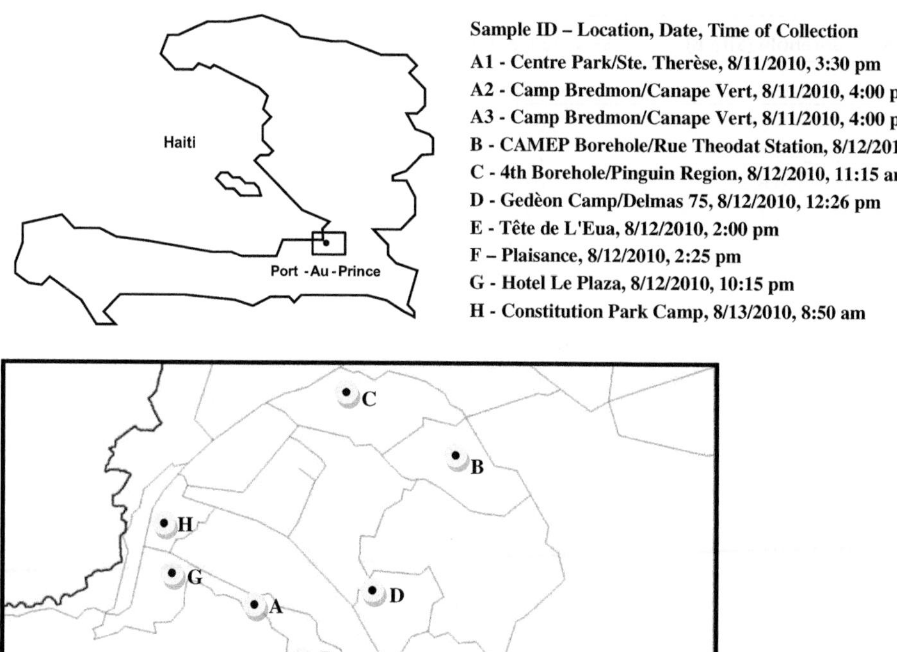

FIGURE 10-1 Study area map showing the specific location, date, and time of the sample collection.

10.2.2 Water Quality Measurements

A comprehensive analysis of 22 elements (sodium, calcium, magnesium, potassium, strontium, Pb, Cr, Ni, titanium, aluminum, antimony, boron (B), Cu, Co, molybdenum, uranium, vanadium, barium (Ba), thallium, Cd, selenium (Se), As, and iron (Fe)) was performed on acidified water samples from Port-Au-Price. To obtain accurate analytical data, the samples were analyzed three times with an inductively coupled plasma mass spectrometer (ICP–MS). A certified reference material (NIST 1640) was used to check instrument accuracy and laboratory efficiency (Table 10-1). Calibration standards were prepared from stock solutions diluted from high-purity standard solutions (1000 mg/L).

10.2.3 Mucilage Extraction

Gelling and nongelling extracts (GE and NE) were obtained from mucilage from the pads of *O. ficus-indica* cactus. The pads were processed according to the procedure described in Ref. [12] with some modifications [13,14,17]. Figure 10-2 shows some of the simple steps used to obtain mucilage powder. The pads were diced and stirred in deionized (DI)

Table 10-1 Comparison of the SRM 1640a Values Certified by the Supplier and Trace-Metal Values Found in This Study

	CAMEP Borehole (Site B)	SRM 1640a (Found)	SRM 1640a (Certified)	Recovery
		mg/L		%
Na	50.3 ± 1.8*	3.04 ± 0.05	3.14 ± 0.031	97.0
Ca	88.6 ± 3.7	5.72 ± 0.09	5.62 ± 0.021	101.8
Mg	18.3 ± 0.6	1.03 ± 0.02	1.06 ± 0.0041	97.3
K	4.2 ± 0.1	0.57 ± 0.01	0.58 ± 0.0023	97.9
		μg/L		%
Sr	1800 ± 100	121.93 ± 0.49	126.03 ± 0.91	96.7
B	1400 ± 100	284.57 ± 4.37	303.1 ± 3.1	93.9
Ba	4700 ± 200	148.6 ± 0.63	151.8 ± 0.83	97.9
Pb	0.4 ± 0.03	11.71 ± 0.16	12.10 ± 0.050	96.8
Cr	10.2 ± 0.4	39.63 ± 0.26	40.54 ± 0.30	97.7
Al	300 ± 10	51.92 ± 1.09	53.00 ± 1.8	98.0
Sb	0.00 ± 0.01	4.86 ± 0.12	5.11 ± 0.046	95.2
Cu	0.9 ± 0.0	84.40 ± 3.88	85.75 ± 0.51	98.4
Co	0.4 ± 0.02	19.08 ± 0.08	20.24 ± 0.24	94.3
Mo	0.00 ± 0.03	46.99 ± 4.11	45.60 ± 0.61	103.1
U	0.9 ± 0.04	24.17 ± 0.47	25.35 ± 0.27	95.3
V	46.3 ± 3.5	15.85 ± 1.22	15.05 ± 0.25	105.3
Tl	0.0 ± 0.0	1.56 ± 0.02	1.62 ± 0.016	96.1
Cd	0.01 ± 0.00	4.02 ± 0.06	3.99 ± 0.074	100.8
Se	53.5 ± 0.8	20.74 ± 0.53	20.13 ± 0.17	103.0
As	1.1 ± 0.1	7.56 ± 0.81	8.08 ± 0.070	93.7
Fe	25.5 ± 0.8	36.64 ± 1.81	36.80 ± 1.8	99.6

*Average ± standard deviation ($n = 3$)

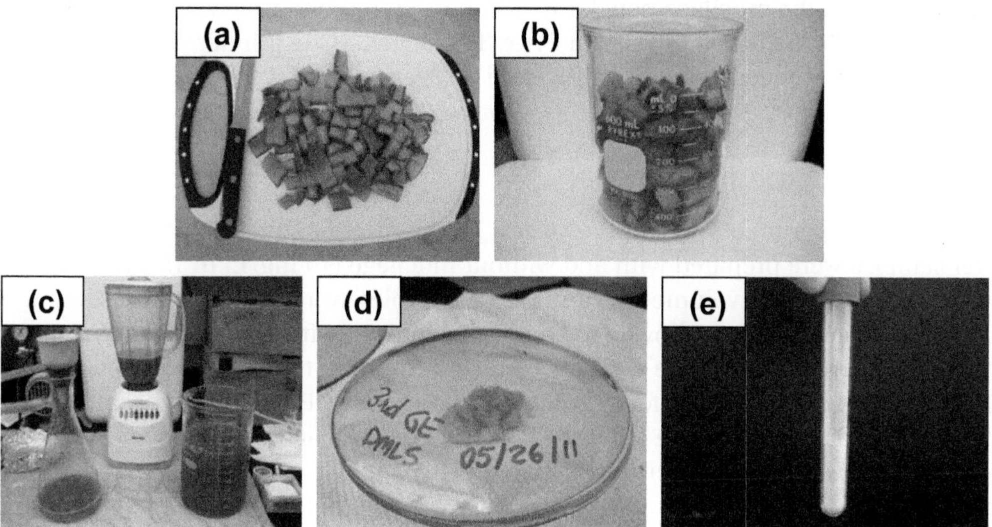

FIGURE 10-2 Some of the steps to extract the mucilage from *Opuntia ficus-indica*: (a) The pads are diced; (b) pads are stirred in diluted NaCl (1% m/v) for 20 min at 60 °C; (c) pads are blended and mucilage extracted; (d) mucilage is allowed to dry at room temperature; and (e) mucilage is stored in powder form. (For color version of this figure, the reader is referred to the online version of this book.)

water with NaCl (1% m/v) at 60 °C for 20 min and then liquidized in a blender. The pH of the suspension was then adjusted to 7 by using 1 M NaOH solution. The solid phase was separated from the liquid phase by centrifugation using an Accusing 400 centrifuge (Fisher Scientific) with a centripetal force of 2522× *g* for 10 min. The precipitate was processed to obtain the GE, and the supernatant was reserved for extraction of the NE.

A solution with sodium hydroxide (50 mM) and sodium hexametaphosphate (0.75% m/v) was added to the precipitate until it was covered, and the suspension was then stirred for 1 h. The pH was adjusted to 2 by using 1 M HCl solution, and the suspension was stirred for 10 min more and left overnight to allow for equilibration and to provide a higher extraction rate. The next day, the mixture was centrifuged, and the supernatant was discarded. The precipitate was resuspended in DI water, and the pH was increased to 8 by using 1 M NaOH solution. The suspension was then filtered by using vacuum filtration with a #41 Whatman filter paper (thickness 1.2 μm).

The supernatant portion of the original suspension, reserved for the extraction of NE, was filtered using a #41 Whatman filter paper in a vacuum filtration system. The filtrates of both the NE and GE were separately mixed with acetone (1:1 v/v) and centrifuged with a centripetal force of 2522× *g* for 10 min. The precipitate was washed in sequential solutions of 70%, 80%, 90%, 95%, and 100% ethyl alcohol (1:1 v/v). The solutions with the mucilage were centrifuged, the supernatant was discharged, and the precipitated mucilage was spread on sterile Petri dishes, dried at room temperature, ground using a mortar and pestle, and was then stored in sealed containers at room temperature. Before experimentation, the mucilage powder was dispersed in the sample water from Port-au-Prince that was being tested by using a Pyrex Tenbroeck tissue grinder (Fisher Scientific) with final stock concentrations of 100 and 1000 mg/L, which were diluted during experimental assays.

10.2.4 Batch Experiments

Simple batch experiments were used to test element reduction using the GE and NE. Water samples were prepared with and without mucilage (10 mg/L) in 15-mL centrifuge tubes to obtain a final volume of 10 mL. The samples were vortexed until they were thoroughly mixed and then allowed to equilibrate for 24 h, after which the topmost 1 mL of the liquid was removed. The samples extracted from the tubes were filtered by using a 0.45-μm syringe top filter before characterization with ICP–MS. All the samples were analyzed in triplicate to ensure repeatability and statistical accuracy.

The accuracy and the precision of the ICP–MS assay were verified by analysis of Standard Reference Material (SRM) 1640a. Calibration solutions were prepared on the same day as of the analyses from dilutions of high-purity standard stock solutions (1000 mg/L) and DI water with a resistivity >17.9 mΩ.

The relative removal (RR) percentage for experimental batch tests was calculated using the following equation:

$$RR\% = 100 - [(C_S \times 100)/C_C]$$

where C_S is the residual concentration (μg/L or mg/L depending on the element) in the sample after treatment with the GE or NE and C_C is the concentration in the control with no mucilage (μg/L or mg/L).

10.3 Results and Discussion

10.3.1 Elemental Composition of Water Samples from Port-au-Prince

Concentrations of the 22 elements analyzed for site B are shown in Table 10-1. Similar results were obtained for all the samples collected from Haiti. For many of the major and trace elements evaluated in this study, acceptable concentration levels were found. In samples from sites B–D, and H (Figure 10-1), however, concentrations of B, Ba, Fe, and Se were determined to be above the maximum contaminant level (MCL) established by the United States Environmental Protection Agency (USEPA) [19], as shown in Table 10-2. These elements could potentially pose serious health risks if consumed.

The guideline for B in drinking water recently increased from 0.5 mg/L [20] to 2.4 mg/L in 2011 [21]. The samples from sites B–D and H were determined to contain B levels ranging from 1.3 to 1.9 mg/L (Table 10-2). Although these levels are lower than the guideline levels established in 2011, side effects resulting from long-term exposure could include loss of fertility, reproductive tract and testicular lesions, and prenatal and general mortality [20].

Table 10-2 Concentration of B, Ba, Fe, and Se in Water Samples from Four Sites in Port-Au-Prince

Element	MCL*	Site B	Site C	Site D	Site H
			mg/L		
B	0.5	1.43 ± 0.06	1.88 ± 0.04	1.27 ± 0.02	1.36 ± 0.10
Ba	2	4.68 ± 0.19	1.27 ± 0.01	3.91 ± 0.01	0.99 ± 0.01
			μg/L		
Fe	300	360.9 ± 1.9	495.2 ± 1.1	323.5 ± 2.3	341.9 ± 2.5
Se	50	53.48 ± 0.77	65.41 ± 2.40	57.53 ± 1.17	64.99 ± 0.97

*Maximum contaminant level (MCL) in [19]

The National Primary Drinking Water Regulation of the USEPA established 2 mg/L as the MCL for Ba in drinking water; the World Health Organization's (WHO) guideline value for Ba in drinking water is 0.7 mg/L [21]. Ba concentrations in water samples from sites B and D were determined to be higher than the USEPA and the WHO's established values (4.7 and 3.9 mg/L). Sites C and H had concentrations that were below the regulations established by the USEPA although they were slightly above the regulations set by the WHO (1.3 and 1.0 mg/L; Table 10-2). Association between Ba content in drinking water and the increase in mortality from cardiovascular disease has been observed [22].

Concentration levels of Fe in drinking water are generally well under the MCL (300 μg/L); however, concentrations can vary depending on whether Fe salts are used in the water treatment process and/or whether iron pipes are used for water distribution. High levels of Fe (0–495 μg/L) were detected in the first evaluation of the samples from Haiti. At these levels, Fe could change the organoleptic characteristics of the water including its flavor and smell [21]. Consecutive analysis did not show the same high concentration, which could have been due to the high reactivity of Fe in oxygenated water, which would lead to the oxidation and subsequent precipitation of iron hydroxides. Wide variations in toxicity have been reported for different Fe salts. Some effects of toxic doses of Fe include depression, rapid and shallow respiration, coma, convulsions, respiratory failure, and cardiac arrest [21]. The iron dextran complex is considered to be carcinogenic to animals by the International Agency for Research on Cancer [23].

Although it is a trace essential nutrient for humans, excessive levels of Se can cause adverse effects and damage to health. Therefore, the MCL for Se in drinking water has been established as 50 μg/L by the USEPA and 40 μg/L as a provisional guideline established by the WHO [21]. Samples from sites B–D, and H contained 50.0–65.4 μg/L of Se (Table 10-2).

10.3.2 Water Samples Treated with Mucilage Extracts (GE and NE)

All mucilage tests focused on samples extracted from sites B–D, and H, and on the elements discussed in the previous section that were determined to have higher levels than the MCL (B, Ba, Fe, and Se). Additionally, the mucilage tests were also used to study

FIGURE 10-3 Left tube: untreated water samples from the CAMEP Borehole, Port-Au-Prince (no mucilage). Right tube: water sample from CAMEP Borehole treated with 10 mg/L of the NE. The cactus mucilage aids the separation of contaminants by forming a dense precipitate at the bottom of the water column. (For color version of this figure, the reader is referred to the online version of this book.)

the removal of five other dangerous elements determined to be in the water (although not necessarily at higher concentrations than the MCL); these included Cr, As, Ni, Zn, and Pb. After 24 h, water samples treated with mucilage were observed to form precipitates that settled to the bottom of the centrifuge tubes, which indicated potential element removal from the water (Figure 10-3).

Water samples from site B were treated with 10 mg/L GE, which demonstrated highly reduced concentrations after 24 h compared to control levels (Figure 10-4). All elements were observed to decrease significantly after treatment with GE, and the final concentrations were below the established MCL. Figure 10-5 shows the removal resulting from mucilage treatment with respect to the control. In all cases, the RR of elements in the treated water compared to the control was very high (from 55 to 100%), indicating that significant removal is achieved with 10 mg/L of GE mucilage.

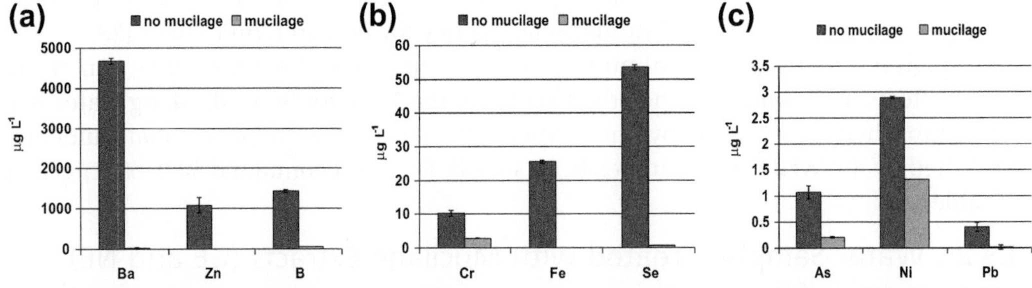

FIGURE 10-4 Concentration of elements naturally found in water samples from the CAMEP borehole (site B). (a) Ba, Zn, and B; (b) Cr, Fe, and Se; and (c) As, Ni, and Pb. Red columns represent the samples before mucilage treatment and green columns indicate samples after mucilage treatment with 10 mg/L of GE extract. (For interpretation of the references to color in this figure legend, the reader is referred to the online version of this book.)

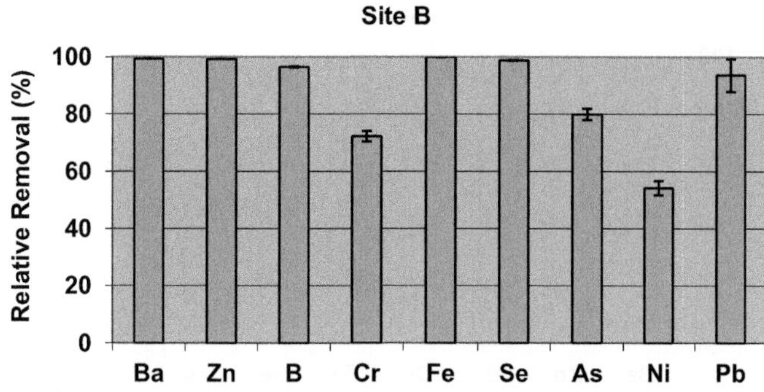

FIGURE 10-5 RR (compared to control batch tests) of elements naturally found in water samples from the CAMEP borehole (site B) treated with 10 mg/L of GE from *Opuntia ficus-indica*. RR rates close to 100% were found for most of the elements. (For color version of this figure, the reader is referred to the online version of this book.)

Water samples from site C were treated with 10 mg/L NE, and results similar to those with GE treatment were obtained. The concentrations of many elements were determined to decrease significantly when compared to that of untreated water samples (controls), and after 24 h, all were below the USEPA established guideline levels for drinking water (Figure 10-6). By studying the RR of the treated water compared to that in control batch tests, we observed that, for 7 of the 9 elements, there was a significant reduction in concentrations on using NE (80–97%; Figure 10-7). Although Ni removal with NE was observed to be slightly lower compared to that of other elements, reduction was still observed and as concentrations were initially <70 μg/L, the WHO recommended concentration level, this was not addressed as a serious problem.

Although both mucilage fractions (GE and NE) were observed to reduce the contaminants studied to levels acceptable by water quality agencies, GE was observed to have a more significant effect on treatment than NE had. This is most likely due to chemical differences (Table 10-3) that exist between mucilage types. Although both NE and GE contain neutral sugars, published results suggest high concentrations of uronic

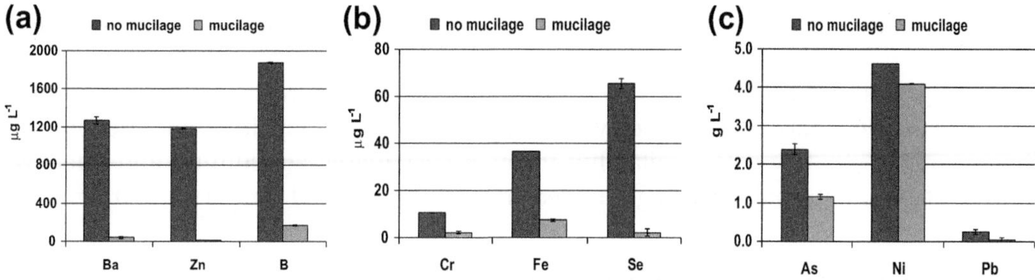

FIGURE 10-6 Concentration of some elements naturally found in water samples from the Pinguin region (site C). (a) Ba, Zn, and B; (b) Cr; Fe, and Se; and (c) As, Ni, and Pb. Red bars represent the samples with no mucilage, and green bars indicate samples after the treatment with 10 mg/L of the NE extract. (For interpretation of the references to color in this figure legend, the reader is referred to the online version of this book.)

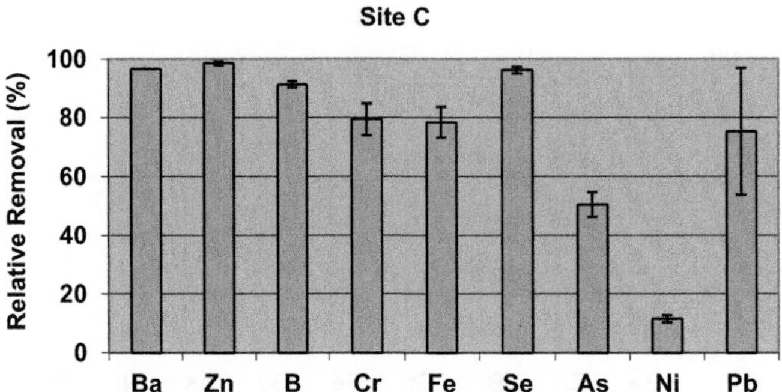

FIGURE 10-7 RR (compared to that in control batch tests) of elements naturally found in water samples from the Pinguin region (site C) treated with 10 mg/L of NE from *Opuntia ficus-indica*. RRs >80% were found for most of the elements. (For color version of this figure, the reader is referred to the online version of this book.)

Table 10-3 Monomeric Sugar Composition of *Opuntia ficus-indica* GE and NE

Extract	Glucose	Xylose	Galactose	Arabinose	Total
			%		
GE	9.89	3.95	7.36	8.90	30.10
NE	1.03	10.80	14.66	24.38	50.86

GE = Gelling extracts
NE = Nongelling extracts

acids, particularly in GE, which has been studied for its pectin qualities [12]. As the literature suggests, significantly more uronic acid groups are present in GE than in NE, which could potentially account for the different removal efficiencies observed in the experimental data provided in Figures 10-4–10-7.

Both NE and GE contain water-soluble pectic polysaccharides with a high content of ionizable negative groups such as hydroxyl, carboxyl, and carbonyl [15]. Those chemical groups are widely known to interact with metals, especially with cations. The interaction between mucilage and anions or oxyanions in solution may occur through bonding by polyvalent cation bridging, hydrogen bond bridging, or formation of chelates and complexes.

Mucilage extracted from *O. ficus-indica* has demonstrated great potential in a water treatment method that could be used by communities in emergency situations after natural disasters occur. Its low cost, availability through sustainable agriculture, and effective removal of pollutants make mucilage an ideal material for fast and simple water treatment.

10.4 Summary and Conclusions

Our main objective was to check the ability of two mucilage extracts from *O. ficus-indica* (NE and GE) to clean contaminated water in a post-earthquake scenario.

We collected water samples from 10 locations of Port-Au-Prince, Haiti, and tested them after mucilage treatment. The results are listed as follows:

- High concentrations of Fe, Ba, Se, and B were found in drinking water samples from Port-Au-Prince. The mucilage extracts (GE and NE) were able to decrease the concentration of those elements to levels acceptable to regulatory agencies.
- Water samples from the topmost 1 ml in the batch test treated with the mucilage reveal that high metal removal rates were achieved by using both NE and GE. Cactus mucilage effectively decreased the hazardous metal content in water samples from Haiti.
- GE removed 80–100% of Ba, Zn, B, Fe, Se, and Pb; 60–80% of Cr and As; and approximately 50% of Ni.
- NE removed 80–100% of Ba, Zn, B, Fe, Se, Pb, and Cr; 50% of As; and approximately 10% of Ni.
- Mucilage (10 mg/L) was able to remove most of the evaluated elements from the water.

This study confirms earlier findings and suggests that cactus-mucilage extracts interact with cations and anions dissolved in water. This property makes this biomaterial an excellent option to be used in emergencies, where a wide range of different contaminants may be present. The cactus mucilage is extracted by an affordable process, and it can easily be used by local communities.

Acknowledgments

The authors wish to acknowledge the full support provided by the National Science Foundation grant: CBET-1034849, and partial support from the Gulf of Mexico Research Initiative to complete this work. They also wish to acknowledge Dr Thomas Pichler for providing analytical instrumentation and Dr. Zachary Atlas for the insightful help with the measurements.

References

[1] Pardue JH, Moe WM, et al. Chemical and microbiological parameters in New Orleans floodwater following Hurricane Katrina. Environ Sci Technol 2005;39(22):8591–9.

[2] Belin B, Yalcin T, et al. Earthquake-related chemical and radioactivity changes of thermal water in Kuzuluk–Adapazari, Turkey. J Environ Radioact 2002;63(3):239–49.

[3] Fujimoto T, Tsuchiya Y, et al. Contamination of the Shinano river water with mutagenic substances after the Niigata Chuetsu earthquake. Tohoku J Exp Med 2007;211(2):171–80.

[4] Krausmann E, Renni E, et al. Industrial accidents triggered by earthquakes, floods and lightning: lessons learned from a database analysis. Nat Hazards 2011;59(1):285–300.

[5] Srinivasalu S, Thangadurai N, et al. Evaluation of trace-metal enrichments from the 26 December 2004 tsunami sediments along the southeast coast of India. Environ Geol 2008;53(8):1711–21.

[6] Young S, Balluz L, et al. Natural and technologic hazardous material releases during and after natural disasters: a review. Sci Total Environ 2004;322(1–3):3–20.

[7] Gupta SK, Suantio A, et al. Factors associated with *E-coli* contamination of household drinking water among tsunami and earthquake survivors, Indonesia. Am J Trop Med Hyg 2007;76(6): 1158–62.

[8] Johnson W, Kimbrough K, et al. Chemical contamination assessment of Gulf of Mexico oysters in response to hurricanes Katrina and Rita. Environ Monit Assess 2009;150(1):211–25.

[9] Karmakar S, Rathore AS, et al. Post-earthquake outbreak of rotavirus gastroenteritis in Kashmir (India): an epidemiological analysis. Public Health 2008;122(10):981–9.

[10] Morkoc E, Tarzan L, et al. Changes of oceanographic characteristics and the state of pollution in the Izmit Bay following the earthquake of 1999. Environ Geol 2007;53(1):103–12.

[11] Dortch MS, Zakikhani M, et al. Modeling water and sediment contamination of Lake Pontchartrain following pump-out of Hurricane Katrina floodwater. J Environ Manag 2008;87(3):429–42.

[12] Goycoolea FM, Cardenas A. Pectins from *Opuntia* spp.: a short review. J Prof Assoc Cactus Dev 2003;5:17–29.

[13] Buttice AL, Stroot JM, et al. Removal of sediment and bacteria from water using green chemistry. Environ Sci Technol 2010;44(9):3514–9.

[14] Fox D. Cactus mucilage-assisted heavy metal separation: design and implementation. Ph.D. dissertation, Chemical and biomedical engineering. Tampa: University of South Florida; 2011.

[15] Fox DI, Pichler T, et al. Removing heavy metals in water: the interaction of cactus mucilage and arsenate (As (V)). Environ Sci Technol 2012;46(8):4553–9.

[16] Pichler T, Young K, Alcantar N. Eliminating turbidity in drinking water using the mucilage of a common cactus. Water Sci Technol Water Supp 2012;12(2):179–86.

[17] Young K. The mucilage of *Opuntia ficus indica*: a natural, sustainable, and viable water treatment technology for use in rural Mexico for reducing turbidity and arsenic contamination in drinking water. Master's thesis. Tampa: Department of Chemical Engineering, University of South Florida; 2006.

[18] Rathje EM, Bachhuber J, et al. Damage patterns in Port-au-Prince during the 2010 Haiti earthquake. Earthq Spectra 2011;27:S117–36.

[19] USEPA. National primary drinking water regulations. Available from: http://water.epa.gov/drink/contaminants/upload/mcl-2.pdf; 2009 [accessed 23.01.11].

[20] WHO. Boron in drinking-water. Available from: http://www.who.int/water_sanitation_health/dwq/boron.pdf; 2003.

[21] WHO. Guidelines for drinking-water quality. 4th ed. Available from: http://www.who.int/water_sanitation_health/publications/2011/dwq_guidelines/en; 2011.

[22] WHO. Barium in drinking water. Available from: http://www.who.int/water_sanitation_health/dwq/chemicals/barium.pdf; 2004.

[23] IARC. Some inorganic and organometallic compounds. Available from: http://monographs.iarc.fr/ENG/Monographs/vol2/volume2.pdf; 1973, 16 March 1998 [accessed 23.01.11].

[24] Greenwood NN, Earnshaw A. Chemistry of the elements. Oxford: Butterworth-Heinemann; 2002.

Potable Water: Nature and Purification

Abul Hussam

CENTER FOR CLEAN WATER AND SUSTAINABLE TECHNOLOGIES, DEPARTMENT OF CHEMISTRY AND BIOCHEMISTRY, GEORGE MASON UNIVERSITY, FAIRFAX, VA, USA

CHAPTER OUTLINE

11.1 Introduction

Fresh water is the primary source of human health, prosperity, and security. By around 2050 the world's population is expected to reach about nine billion. Assuming that standards of living continue to rise, the requirement of potable water for human consumption will amount to the resources of about three planet Earths [1]. A key United Nations report indicates that water shortages will affect 2.3 billion people or 30% of the world's population in four dozen nations by 2025 [2]. Already, the crisis of potable water in most developing countries is creating public health emergencies of staggering

Monitoring Water Quality, http://dx.doi.org/10.1016/B978-0-444-59395-5.00011-X

proportions. In Bangladesh, for example, it is officially recognized by the government of Bangladesh that 50% of the country's approximately 150 million people, are at risk of arsenic poisoning from groundwater used for drinking. Recently, the government of Bangladesh, in its *Action Plan for Poverty Reduction*, stated its desire to ensure 100% access to pure drinking water across the region within the shortest possible time frame [3]. This is also consistent with key goals of the Millennium Development Goal "Eradication of extreme poverty and hunger" and "Halving by 2015, the proportion of people without sustainable access to safe drinking water" [4]. Whether this is achievable within the stated time is debatable, but it clearly delineates the state of the world we live in.

The problem is magnified by anthropogenic and geogenic pollutants rendering significant amounts of water unfit for human use. For example, more than six million tons of petroleum hydrocarbons enter into the aquatic food web of the oceans yearly. This is equivalent to indiscriminate dumping of many persistent toxic organic pollutants into the sea. The pollution of our coastal and inland waters by inorganics containing Cu, Pb, Cd, Zn, Hg, As, and Sn started in 1820, as evidenced by the detection of these metals in the Greenland ice sheet [5]. At present the surface water of ocean is oversaturated with respect to $CaCO_3$ (s) in the forms of calcite and aragonite. With global warming and increase in CO_2 (g), biotic $CaCO_3$ (s) could begin to dissolve because of undersaturation. A doubling of the partial pressure of carbon dioxide (p_{CO_2}) will double the $[H^+]$ and thus increase chemical weathering and enhance the release of phosphate and trace metals [6] (controlled by $=2XOH + M^{2+} \leftrightarrow (XO)_2M + 2H^+$, where $=XO$ is the metal oxide surface). This is an example of how the physicochemical principles are central to the understanding of clean and potable water. In this paper, I will focus on some of the relevant aquatic chemical principles and models, some recent methods for the mitigation of water pollutants, and emphasize the development of filtration technologies and the need for reliable analytical techniques [7].

11.2 Water Pollution: Nature and Quality

There is a qualitative relation between water quality and the influent and effluent water and their recycling for a sustainable biodiversity, where human water consumption is a part of this cycle [8]. This is depicted in Figure 11-1. Water quality is a process objective (or transfer function) between influent and effluent water. These processes are based on physicochemical principles, such as sorption, surface complexation, precipitation, sedimentation, redox reactions, and bioaccumulation coupled with transport processes such as convection, diffusion, dispersion, etc. These processes have been known for a long time; however, the complex interplay of various chemical processes, kinetics, and reactive mass transport is just beginning to emerge.

Table 11-1 shows a portion of the inorganic water quality parameters for potable water. These numbers were obtained after many years of research on the human toxicity of some selected chemical species. This is just one aspect; there are also maximum contamination limit (MCL) values for organics, microbes, and other esthetic parameters.

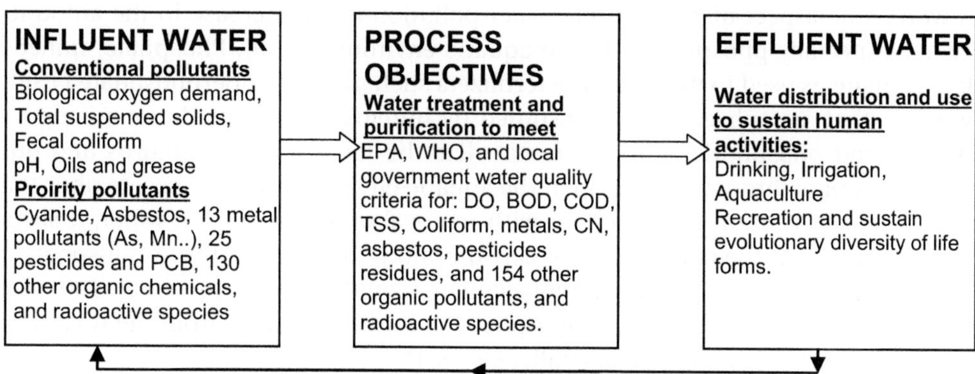

FIGURE 11-1 Depiction of clean water as a process cycle.

At the top of this chart (left box of Figure 11-1) is the biological oxygen demand (BOD), because BOD is nearly directly proportional to local population density and signifies human activity. This is especially true for developing nations with high population densities where the major sources of water are surface water or groundwater, contaminated in the process of distribution. There is, unfortunately, very little information on the water quality parameters of rivers, estuaries, ponds, lakes, streams, and drains in these countries. Some of these natural water bodies are also on the verge of extinction.

Table 11-1 Water Quality Parameters for Inorganic Species Shown as Maximum Contamination Limit (MCL) by USEPA, World Health Organization (WHO) Guideline Values, and Bangladesh Standard Values in Comparison to Typical Groundwater Quality in Bangladesh

Constituent	USEPA (MCL)	WHO Guideline	Bangladesh Standard*	Typical Groundwater Composition
Arsenic (total) (μg/L)	10	10	50	5–4000
Arsenic (III) (μg/L)				5–2000*
Iron (total) (mg/L)	0.3	0.3	0.3 (1.0)	0.2–20.7
pH	6.5–8.5	6.5–8.5	6.5–8.5	6.5–7.5
Sodium (mg/L)		200		<20.0
Calcium (mg/L)			75 (200)	120 ± 16
Manganese (mg/L)	0.5	0.1–0.5	0.1 (0.5)	0.04–2.00
Aluminum (mg/L)	0.05–0.2	0.2	0.1 (0.2)	0.015–0.15
Barium (mg/L)	2.0	0.7	1.0	<0.30
Chloride (mg/L)	250	250	200 (600)	3–12
Phosphate (mg/L)			6	<12.0
Sulfate (mg/L)			100	0.3–12.0
Silicate (mg/L)			-	10–26

*The Bangladesh Standard Values are Given as Maximum Desirable Concentration with Maximum Permissible Concentration in Parentheses. In Some Wells, As (III) Concentrations Could Exceed 90% of the As (total). (1 mg/L = 1000 μg/L).

A worrisome aspect of present-day water pollution is the increase in the amount of nutrients, especially phosphorus, in the aquatic environment [9]. Phosphate, mined as rock phosphate, is used to manufacture fertilizers such as triple superphosphate (TSP) to grow crops that are subsequently consumed by humans. TSP is made by mixing rock phosphate and phosphoric acid. We metabolize the food, and phosphorus and other nutrients that pass through us end up in the aquatic environment. The excess phosphate can cause eutrophication and interfere with removal of arsenate from water, if seeped into groundwater. At the current rate of consumption, the supply of rock phosphate, an essential nutrient with no known replacement, could exhaust in about 100 years. Therefore, the recovery and reuse of phosphate and other nutrients from the wastewater stream, is a scientific and technical challenge yet-to-be solved.

In developed nations, industry is the single largest consumer of water, whereas it is agriculture in yet-to-be developed nations. These are the primary sources of pollutants in water. However, this is changing because of the extraction, use, and depletion of non-renewable groundwater resources. In developing countries, a critical source of water pollution is chlorine, originating from chlorinated hydrocarbons and industrial effluents from water treatment, pulp and paper bleaching, metal processing, pharmaceutical manufacturing, textile dyeing and cleaning, corrosion control, and other processes as varied as photography and photolithography of anthropogenic origin. In most other countries, the critical pollutants in groundwater are pathogenic bacteria, agriculture runoff (fertilizers and pesticides), and toxic trace metals (such as arsenic, manganese) of geogenic origin. The disparate nature of water contamination and of potable water in particular, requires new solutions and approaches developed locally with appropriate technologies. This is especially true considering the financial constraints of these countries. In the following sections some physiochemical aspects are described, to provide an understanding of the nature of pollutants and approaches for their mitigation, with emphasis on some recent potable water crises in Bangladesh.

11.3 Chemical Models in Aquatic Chemistry

To understand the underlying processes by which chemical species are evolved (produced-transformed-cycled) in the aquatic system, irrespective of their origins (anthropogenic or geogenic), we need to understand some basic physicochemical principles. These principles include chemical speciation, thermodynamic equilibrium (sorption, precipitation, mineralization, surface-complexation reactions [SCRs], etc.), redox chemistry of species, the kinetics of reaction, and the kinetics of diffusion, to ascertain the rate-limiting steps. Chemists have come a long way in understanding these processes even in very complex systems, through chemical–mathematical models and computational techniques. However, weakness still remains when we are asked to input the biological factor into chemical speciation. In this chapter, I will use some computational models to understand speciation, natural attenuation, and the study of arsenic mitigation from groundwater, by using iron-based sorbents as an example. This is a limited view of the subject but it covers some principles yet-to-be included in our educational curriculum.

11.3.1 Chemical Equilibrium Model: Surface-Complexation Reactions

A very large number of chemical species interact with the soil surface and with different mineral surfaces (activated alumina and hydrous ferric oxide [HFO], in particular) and determine the fate and transport of such species. While the knowledge of speciation in aqueous solution is well understood, understanding of the interaction between solute and surface is evolving. A surface complexation model (SCR) involves three steps: surface ionization of a solid surface, formation of an electrical double-layer on the solid surface, and formation of a complex between surface and the ionic species (e.g., anionic arsenate). In SCR, the double-layer is assumed to consist of an inner compact layer and an outer plane where the diffuse layer starts. The inner layer is where the ions are located to form surface complexes. This model is appropriate for low-ionic strength solutions, such as groundwater or surface water. The surface potential term ψ can be obtained from the Boltzmann distribution [10] of ionic species. The activity change of an ion (as chemical component) moved from bulk on to the surface is given by

$$X_S^Z = X^z[\exp(-\psi F/RT)]^z$$

Where z is the charge of the ion X, X_S^Z is the activity of ion X on surface, and the exponential term is the Boltzmann factor. In a complex chemical system with many cations, anions, and active surfaces, the first step is to solve the speciation problem by using computational models, such as MINTEQA [11] and then to modify the speciation calculation based on SCR, when the surface area of the solid sorbent, the double-layer capacity, site types, and site densities are known. The basic principle of MINTEQA is shown in Figure 11-2. The applications of such models require reliable analytical measurements of all possible cations and anions, so that charge imbalance does not exceed 5%. Recent refinements in the model include an extended database of many organic acids, bases, humic and fulvic acids and dissolved organic matters [12].

Another example application of MINTEQA, using SCRs is the development of sorbents for arsenic removal from groundwater. The use of MINTEQA in understanding the breakthrough capacity of HFOs (s) as solid sorbentsfor the removal of arsenate is shown in Figure 11-3. Three sorption isotherms were calculated from experimental and theoretical parameters for goethite (most similar to the composite iron matrix [CIM] surface) for three different surface areas as indicated. The experimental data points collected for over a month and superimposed on the curves, show that the CIM surface area has increased. The experimental BET surface area of CIM is also in the range of 40–60 m/g [2]. Although the experimental data fits a classical Freundlich isotherm ($r^2 = 0.97$), there is no provision for including surface area and other surface-specific properties in such isotherms. Classical isotherms are therefore inadequate in predicting and optimizing the nature of surface interactions of ionic species in solution. In a similar calculation, we have shown that the system open to air CO_2, does not change

MINTEQA
A Chemical Equilibrium Model
I
Selection of Chemical Components
Entities to create species by chemical reaction
I
Search Reaction Data Base
Get thermodynamic equilibrium constant and enthalpy of reaction
I
Use Input Data
Total concentrations of components, temp, sorption model etc.
I
Setup Tables Using
Mass balance, charge balance, surface complexation parameters
I
Solve Nonlinear Algebraic Problems
by Newton - Raphson Method

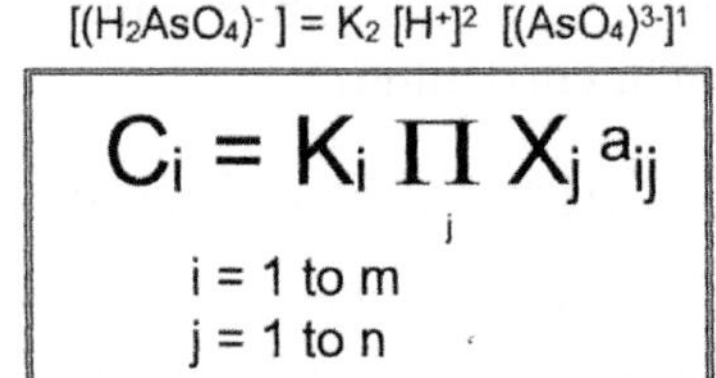

$[(H_2AsO_4)^-] = K_2 [H^+]^2 [(AsO_4)^{3-}]^1$
$C_i = K_i \prod_j X_j^{a_{ij}}$
i = 1 to m
j = 1 to n

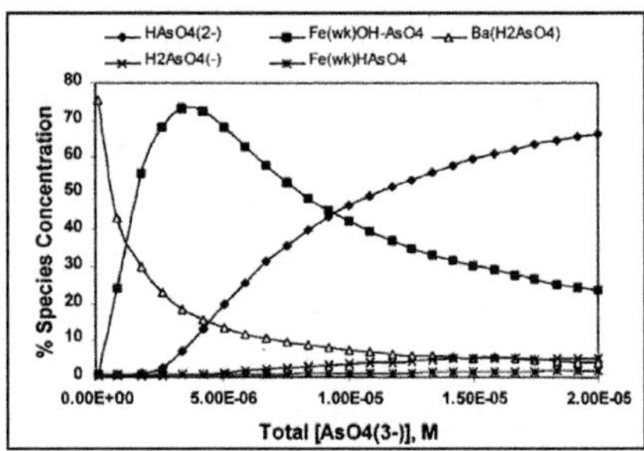

FIGURE 11-2 Left: Basic principle for MINTEQA based speciation calculation. Top-right: The nonlinear equation in the box, where, C_i is the chemical species i ($H_2AsO_4^-$), K_i the equilibrium constant for species (K_2), X_j is the equilibrium concentration of components j (H^+, AsO_4^{3-}), and a_{ij} is the stoichiometric coefficient. Bottom right: Percent distribution of arsenic species as a function of total AsO_4^{3-} calculated using the MINTEQA model for the natural attenuation of arsenic inorganic arsenic species in presence of hydrous ferric oxide (HFO). ▽ – Fe(wk)OH–AsO_4; o – $Ba(H_2AsO_4)$ (s); △ – Fe(wk)$HAsO_4$; ◆ – $HAsO_4^{2-}$; X – $H_2AsO_4^-$. Total Fe = 1.21×10^{-4} M or 6.77 mg/L) [14].

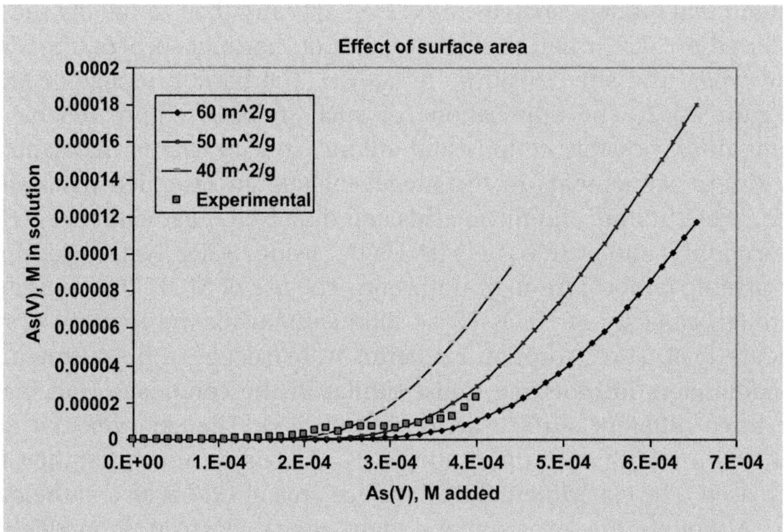

FIGURE 11-3 Figure shows the sorption isotherm calculated based on the surface-complexation reaction As(V) on Goethite using MINTEQA. The experimental data points are also superimposed on the family of curves with different surface areas as indicated. Here the composite iron matrix (CIM) is considered as Goethite with plane geometry, measured surface area 50 m^2/g, one surface, four sites, inner capacitance 1 $\mu F/cm^2$, outer capacitance 5 $\mu F/cm^2$, counter ion accumulation allowed at measured water pH = 7.6 and ionic strength 0.01 M.

the arsenic removal capacity [13]. These calculations are based on the existing thermodynamic database in the program. Despite this limitation in the database and in the assumed values of site density and double-layer capacity when using such models, one can gain significant insight on sorbent capacity in complex aquatic systems. It is also possible to model interferences from ions such as phosphate, silicate, carbonate, sulfate, humic and fulvic acids, and other dissolved organic matters (DOM) on sorption capacity. The MINTEQA model also allows one to obtain natural attenuation of arsenic species from groundwater, which is critical in formulating synthetic water for filter testing, identifying possible mineral phases in the filter, and leaching species from mineral-saturated sorbents [14].

11.3.2 Continuous Proton-Binding Model and pK_a-Spectrum

In the natural aquatic environment, minerals, high molecular weight DOM (e.g., humic and fulvic acids), and bacteria are important substrates for sorption of many chemical species. There have been numerous studies on metal binding on these surfaces [15]. The most fundamental acid–base properties of these materials are the dissociation constant (K_a) for the proton-binding sites and their concentrations. A simple and carefully performed acid–base titration is all that is needed to find these fundamental parameters. First the data are transformed into excess charge, b_i, for the i-th addition of titrant and then the data are fitted in a pK_a spectrum:

$$b_i = (C_b - C_a + [H^+] - [OH^-])_i / C_s$$

$$b_i = \sum (L_j K_j)/(K_j + [H^+]_i) \text{ for } i = 1 \ldots n,\ j = 1 \ldots m$$

where, C_b and C_a are the concentrations of base and acid added, respectively; C_s is the solid sorbent concentration in g/L, L_j is the concentration of j-th binding site; and K_j is the corresponding binding constant. The nonlinear data fitting is performed on calculated b_i vs. $[H^+]_i$ assuming reasonable initial values for L_j and K_j within some limits until optimized. A typical titration of HFO and HFO coated with anaerobic bacteria is shown in Figure 11-4. The pK_a spectrum represents the overall noninteracting proton binding of the surface as the summation of discrete monoprotic binding reaction, each reaction being defined by a unique pK_a and the corresponding site concentration, L_j [16]. Although this model has significant simplifying assumptions, the most interesting aspect of this work is that it requires only simple acid–base pH titration instruments.

11.3.3 Octanol–Water Partitioning Model and Bio-Concentration Factors (BCF)

The chemical abstracts service (CAS) registers over 67.72 million commercially available chemicals [17]. This is an alarming number and many of these are polluting our environment in various ways. For example, the drugs we use for ourselves and animals are being flushed directly into wastewater, which then becomes a drinking water source

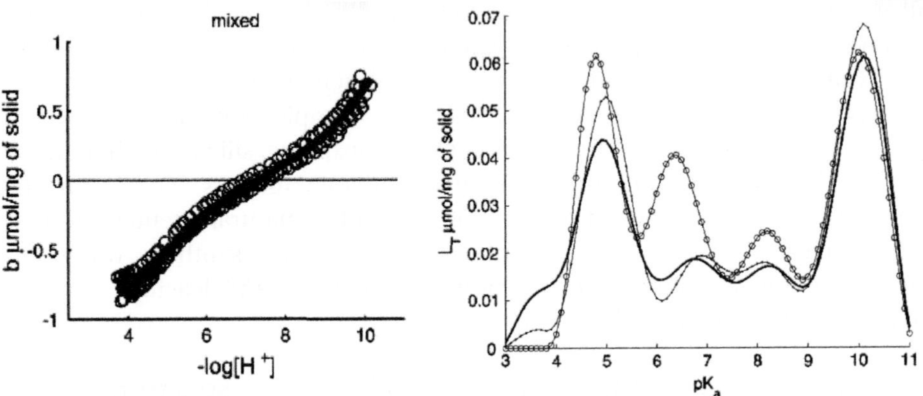

FIGURE 11-4 Left: Acid–base titration of hydrous ferric oxide. Right: pK_a spectra of HFO coated with *Shewanella putrefaciens* strain CN32 anaerobic bacteria (o), 50/50 mix of HFO:CN32 (.). The heavy solid line is a theoretical spectrum assuming the latter. The pK_a spectra show five proton-binding sites and their relative concentrations (in ordinate). Typical pK_j range 1–10, and corresponding L_j 0.1–0.6 µmol/mg solid with 4 primary sites, $i = 4$ [16] (by permission).

downstream. Researchers have found antidepressants, antacids, synthetic hormones from birth control pills, and many other human and animal medicines along with traces of caffeine, in the drinking water supplies of 24 of 28 US [18]. Many of these organic compounds, including pharmaceuticals and endocrine-disrupting chemicals, have already accumulated in the aquatic environment and have been propagated into our food chain (fish, mussels, etc.) through bioaccumulation in the fatty tissues. Some of these are persistent organic pollutants (POP) with good and bad features. The beneficial aspect is that some POPs are less mobile and may not find their way readily into the groundwater. The negative aspect is that their in situ remediation through biodegradation becomes difficult.

One quantitative measure of POP bioaccumulation, lipophilic storage capacity, biomagnification, and sorption tendency is related to the octanol–water partition coefficient ($K_{OW} = [\text{Solute}]_{\text{Octanol}}/[\text{Solute}]_{\text{Water}}$). The K_{OW} is a direct measure of the relative solubility of organic pollutants in fat tissue with respect to that in water. In a simple shake-in-a-flask experiment containing immiscible layers of octanol and water, the equilibrium concentrations of both species can be measured [19]. Recent re-examination of more advanced solvatochromic parameters [20] to estimate solubility, shows that K_{OW} is a far superior predictor of lipophilic solubility [21]. For a large number of pesticides, the relation between K_{OC} (OC—organic carbon based equilibrium) and the BCF was found to be [22]:

$$\log K_{OC} = 1.886 + 0.681\ \log(\text{BCF}) \quad (r^2 = 0.83)$$

The toxic effect of these chemicals can be reduced by making them biodegradable by incorporating functional groups (degradaphores) in their structure-sensitive microbial oxidation processes. One classical example is the DDT-type analog, where two –Cl is replaced with two $-CH_3$, resulting in a 600-fold decrease in the biomagnification factor (=ppm in fish/ppm in water). K_{OW} is further related to biological responses through the Hansch quantitative structure activity relationship (QSAR):

$$\log \text{BR} = -a(\log K_{OW})^2 + b \log K_{OW} + \sigma + cE_s + d$$

where BR is a measure of biological response, σ is the Hammet constant [23] of substitutients, E_s is summation of the Taft steric parameters, and *a, b, c, d* are multiple linear regression coefficients.

The transport retardation of most organic compounds in aquifers is also related to K_{OW}. Generally, if the fraction of organic carbon (DOM) in water is known, the distribution coefficient in the sediment can be evaluated. In many countries, where malaria is prevalent and insecticide/pesticide use is extensive, it is necessary to learn how these chemicals accumulate in sediment–water interfaces and are transported into the biota. This exercise may well be too late for some of the aquatic species (fish) from our streams and rivers that have become extinct because of the toxicity of these chemicals. However, it is absolutely essential to understand this process for the preservation of existing biodiversity.

11.4 Development of Mitigation Technologies

The purification of raw water into potable water is based exclusively on a few physicochemical processes. Over the years, the basic process has not changed very much, except for the engineering scale-up to meet increasing demand. However, significant developments have taken place in new sorbent design, high efficiency reverse-osmosis (RO) membranes, and nanofiltration (using nanomaterials). The appearance of arsenic and other trace metals in our potable water has also prompted the development of iron-based sorbents and filters for countries like Bangladesh, India, and Nepal. The development of mitigation measures is further challenged by the reduction of the maximum contamination limit (MCL) for many toxic species. This section describes some critical aspects of water purification using chemicals and the science of water filter development.

11.4.1 Chemical Purification of Water: Chlorine and Derivatives

Chlorine or hypochlorite is the classic surface water purification agent for removing bacteria and viruses. It was chlorine that drastically reduced the outbreak of waterborne diseases in the early 1900s. In 1997, Life magazine declared that "the filtration of drinking water plus the use of chlorine is probably the most significant public health advancement of the millennium." However, chlorine and reaction products find their way into aquatic ecosystems and affect organisms that are integral to food chains. Once present, they interact with organic carbon and lead to the formation of carcinogenic chlorinated hydrocarbons, which bioaccumulate within the food chain. An alternative to chlorine is the monochloramine, NH_2Cl, which oxidizes sulfhydryls and disulfides in the same manner as HClO [3], but only possesses 0.4% of the biocidal effect of HClO [4]. A switch from free chlorine to chloramine disinfectant, triggered problems with excessive Pb^{2+} in drinking-water service lines with lead solder or brass plumbing materials in Washington, DC. Laboratory tests proved that chlorine reacts with soluble Pb^{2+} to rapidly precipitate a red-brown colored lead solid that was insoluble even at pH 1.9 for 12 weeks. This solid did not form in the presence of chloramine [24]. Chloramine breakdown products with no disinfectant capacity can corrode plumbing as water stagnates and bacteria multiply. This example clearly demonstrates the need for deeper chemical understanding of disinfectant behavior under field conditions. A review article envisaged the involvement of some new aspects of molecular biology in the purification of water [25].

Industrial wastewater treatments also have several problems. Chlorine and bromine introduced into industrial water systems to address microbial contamination, react with the corrosion inhibitor TTA (tolyltriazole) to produce a chlorinated by-product that does not protect copper alloy piping. At least 200 million lb of acrylamide-based polymers are used annually in a variety of industrial and wastewater facilities. In order to make these polymers water soluble, an emulsion of water, oil, and surfactants in roughly equal amounts is used. Over 100 billion gallons of industrial wastewater have been treated with STABREX antimicrobial (*N*-bromoaminoethanesulfonic acids) since 1997, replacing 20 million lb of chlorine. Such "green chemistry" solutions are yet-to-be validated for environmental impact. In developed countries, in addition to activated carbon, advanced oxidation processes using combinations of ozone, ultraviolet (UV) light, and hydrogen peroxide to create the highly reactive hydroxyl radical ($OH^{\bullet}$), are becoming popular. However, ozone is found to form toxic perchlorate (MCL 6 μg/L) [26] and perbromate anions.

11.4.2 Membrane Filtration Systems

Membrane technologies for removing particulate matter by micro- and ultrafiltration, and dissolved substances by nanofiltration and reverse osmosis (RO) are increasingly being used for wastewater reclamation in modern, urban water systems. Immersed micro- and ultrafiltration membranes provide excellent pretreatment for RO, which can remove a wide range of dissolved constituents. Figure 11-5 shows a membrane filtration cartridge, which is the workhorse for modern membrane filtration systems. For the reclamation of potable water, membrane technology must be followed by UV-oxidation treatment for disinfection [27]. Here, nanotechnology concepts are being investigated for higher-performing membranes with fewer fouling characteristics, improved hydraulic conductivity, and more selective rejection/transport characteristics. Membrane clogging and fouling are the primary causes for failure of these technologies.

11.4.3 Development of Sorbent Based Water Filters

The development of household water filters has huge potential both scientifically and commercially. One can see the potential in the usage of commercial water filters and fancy plastic containers or water jugs with charcoal or stones, sold in the market by the

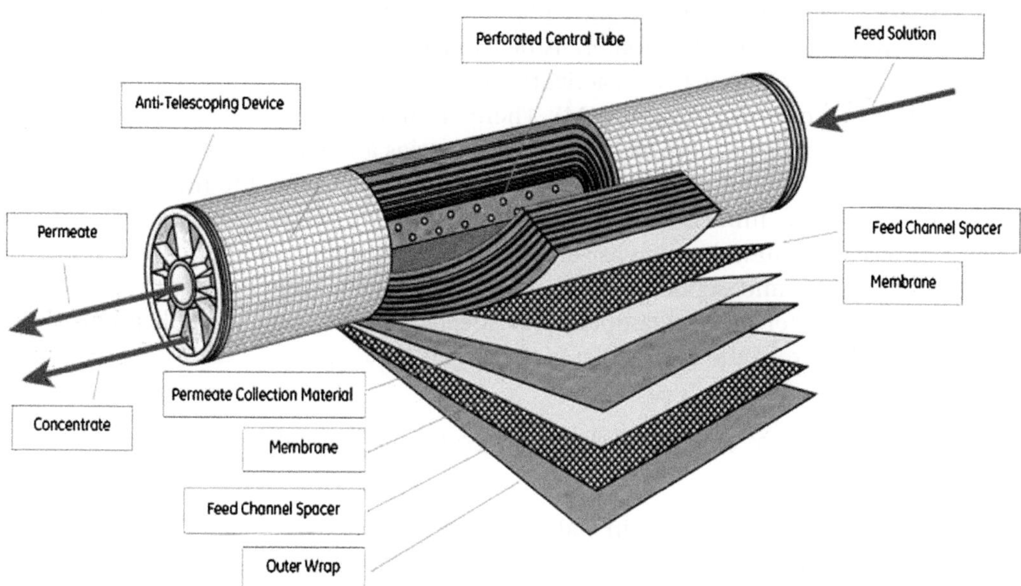

FIGURE 11-5 General Electric reverse-osmosis membrane filtration cartridge. *http://www.gewater.com/products/consumables/index.jsp*

million. It is questionable whether some of these products are filtering any toxic species to make potable water. There are no regulations to control the quality of these products, although they are supposed to purify drinking water. The demand for such filters is on the rise because of the scarcity of potable water even in cities with piped water supplies. At one extreme, people do not trust city water supplies even in the United States. At another extreme, at least 50 million people in Bangladesh are drinking groundwater containing toxic levels of arsenic and other minerals. There is a market for at least 10 million household filters valued at $300 million in Bangladesh alone. The worldwide demand and commercial potential can only be addressed through science and technology dedicated to the development of the small and affordable filter.

Real water filters must contain active material for the separation and removal of toxic species from the influent water. Therefore, material development is crucial. Once a new sorbent material is developed, there are three scientific aspects to be understood: (1) the sorption equilibrium thermodynamics (SET), (2) sorption kinetics (SK), and (3) flow and mass transport dynamics (FMTD). The SET determines the overall capacity of the sorbent and the sorbent mass needed to manufacture filters with a specific capacity. The sorbent SK determines the timescale of sorption and must not be a rate-limiting factor. A fast SK is essential for practical filters. The FMTD is determined by the physical design and determines the final commercial design. Modern computational tools such as finite element methods (FEM) for the solution of mass-transport equations coupled with SET and SK are now essential for design optimization.

Real water filters require active sorbents to sorb unwanted chemical and biological species. There are only very few materials developed for water filtration. These are activated charcoal, ion-exchange resins, activated alumina, ceria, zirconia, and hydrous ferric oxide matrix (iron oxyhydroxide or CIM). There are many physical and chemical variations of these materials with affinity for specific species and capacity. Numerous studies have been made of these materials but activated carbon remains the most popular and inexpensive material, limited to the removal of organics. Filters may often contain inactive or passive materials, such as sand, stones, brick chips, or filter cloths to impart mechanical stability and separate suspended particulates. Here we describe only the CIM-based water filter as an example, with special reference to its application in the removal of arsenic from the groundwater of Bangladesh.

11.4.3.1 Development of Iron Oxides and Iron Oxide–Amended Materials for Water Filters

Iron and its various compounds in nature (oxides and oxyhydroxides) are known as nature's "cleansing agent." There have been renewed efforts to understand and use iron-based materials because they are environmentally benign. Zero-valent iron has been used in the past to mitigate chlorinated hydrocarbons, arsenic, and other toxic species in the environment [28]. Significant research has been done on the development of sorbents and other materials for removing arsenic and other toxic species from water [29]. These materials include: oxides of granular metals, amorphous iron hydroxide, HFO, granular ferric hydroxide, ferrihydrite, red mud, iron oxide-coated polymeric materials, iron oxide-coated sand, Fe(III)–Si binary oxide, iron oxide-impregnated activated alumina, blast-furnace slug, iron-cerium bimetal oxide, iron-coated sponge, nanoscale zero-valent iron, sulfate-modified iron oxide-coated sand, hydrous ferric oxide incorporated into naturally occurring porous diatomite, crystalline hydrous ferric oxide, crystalline hydrous titanium oxide, granular hydrous zirconium oxide, and iron(III)–tin(IV) binary mixed oxide. Except for the classic method based on the coprecipitation–coagulation of toxic impurities by flocculent iron hydroxide precipitate, most of these studies were confined to the determination of removal capacity and kinetics. None of these materials was extensively tested in the field or passed through rigorous environmental technology verification.

The list of materials above shows that iron is the key component. However, iron-based technologies are subject to many problems: uncontrolled leaching and rusting, which can clog the filter media and filter outlets and render the filter useless; low capacity to remove As(III) species; and the complexity of regenerating and reusing the material in practical filtration systems.

In the process of our research, we invented a technique for processing easily available surplus iron into composite iron granules (CIGs) and then into a CIM [30]. It was found that readily available iron turnings (cast iron, low- and high-carbon steel) can be processed into CIG by a controlled rusting process to produce mixed oxides of iron and other metals, (e.g., manganese) and then into a CIM. CIM is different from granular metal oxides in that the

FIGURE 11-6 Left: Picture of the composite iron matrix (CIM) recovered from a used SONO filter. Right: Fine CIM particles dispersed and embedded in fabrics as filter media. The fabrics are attracted to magnet as shown.

active substrate is made from CIG into a solid, porous matrix through in situ processing inside the filter. Figure 11-6 shows a typical round-shaped CIM formed in situ in a SONO filter and the same dispersed and embedded in fabrics. CIG can be used to make a CIM of any shape, including a thin disc containing 1.0 g of CIG inside a tube to build experimental filters. The purpose of such a small filter is to study the breakthrough in the laboratory timescale, (discussed later). The active material in the filter removes inorganic arsenic species quantitatively by generating new complexation sites on the CIM through iron oxidation and SCRs. Chemical species diffusion through iron-corrosion products determines the rates and the removal kinetics ranged between zero and pseudo-second order in influent As(V) concentration. It was argued that excess Fe^{2+}, Fe^{3+}, and Ca^{2+} in groundwater could increase positive charge density of the inner Helmholtz plane of the electrical double-layer and enhance binding of anionic species, such as arsenate.

Surface complexation of metal species is the primary mode of metal ion removal with charged solid oxide sorbents. The primary active material in iron-based filters is made from hydrolysis of iron salts or from partial rusting of FeO or Fe-amended materials, sometimes through proprietary processes. These sorbents can remove inorganic arsenic species quantitatively through possible reactions shown in Table 11-2. It is known that arsenate and arsenite form bidentate, binuclear surface complexes with =FeOH (or =FeOOH or hydrous ferric oxide), and HFO as the predominant species, tightly immobilized on the iron surface. The primary surface reactions are: $=FeOH\ (s) + H_2AsO_4^- \rightarrow =FeHAsO_4^-\ (s) + H_2O$ ($K = 10^{24}$) and $=FeOH\ (s) + HAsO_4^{2-} \rightarrow =FeAsO_4^{2-}\ (s) + 2H_2O$ ($K = 10^{29}$). These intrinsic equilibrium constants indicate very strong complexation and immobilzation of inorganic arsenic species.

Table 11-2 illustrates the possible reaction zones in a SONO filter. The two-stage filtration system also enhances the amount of arsenic removed through flocculation and precipitation of naturally-occurring iron in groundwater. It shows that groundwater containing Fe(II) plays a role where inorganic As(III) species are oxidized to As(V) species by the active $O_2^{-\cdot}$, which is produced by the oxidation of soluble Fe(II) with dissolved O_2. The presence of manganese in most raw-iron turnings can catalyze oxidation of As(III) to As(V). Therefore, the process does not require chemical pretreatment of water with

Table 11-2 Chemical and Surface-Complexation Reactions in an Iron-Based Filter

Description	Reactions	Example Filter
Oxidation of soluble iron Oxidation of ferrous to ferric through active oxygen species **Oxidation of As(III)** (Equations are balanced for reactive species only)	$Fe(II) + O_2 \rightarrow O_2^{\cdot -} + Fe(III)OH_2^+$ $Fe(II) + O_2^{\cdot -} \rightarrow Fe(III) + H_2O_2$ $Fe(II) + CO_3^{\cdot -} \rightarrow Fe(III) + HCO_3^-$ $As(III) + O_2^{\cdot -} \rightarrow As(IV) + H_2O_2$ $As(III) + CO_3^{\cdot -} \rightarrow As(IV) + HCO_3^-$ $As(III)OH^{\cdot} \rightarrow As(IV)$ $As(IV) + O_2^{\cdot -} \rightarrow As(V) + O_2^{\cdot -}$	**SONO filter:** Cleaning arsenic from well water Cover Empty space Tap fitted with flow control junction Coarse sand CIM* Coarse sand Brick chips Cover Plastic pipe Coarse sand Wood charcoal Fine sand Brick chips Tap fitted with flow control junction *Composite iron matrix. Source. Dr. Abul Hassam Post-Gazette
Formation of HFO in presence of Fe(III) Fe(III) complexation and precipitation.	$=FeOH + Fe(III) + 3\,H_2O \rightarrow Fe(OH)_3$ **(s, HFO)** $+ =FeOH + 3H^+$ (=FeOH is surface of hydrated iron)	
Surface complexation of arsenates Surface complexation and precipitation of anionic species As(V) on HFO. log K values are shown in ( ). Ψ is the surface potential	$=FeOH + AsO_4^{3-} + 3\,H^+ \rightarrow =FeH_2AsO_4 + H_2O$ (29.31) $=FeOH + AsO_4^{3-} + 2\,H^+ - \exp(-F\psi/RT) \rightarrow =FeHAsO_4^- + H_2O$ (23.51) $=FeOH + AsO_4^{3-} + H^+ - 2\exp(-F\psi/RT) \rightarrow =FeAsO_4^{2-} + H_2O$ $=FeOH + AsO_4^{3-} - 3\exp(-F\psi/RT) \rightarrow =FeOHAsO_4^{3-} + H_2O$ (10.58)	
Surface complexation of silicate species Reactions with iron surface and silicates can produce a porous solid matrix with extremely good mechanical stability for long term use.	$=FeOH + Si(OH)_4 \rightarrow =FeSiO(OH)_3\,(s) + H_2O$ $=FeOH + Si_2O_2(OH)_5^- + H^+ \rightarrow =FeSi_2O_2(OH)_5\,(s) + H_2O$ $=FeOH + Si_2O_2(OH)_5^- \rightarrow =FeSi_2O_3(OH)_4^-\,(s) + H_2O$	
Precipitation of other metals Surface precipitation of arsenate with soluble metal ions if surface concentrations exceed solubility limits. Many metal ions are also quantitatively removed this way.	$=FeOHAsO_4^{3-} + Al(III) \rightarrow =FeOHAsO_4Al\,(s)$ $=FeOHAsO_4^{3-} + Fe(III) \rightarrow =FeOHAsO_4Fe(s)$ $=FeOH.HAsO_4^{2-} + Ca(II) \rightarrow =FeOH.HAsO_4Ca\,(s)$ $M(III) + HAsO_4^{2-} \rightarrow M_2(HAsO_4)_3\,(s)$, M= Fe, Al, $M(II) + HAsO_4^{2-} \rightarrow M(HAsO_4)\,(s)$ and other arsenates M = Ba, Ca, Cd, Pb, Cu, Zn and other trace metals	

All Surface Species are Indicated by ═FeOH

oxidizing agents such as hypochlorite or potassium permanganate. More than 290,000 SONO filters based on CIM are functioning in Bangladesh, Nepal, India, and Pakistan. Experimental filters installed in the field at different times have functioned for eight years in Bangladesh without a break through, even with high-soluble iron (21 mg/L in one case), high phosphate content (50 mg/L), and varied water chemistries. Figure 11-7 shows some results from the use of SONO filters in removing arsenic and manganese from groundwater in Bangladesh.

11.4.3.2 Nanoiron

There has been significant interest in nanoiron particles (<100 nm in dia.) because they are easy and inexpensive to prepare even in large quantities. Although Fe-nanoparticles in aqueous suspension are inherently unstable, the higher surface-to-volume ratio allows

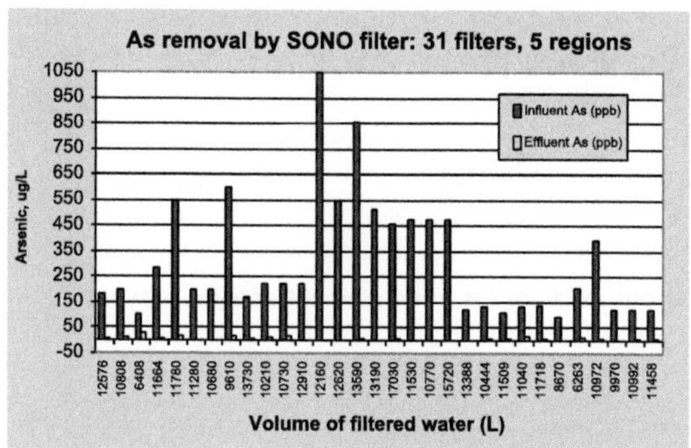

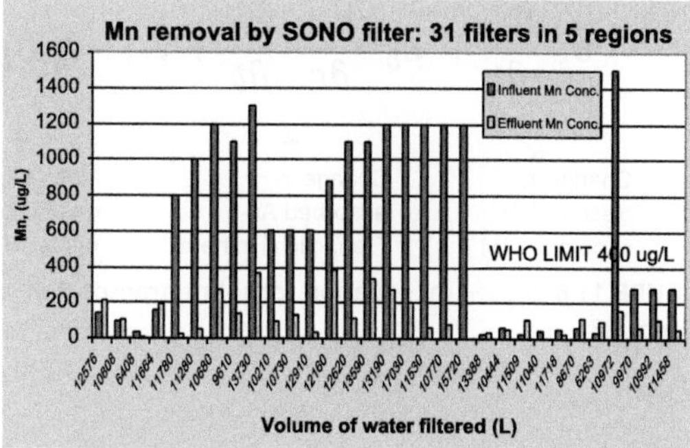

FIGURE 11-7 Left: Groundwater arsenic removal capacity by 31 SONO filters used in five regions of Bangladesh, each filtering 10,000 L water. The maximum influent As (total) is shown as red bars (1050 μg/L As (total). The filtered water (blue bar) had an average of 7.3 μg/L As (total) (Max 32 μg/L). Right: Manganese (a neurotoxin) is also removed by SONO filters. All 31 filters passed WHO limit 400 μg/L Mn. (For interpretation of the references to color in this figure legend, the reader is referred to the web version of this book.)

for more efficient sorption. Experiments by a Rice University team showed that 20-nm magnetite nanoparticles could treat 20 L of water a day for a year, yielding 3.65 kg of magnetite bound with arsenic—the waste that would have to be removed or recycled. Only 1.8 kg of waste would be produced with 12 nm magnetite particles, the most efficient size for removal. About 200–500 mg of the magnetite nanoparticles would be necessary to treat a liter of water [31]. A 6-year test of such a technology in Bangladesh did not, however, result in a practical product.

11.4.3.3 Modeling Sorption Breakthrough of Water Filters: Finite Element Analysis

In our previous work we discussed the use of an equation to calculate the solute breakthrough from a fixed-bed column experimental data, which can be used to determine the scale-up size for a known breakthrough volume [14]. This approach considers surface diffusion to the inside of the adsorbent pore as the rate-controlling step. If the volumetric flow is low and an instantaneous equilibrium is assumed, a good approximation for the breakthrough curve can be obtained [32]. This approach is only approximate for the one-dimensional column, does not consider the porosity of the sorbent materials, or any dynamic reaction terms. In order to understand and visualize the concentration gradient profile of solute in solution and in the solid phase during filtration, one should use the reactive transport model solved through FEM in real filter geometry as described here.

The dynamic sorption of soluble arsenic species as solute in a model column was studied experimentally and theoretically. The theoretical modeling was accomplished by solving coupled differential equations containing diffusion, sorption, advection or dispersion by COMSOL [33], which is the most comprehensive model for solving such problems. The general equation for solute transport in a saturated porous media is shown Figure 11-8. The first term is the rate of change in dissolved solute concentration,

$$\theta_S \frac{\partial c_i}{\partial t} + \rho_b \frac{\partial c_{Pi}}{\partial c}\frac{\partial c_i}{\partial t} + \nabla.[-\theta_s D_{Li} \nabla c_i + \boldsymbol{u} c_i] = R_{Li} + R_{Pi} + S_{ci}$$

Change in dissolved As in porous CIM

Change in adsorbed As- Freundlich isotherm

Mechanical dispersion & diffusion -dispersivity

Active flux Linear velocity

Reaction terms =0

FIGURE 11-8 A general equation for solute transport in a porous medium for finite element analysis of water filters [33].

c_i, in the aqueous media over time. The second term explains the rate of change of sorbed species, c_{Pi}, attached to the sorbent. The sorption is dictated by typical sorption isotherms like the Langmuir Freundlich isotherms. The third term inside [ ] contains mechanical dispersion because of local concentration gradient coupled with molecular diffusion and advective flux coupled with linear velocity. The terms on the right-hand side involve chemical reactions and solute source terms, which could also change the solute concentration. They could be source or sink terms inside the column. These terms are generally absent when the reactive transport of the solute is expressed by a suitable isotherm, except where the sorption is coupled with post-chemical transformation of solute, where the R_{Pi} term is invoked. We consider all reaction terms to be zero. Other required parameters in COMSOL are explained in Table 11-3 with some experimental values.

The utility of COMSOL is illustrated here with an experimental arsenic filter column (9 cm in length and 1.5 cm inside diameter), in which a 0.4 cm thick 1.0 g CIM plug was

Table 11-3 Parameters Defined for COMSOL Solution and Visualization

Parameter	Value	Description
D	6.0×10^{-6}	Diffusion coefficient of solute (cm^2/s)
θ-sand	0.35	Porosity of sand
θ-cim	0.37	Porosity of sorbent, CIM
K	0.05965	Freundlich isotherm, K (cm^3/g) CIM
N	0.3822	Freundlich constant
C_{in}	3.0×10^{-7}	Inlet solute concentration (g/mL or 300 μg/L)
D_{ts}	1/5000	Time scaling (5000 s run time)
u-expt	0.05	Experimental linear velocity = length of column/time for travel (cm/s)
u_l	0.14	Longitudinal velocity corrected for porosity, (cm/s)
u_t	0.023	Transverse velocity corrected for porosity (calculated), (cm/s)
ρ-sand	1.328	Density of sand (g/mL)
ρ-cim	1.849	Density of CIM (g/mL)
αL	5.6	Longitudinal dispersivity (cm)
αD	1.0	Transverse dispersivity (calculated) cm = αL/6

Parameters Were Experimentally Obtained Unless Mentioned as Calculated.

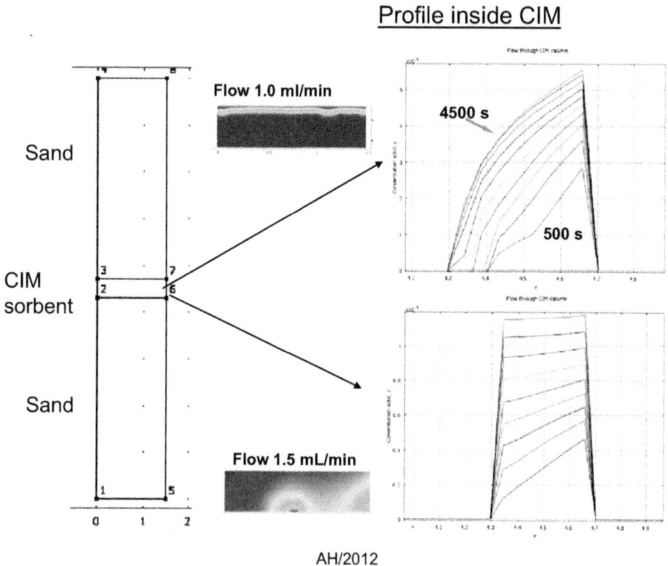

FIGURE 11-9 Figure shows column geometry, longitudinal and cross-sectional concentration profile of arsenic at different times, and 2 flow rates. The right-hand plot shows the middle longitudinal sorbed solute concentration gradient in the sorbent phase at 500–4500 s in 500 s increments at 1 mL/min (top-right) and at 1.5 mL/min (bottom right) flow rates. The colored cross-sections show the sorbed concentration profile with red as the highest and blue as the lowest. (For interpretation of the references to color in this figure legend, the reader is referred to the web version of this book.)

prepared in the middle between two sand columns as shown in Figure 11-9. The bottom of the column is taken as 0 cm. Aqueous solution containing 300 μg/L As(III) was pumped at a constant rate from the top and the effluent was collected at the bottom outlet. The dynamics of sorption are solved by the model described, using the parameters in Table 11-3. The appropriate boundary conditions are: constant inlet concentration (i.e., 300 μg/L As(III)); advective flux at outlet; continuity at interface between sand and sorbent pair boundaries. All other boundaries have zero fluxes. The longitudinal dispersivity was measured by monitoring the elution profile of a non-interacting solute (such as sodium nitrate) in a sand column of exact dimension. The transverse dispersion was scaled by diameter/length of the column. Here, we show that COMSOL finite element analysis allows the best possible prediction and visualization of solute profile in solution and in sorbent phases. The utility of such an exercise is to predict the solute breakthrough as a function of time and flow rates as shown in Figure 11-9. Figure 11-10 shows the solute flow patterns inside the column. It clearly demonstrates that the solute gradient drops at the sorbent interface almost linearly and then increases with a zero gradient as effluent. The effluent solute concentration increases from 5- to 60 μg/L as a function of time. The plot shows that solute concentration increases immediately after the sorbent phase and the effect of sand is negligible. Figure 11-11 shows a reasonable agreement between experimental and COMSOL calculated values. We observe that column breakthrough

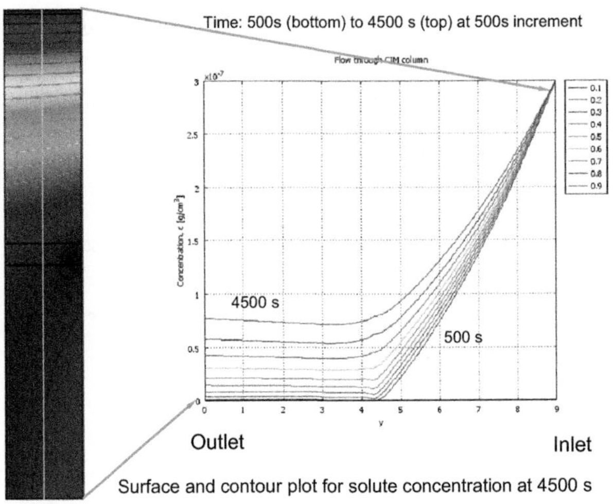

FIGURE 11-10 Left: Concentration profile and contour plot of solute in the aqueous media in the column at 4500 s. Right: Plot shows the middle longitudinal concentration gradient from inlet to the outlet at 500–4500 s in 500 s increments.

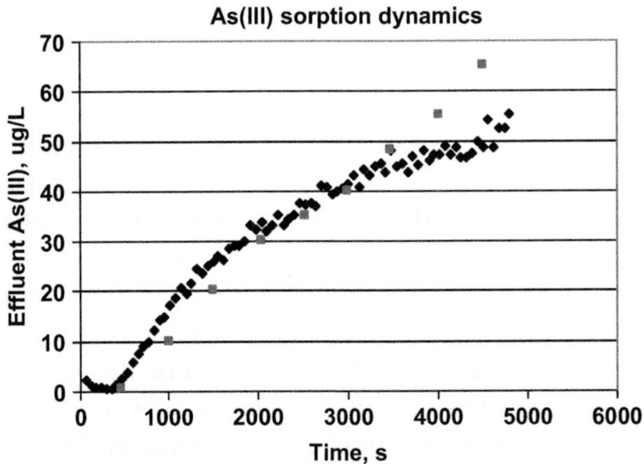

FIGURE 11-11 COMSOL prediction (square) and experimental values (diamond) of effluent As(III) breakthrough. Experimental conditions are shown in Table 11-3. Effluent As(III) concentrations were measured by a continuous flow hydride generation atomic fluorescence spectrometry. Influent 300 μg/L As(III) at a flow rate of 1.5 mL/min. (For interpretation of the references to color in this figure legend, the reader is referred to the web version of this book.)

(50 μg/L As(III) MCL breakthrough) can be sensitive to flow rates. Figure 11-9 shows the concentration gradient of sorbed solute at two different flow rates. At low flow rate (top-right graph) the sorbed solute concentration is the highest at the top of the sorbent and steadily decreases to zero near the bottom interface, without a solute breakthrough even at 4500 s. The situation changed at 1.5 mL/min when sorbent breakthrough occurred at

the 3500 s. The sorbed solute profile shows uneven distribution of solute. The colored profile also shows the highest and the lowest concentration of arsenic in the column. In order to validate this calculation the cell Peclet number (Pe) must be less than unity, which assures hydrodynamic rather than numeric dispersion [34]. In our case the cell Pe was 0.06–0.1. In the absence of experimental data, one could still use reasonable values satisfying the cell Pe limits and obtain filter design insights. Clearly, these tools are extremely useful in concentration profile visualization beyond one dimension and in optimizing practical filter design.

11.5 The Role of Analytical Chemistry

Accurate and precise measurements are the hallmarks of water-quality evaluation and validation because "Poor water quality intensifies water scarcity." The best example is the discovery of arsenic in water in 1997, by Professor Dipankar Chakraborti, an analytical chemist, who continues to expose the arsenic crisis in the Indian subcontinent through reliable measurement of arsenic in the environment. It should be emphasized that all developments and decisions are based on laboratory and field measurements of water-quality parameters. Khan proposed the following first order (in bold) and second order parameters to assess short- and long-term trends in water quality from different sources [35].

- Salinity: **Cl^-**, SEC (specific electrical conductance), TDS (total dissolved solids), **SO_4^{2-}**, Br^-, **SO_4^{2-}** Mg/Ca, δ18O, δ2H, F^-
- Acidity and redox status: **pH**, **HCO_3^-**, **E_h**, dissolved oxygen, Fe(II)/Fe(III), Mn(II)/Mn(III), As(III)/As(V)
- Radioactivity: **^3H**, ^{36}Cl, ^{222}Rn
- Agricultural pollution: **NO_3^-**, **SO_4^{2-}**, **DOC** (dissolved organic carbon), K/Na, P, pesticides, and herbicides
- Mining pollution: **SO_4^{2-}, pH**, Fe, As, F^-, Sr, and other metals,
- Urban pollution: Pb, **Cl^-, HCO_3^-, DOC**, B, hydrocarbons, organic solvents

Reliable and sustainable laboratory measurement takes time and resources— often ignored by long-term planners. The quality control of analytical measurements involves instrument tuning, calibration, calibration verification, precision and recovery tests, matrix spike and surrogate recovery, labeled compound recovery, and blank results. Two categories of measurement, which are the workhorses in a modern analytical laboratory, are atomic spectroscopy (absorption, emission and fluorescence) for trace element analysis and separation techniques (gas or liquid chromatography, GC or LC) interfaced to a mass spectrometer (MS) for trace organics. Instrument-interface development is also necessary for specialized applications. One such instrument interface, developed using the hydride generation atomic fluorescence spectrometer (HGAFS), is shown in Figure 11-12. In order to study the separation efficiency of a small amount of sorbent material in a column, we have developed a flow manifold interfaced to an HGAFS for

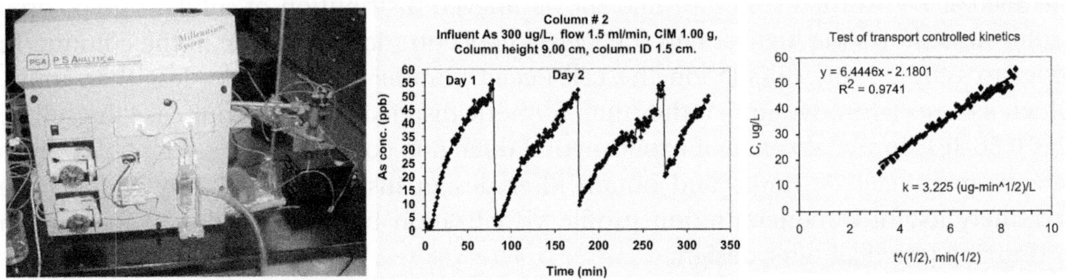

FIGURE 11-12 Left: Hydride generation atomic fluorescence spectrometer (HGAFS) with the sorbent column and continuous flow manifold. The iron-based CIM sorbent (1.0 g) is sandwiched between two sand layers and is the As(III) removal column. Influent As(III) 300 ug/L was at 1.5 mL/min. Middle: Continuous measurement of effluent As(III) over 4 days with about 12 h aging of the sorbent between days. The sharp line shows recovery of As(III) sorption capacity after each aging. Right: Plot of effluent [As(III)] vs. $t^{1/2}$ showing that sorption is a diffusion-controlled process.

arsenic measurement [36]. The day-1 data are used for COMSOL validation as described earlier. The graph in the middle shows that As(III) sorbent capacity increased in the next day's run (sharp drop) and 400 arsenic measurements were done automatically in the μg/L concentration level. The graph on the right shows that the sorption is a diffusion-controlled process ([As(III)] vs. $t^{1/2}$ is linear) and the sorption capacity recovered to an equilibrium value after the sorbent was allowed to age— a finding consistent with field observations. Another example of an automated pH and conductivity/TDS measurement for continuous in-line water-filter testing is shown in Figure 11-13. These measurements are now routinely performed in many laboratories in developing countries.

11.6 Conclusions and Future Outlook

Naturally clean water is now becoming a thing of the past, primarily because of human activities. Cleaning water for human consumption is a difficult task because of the huge

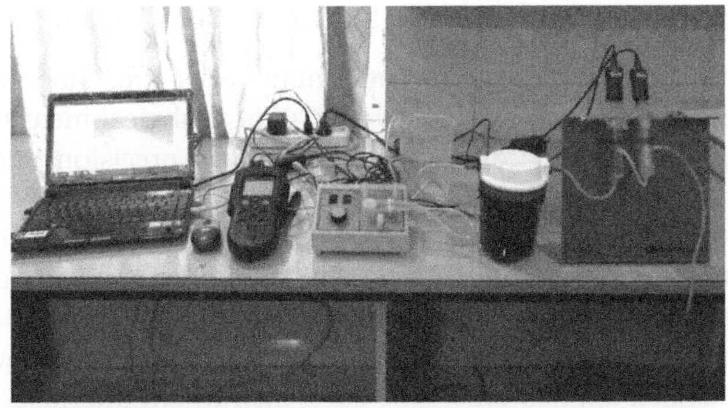

FIGURE 11-13 A setup for testing inline CIM-based filters in Bangladesh. The setup is used to monitor priming of CIM filters by recording pH, E_h, TDS, and intermittent collection of water for analysis of As(III), Fe (total), and other species as needed. The PC control of instruments allows large data gathering and continuous monitoring even in remote places where laptops are available.

FIGURE 11-14 Pollution of the Sityalakkha river near Narayangong, Bangladesh, seen from satellite. Photograph at left shows the color contrasting water pollution. Picture on the right shows large apparel and other industries on both banks of the river. *Google Geoeye Mapdata 2009, accessed Jan 11, 2009.*

scale of the work. Bangladesh is a case in point where high population density has magnified the water crisis. In fact, all developing and yet-to-be developing nations are going through this crisis. This chapter describes some basic physicochemical principles in understanding and obtaining clean potable water. At the least, in this information age, we should know the relevant facts, understand them scientifically, and act on them. For example, in early 1970, when Bangladesh decided to sink millions of tube wells to avoid waterborne diseases, there was no record of inputs from local scientists and there was a disregard for existing scientific information provided by international development partners, in going ahead with such a massive project. At that time and later, microbiologists stated that "It has been generally believed in Bangladesh that groundwater is relatively free of microorganisms and, therefore, fit for human consumption without treatment. However, the results of this study show clearly that all samples of tube-well water in rural Bangladesh that were examined contained high counts of bacteria and zooplankton, as well as fungi" (1976, 2001) [37]. The outcome could have been different for millions of arsenic-poisoned patients if the significance of this information was known to the country and acted on. In addition to the application of basic sciences as guiding principles, we should also emphasize the power of information. Many countries are at the crossroads of industrial development. Figure 11-14 shows a part of present-day Bangladesh, where significant industrialization is raging on the bank of one of the largest rivers. Does the color of the water demonstrate something to worry about? This is easy to see from any place in the world with access to the internet. Clearly, industrial effluents are contaminating rivers and making them unfit for any use and an environmental catastrophe is waiting to happen.

References

[1] Daigger GT. Tools for future success. Water Environ Technol 2003;15(12):38–45.

[2] (a) Peter Gleick H, Cooley H, Morikawa M. The world's water: the biennial report on freshwater resources 2008–2009. Washington, DC: Island Press; 2008;
(b) Postel S. Last oasis: facing water scarcity. World water development report. New York: Worldwatch Institute, W. W. Norton, www.unesco.org/water/wwap/wwdr/; 1992.

[3] Government of Bangladesh. A strategy for poverty reduction in the lagging regions of Bangladesh. Dhaka: General Economics Division, Planning Commission; 2008. p. 62, 85.

[4] MDG of United Nations signed by heads of 189 states. http://www.un.org/millenniumgoals/.

[5] Stumm W, Morgan JJ. Aquatic chemistry – an introduction emphasizing chemical equilibria in natural waters. John Wiley and Sons Inc; 1981.

[6] Milligan AJ, Morel FMM. A proton buffering role for silica in diatoms. Science 2002;297(5588):1848–50. reference 5, pg. 732–734.

[7] Hussam A, Chemistry for clean water. Keynote presentation, Bangladesh Chemical Congress 2008, Dhaka Bangladesh. January 30, 2009.

[8] (a) Mann D, Droop SJM. Biodiversity, biogeography and conservation of diatoms. Hydrobiologia 1996;336:19–32;
(b) Kübeck C, Hansen C, van Berk W, Zervas A. Development of a transfer function based method assessing long-term trends of solute influxes into the groundwater as a tool for quality management. Geophys Res Abstr 2011;13. EGU2011-5406.

[9] (a) Steen I. Phosphorus availability in the 21st century: management of a non-renewable resource. Phosphorous Potassium 1998;217:25–31;
(b) Wilsenach JA, M. Maurer JA, Larsen TA, van Loosdrecht MC. From waste treatment to integrated resource management. Water Sci Technol 2003;48(1):1–9.

[10] Dzombak DA, Morel FMM. Surface complexation modeling: hydrous ferric oxide. New York: Wiley Interscience; 1990.

[11] MINTEQA2 Model System, Center for Exposure Assessment Modeling, Environmental Protection Agency, 960 College Station Road, Athens, GA 30605–32720, 2001.

[12] Visual MINTEQ ver 2.53, 2012, www.lwr.kth.se/English/OurSoftware/vminteq/.

[13] Ahamed S, Munir AKM, Hussam A. Groundwater arsenic removal technologies based on sorbents: field applications and sustainability. In: Handbook of water quality and water purity. Elsevier Inc.; 2009.

[14] Hussam A, Habibuddowla M, Alauddin M, M., Hossain ZA, Munir AKM, Khan AH. Chemical fate of arsenic and other metals in groundwater of Bangladesh: experimental measurement and chemical equilibrium model. J Environ Sci Health 2003;A38(1):71–86.

[15] (a) Borrok D, Turner BF, Fein JB. A universal surface complexation framework for modeling proton binding onto Bacterial surfaces in geologic settings. Am J Sci 2005;305:826–53;
(b) Jambor JL, Dutrizac JE. Chem Rev 1998;98:2549–85.

[16] Smith DS, Ferris FG. Specific surface chemical interactions between hydrous ferric oxide and iron reducing bacteria determined using pK_a spectra. J Colloid Interface Sci 2003;266:60–7.

[17] http://www.cas.org/cgi-bin/cas/regreport.pl.

[18] Leonning CD. Washington Post, Monday, Page B01, March 10, 2008.

[19] $K_{O/W}$ can be also be found with high performance liquid chromatographic technique using a revered phase column based on capacity factor relation to compounds with known $K_{O/W}$ as a calibrant.

[20] Solute-solvent interactions of nonelectrolytes can be characterized by the linear free energy relation of solute's dispersive, polarity-polarizibility, hydrogen bond acceptor and hydrogen bond donor strengths. These parameters are known as solvatochromic parameters.

[21] Yalkowsky SH, Pinal R, Banerjee S. Water solubility: a critique of the solvatochromic approach. J Pharm Sci 2006;77(1):74–7.

[22] Chiou CT, Freed VH, Schmedding DW, Kohnert RL. Partition coefficient and bioaccumulation of selected organic chemicals. Env Sci Technol 1977;11:475–8.

[23] Hammet parameter (intercept/2.56) is a measure of the substitution effect on the rate of hydrolysis of ethyl benzoate.

[24] Edwards M, Abhijeet D. Role of chlorine and chloramine in corrosion of lead-bearing plumbing materials. J Am Water Works Assoc 2004;96(10):69–81.

[25] Shannon MA, Bohn PW, Elimelech M, Georgladis JG, Mariñas BJ, Mayes AM. Science and technology for water purification in the coming decades. Nature 2008;452(20):301–10.

[26] water.epa.gov/drink/contaminants/unregulated/perchlorate.cfm, www.cdph.ca.gov/certlic/drinking water/pages/Perchlorate.aspx.

[27] Maurer M, Pronk W, Larsen TA. Treatment processes for source-separated urine. Water Res 2006;40(17):3151–66.

[28] (a) Lackovic JA, Nikolaides NP, Dobbs GM. Inorganic arsenic removal by zero-valent iron. Environ Eng Sci 2000;17(1):29–39;
(b) Tratnyek PG, Johnson TL, Schattauer A. Interfacial phenomena affecting contaminant remediation with zero valent iron metal. In: Emerging technologies in hazardous waste management VII. Atlanta, Georgia: American Chemical Society; 1995. p. 589–92.

[29] Hussam A, Ahamed S, Munir AKM. Arsenic filters for Bangladesh: toward a sustainable solution. The bridge—linking engineering and society. Natl Acad Eng 2008;38(3):14–23.

[30] (a) Hussam A, Munir AKM. A simple and effective arsenic filter based on composite iron matrix: development and deployment studies for ground water of Bangladesh. J Environ Sci Health Part A Toxic/Hazardous Subst Environ Eng 2007;42:1869–78;
(b) Arsenic Water Filter, US Provisional Patent Application numbers: 60/886,989 and 60/913,120, January 2007. Removing viruses from drinking water, US Provisional Patent application no.: 61089143, August 15, 2008.
(c) Composite Iron Matrix Embedded Fabrics – A Prototype, GMU-10-024, US Provisional Patent Application No.: 61/285,635, filed 12/11/2009.
(d) Alauddin M, Hussam A, Khan AH, Habibuddowla M, Rasul SB, Munir AKM. Critical evaluation of a simple arsenic removal method for groundwater of Bangladesh. In: Chappell WR, Abernathy CO, Calderon RL, editors. Arsenic exposure and health effects. Elsevier Science B. V.; 2001. p. 439–49.

[31] Yavuz CT, Mayo JT, William Yu W, Prakash A, Falkner JC, Yean S, et al. Low-field magnetic separation of monodisperse Fe_3O_4 nanocrystal. Science 2006;314:964–7.

[32] Noble RD, Terry PA. Principles of chemical separation with environmental applications. Cambridge University Press; 2004. p. 207–212.

[33] www.comsol.com/. COMSOL is a multiphysics engineering, design, and finite element analysis software for the modeling and simulation of any physics-based problem.

[34] Bethke MC, Craig M. Geochemical and biogeochemical reaction modeling. 2nd ed. Cambridge University Press; 2008. p. 298–299.

[35] Khan AH. Mitigation of arsenic contamination of groundwater in Bangladesh. Symposium on: Arsenic calamity of groundwater in Bangladesh: contamination of water, soil and plants, Dhaka, Bangladesh, August 2008.

[36] Hussam A, Munir AKM, Ahamed S. Arsenic filter based on composite iron matrix. Paper presented at International Conference on water scarcity, global changes, and groundwater management responses, UC Irvine, United Nations Educational, Scientific and Cultural Organization (UNESCO) and U.S. Geological Survey (USGS), University of California, Irvine. Dec. 1–5, 2008.

[37] (a) Levine R. Failure of sanitary well to protect against cholera and other diarrhoeas in Bangladesh. Lancet 1976;II:86–9;
(b) Islam MSA, Siddika MNH, Khan MM, Goldar MA, Sadique MA, Kabir ANMH, et al. Microbiological analysis of tube-well water in a rural area of Bangladesh. Appl Environ Microbiol 2001;67(7):3328–30.

Removal and Immobilization of Arsenic in Water and Soil Using Polysaccharide-Modified Magnetite Nanoparticles

Qiqi Liang, Byungryul An, Dongye Zhao*

ENVIRONMENTAL ENGINEERING PROGRAM, DEPARTMENT OF CIVIL ENGINEERING, AUBURN UNIVERSITY, AUBURN, AL, USA

*CORRESPONDING AUTHOR

CHAPTER OUTLINE

12.1 Introduction

The presence of arsenic (As) in soil and water is widespread. The United States Environmental Protection Agency (USEPA) estimates that approximately 2% of the US population receives drinking water containing >10 μg/L of As [1], and the Natural Resources Defense Council estimates that approximately 56 million people in the United States drink water containing As at unsafe levels.

Arsenic has been associated with various cancerous and noncancerous health effects [2]. According to a recent report put forward by the National Academy of Science and

Monitoring Water Quality, http://dx.doi.org/10.1016/B978-0-444-59395-5.00012-1

National Research Council [3], even at 3 μg/L of As, the risk of bladder and lung cancer being caused is between 4 and 7 deaths per 10,000 people. At 10 μg/L, the risk increases to between 12 and 23 deaths per 10,000 people [3].

Triggered by the risk concern, the USEPA announced its ruling in October 2001 to lower the maximum contaminant level from the previous 50 μg/L (established in 1942) to 10 μg/L with effect from January 22, 2006 [2]. This ruling has a tremendous impact on water utilities. Approximately 4100 water utilities serving approximately 13 million people are affected by the law [4]. The compliance cost has been estimated to be approximately $600 million per year using current treatment technologies [5].

In soil and groundwater, As predominantly exists in two oxidation states, As(V) and As(III), with specific forms influenced by pH and redox [6] conditions. It is quite commonly observed that the both species coexist. Consequently, the simultaneous removal of As(V) and As(III) is often required in drinking water treatment.

Adsorption has been one of the most used technologies for the removal of trace levels of As [7]. Numerous studies have shown that various types of iron oxides can effectively adsorb both As(V) and As(III) [8–12]. Moreover, researchers found that reducing the size of the adsorbent particles to the nanoscale level can substantially increase As uptake because of the much gained specific surface area. For instance, a decrease in the particle size from 300 to 12 nm can increase the sorption capacity for both As(III) and As(V) by nearly 200-fold [13].

However, the nanoparticles without a stabilizer or surface modifier tend to agglomerate rapidly into micron-scale or larger aggregates, thereby greatly diminishing the specific surface area and As sorption capacity.

To prevent particle agglomeration, various particle stabilizing techniques have been developed in recent years [13–15]. Of many stabilizers reported, polysaccharides such as starch and carboxymethyl cellulose (CMC, molecular weight = 90,000) have been found to be not only effective in facilitating size control of various metal and metal oxides nanoparticles but to be also cost effective and environmentally benign. The proper use of stabilizers can maintain the high specific surface area and high reactivity of the nanoparticles. Further, stabilizers can facilitate manipulation of the morphology of the nanoparticles suitable for desired environmental cleanup applications. For example, An et al. [16] reported that the application of low concentrations of a water-soluble starch resulted in a new class of bridged magnetite nanoparticles that maintain the specific surface area of the nanoparticles and can settle out of the water under the influence of gravity. These are thus highly suitable for As removal from water. Alternatively, Liang et al. [17] reported that fully stabilized magnetite nanoparticles can be obtained at elevated stabilizer (starch or CMC) concentrations. The stabilized nanoparticles offer the unprecedented advantage that the particles can be delivered into the contaminated subsurface, thereby enabling in situ immobilization of As contaminated aquifers.

This chapter summarizes some of the latest developments in preparing stabilized or surface modified magnetite nanoparticles and illustrates the potential uses for As removal from water or from soils in which As is immobilized. The effects of stabilizers on the surface physical and chemical characteristics of the nanoparticles are also discussed.

12.2 Preparation and Characterization of Polysaccharide Stabilized Magnetite and Fe–Mn Nanoparticles

12.2.1 Synthesis of Polysaccharide-Bridged or Stabilized Magnetite Nanoparticles

For water treatment, two features of nanoparticles are desired: (1) the particles should offer a high sorption capacity; (2) the spent particles should be amenable to easy separation from water and must not cause any harmful effects on the treated water. To this end, the desired form of magnetite would be a water-based suspension, in which magnetite nanoparticles are present as interbridged flocs as described by An et al. [16]. The bridged nanoparticles offer the advantages of high As sorption capacity and easy separation after the desired use. Alternatively, fully stabilized magnetite nanoparticles may also be used for water treatment. However, an external magnetic field will need to be used to separate the spent nanoparticles [18]. For in situ As immobilization, the prerequisite is that the nanoparticles must be deliverable in the soil. Therefore, fully stabilized nanoparticles are required.

Two main schemes exist for preparing magnetite nanoparticles, and they involve (a) decomposition of organic iron precursors at high temperatures and (b) aqueous-phase coprecipitation of ferrous–ferric ions at elevated pH. In the organic decomposition scheme, for instance, Yean et al. demonstrated that at 320 °C, the reaction of FeO(OH) in oleic acid and 1-octadecene resulted in the formation of magnetite nanoparticles with an average size of 11.72 nm [13]. The main drawback of this approach is the use of organic solvents and the high energy input. In contrast, the aqueous coprecipitation scheme is rather straightforward. It is conducted at room temperature and requires only pH adjustment in the aqueous phase, without the involvement of any organic solvents. Yet, without a stabilizer, this method fails to control the particle size and aggregation of the resulting nanoparticles [19].

An et al. [16] and Liang et al. [17] developed the following modified scheme for preparing polysaccharide-bridged or fully stabilized magnetite nanoparticles:

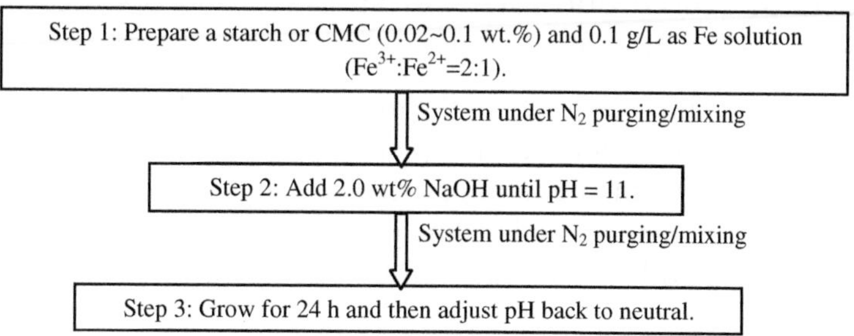

Step 1: Prepare a starch or CMC (0.02~0.1 wt.%) and 0.1 g/L as Fe solution (Fe^{3+}:Fe^{2+}=2:1).
System under N_2 purging/mixing
Step 2: Add 2.0 wt% NaOH until pH = 11.
System under N_2 purging/mixing
Step 3: Grow for 24 h and then adjust pH back to neutral.

As rapid precipitation of $Fe(OH)_2$ occurs after the addition of sodium hydroxide solution to the mixture of Fe^{3+}–Fe^{2+}, the magnetite formation begins with the oxidation of $[Fe(OH)]^+$ (the dissolved form of $Fe(OH)_2$ (solid) in water). Subsequently, the oxidation gives the intermediate product $[Fe_2(OH)_3]^{+3}{}_{(aq)}$, which can combine with another $[Fe(OH)]^+$ to form $[Fe_3O(OH)_4]^{2+}{}_{(aq)}$ that has the same Fe^{2+}/Fe^{3+} ratio as magnetite has. Equations (12.1)–(12.3) depict the key chemical reactions involved [20]:

$$Fe(OH)_{2(solid)} \leftrightarrow [Fe(OH)]^+_{(aq)} + OH^- \tag{12.1}$$

$$2[Fe(OH)]^+_{(aq)} + \frac{1}{2}O_2 + H_2O \rightarrow [Fe_2(OH)_3]^{+3}_{(aq)} + OH^- \tag{12.2}$$

$$[Fe_2(OH)_3]^{+3}_{(aq)} + [Fe(OH)]^+_{(aq)} + 2OH^- \rightarrow [Fe_3O(OH)_4]^{2+}_{(aq)} + H_2O \tag{12.3}$$

At highly oxidizing environments or low pHs, $[Fe_3O(OH)_4]^{2+}{}_{(aq)}$ will further oxidize to goethite or ferric hydroxides. However, at low amounts of dissolved oxygen and high pH, $[Fe_3O(OH)_4]^{2+}{}_{(aq)}$ will nucleate and/or crystallize to form magnetite particles:

$$Fe_3O(OH)^{2+}_{4(aq)} + 2OH^- \rightarrow Fe_3O_{4(solid)} + 3H_2O \tag{12.4}$$

In the absence of a stabilizer, the magnetite particles will agglomerate into micron-scale or larger aggregates that will settle out of the aqueous phase under the influence of gravity. However, in the presence of an effective stabilizer such as starch or CMC, the growth rate and extent, and thus, the size, of the nanoparticles can be controlled [14,21,22]. Researchers at the Auburn University [16,17,22] also demonstrated that the particle size and surface chemistry can be manipulated by means of starch and/or CMC stabilizers at various concentrations and of various molecular weights and degrees of substitution.

Figure 12-1 shows the transmission electron microscopy (TEM) images of bridged and stabilized magnetite nanoparticles. Figure 12-1a shows that the bridged particles are interconnected, yet they remain identifiable individual nanoparticles, that is, the particles did not aggregate into larger solid particles because of the coating of starch on the nanoparticles. Figure 12-2 illustrates how stabilizers affect the stability and morphology of the resulting nanoparticles, and demonstrates that starch can serve as a bridging or flocculating agent at lower concentrations and can also act as a stabilizer at elevated concentrations.

12.2.2 Effects of Starch and CMC on Stability and Surface Properties of Magnetite Nanoparticles

Polysaccharides (starch and cellulose) are the most abundant biopolymers and represent the greatest components of biomass on the planet. The starch and CMC stabilizers used in the preparation of the magnetite nanoparticles are slightly modified polysaccharides, of which starch is a linear, neutral, and gelling polymer, whereas CMC is a linear, anionic, and nongelling polymer with a pK_a value of 4.3 [16]. Because of the difference in the

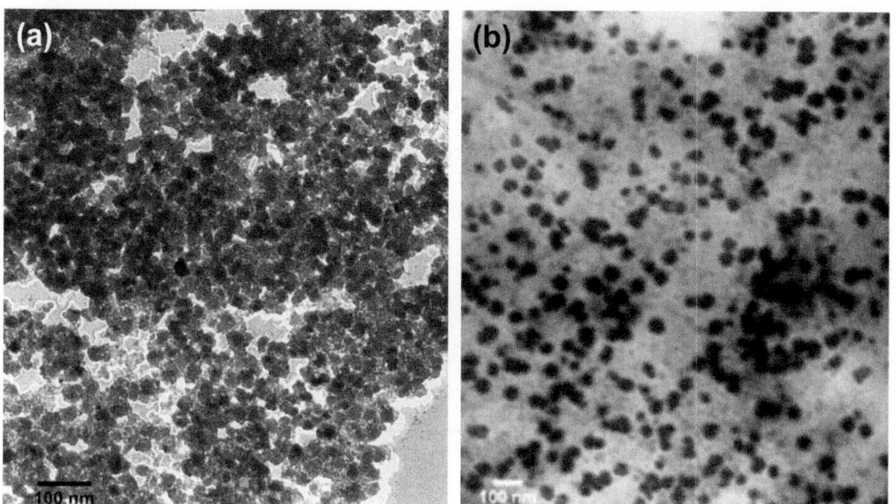

FIGURE 12-1 (a) TEM images of starch-bridged magnetite nanoparticles (26.6 ± 4.8 nm, mean ± standard deviation) prepared at a concentration of 0.057 g/L of Fe in the presence of 0.049 wt% starch and (b) TEM images of starch-stabilized magnetite nanoparticles (75 ± 17 nm) prepared at a concentration of 0.1 g/L of Fe with 0.1 wt% starch [15,16].

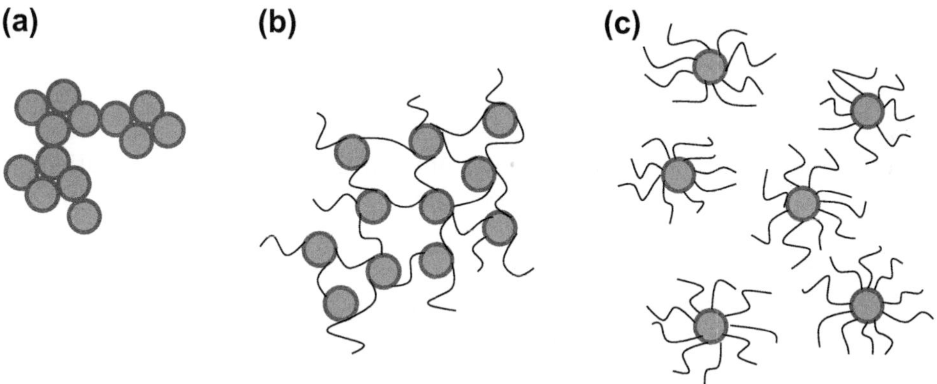

FIGURE 12-2 A conceptualized illustration of the stabilizer effect on particle interactions in aqueous solution: (a) Particles aggregate without a stabilizer; (b) particles are bridged and form loosely packed flocculates in the presence of appropriate concentrations of a polymer stabilizer; and (c) particles remain stable when the surface-attached stabilizer concentration is high enough [15]. (For color version of this figure, the reader is referred to the online version of this book.)

functionality of the two polysaccharides, their stabilizing effectiveness and the surface properties of the resultant nanoparticles are expected to differ.

CMC is a polysaccharide that is modified from cellulose by replacing the native CH_2OH group in the glucose unit with a negatively charged carboxymethyl group. The

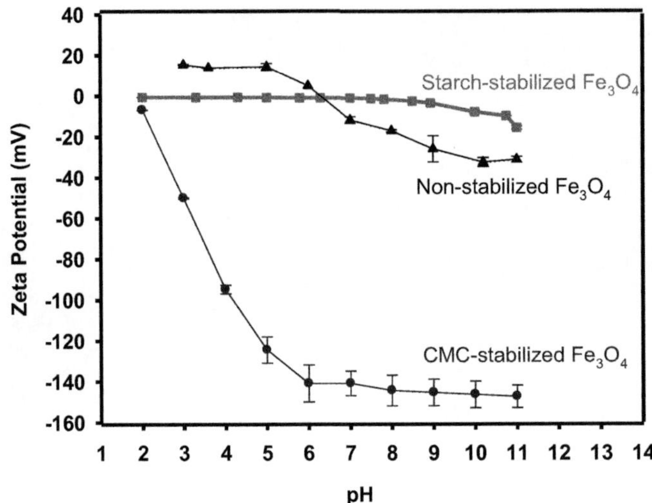

FIGURE 12-3 The ζ potential for bare and 0.1 wt% starch- or CMC-stabilized magnetite nanoparticles at various pH levels of the solution. All particle suspensions were prepared at a concentration of 0.1 g/L (as Fe) [16]. (For color version of this figure, the reader is referred to the online version of this book.)

addition of carboxymethyl groups greatly enhances the water solubility and interaction with metals such as Fe on magnetite surfaces. In contrast, starch is a neutral polymer, and the solubility and interaction with the nanoparticles are expected to be weaker than that for CMC. Liang et al. [17] showed that CMC is a much more powerful stabilizer than starch is. He and Zhao [21,22] indicated that CMC stabilizes nanoparticles through electrosteric repulsions, whereas starch works only through steric hindrance arising from the osmotic force when the individual starch molecular domains overlap as the starch-coated nanoparticles collide. Liang et al. [17] showed that the minimum starch concentration needed to fully stabilize 0.1 g/L as Fe of magnetite nanoparticles is 0.04 wt% of starch because complete particle stabilization was realized at a starch concentration of 0.04 wt%, compared to only 0.005 wt% when CMC is used.

Figure 12-3 compares the zeta (ζ) potential of bare and starch or CMC-stabilized magnetite nanoparticles. The ζ potential is an important parameter that governs the interparticle electrostatic interactions and stability of nanoparticles in water. It can also strongly affect the adsorption characteristics of ionic species such as arsenate. Generally, a high ζ potential (the absolute value of ζ is >60 mV) indicates strong electrostatic stabilization [23]. Figure 12-3 shows that due to the starch coating, a nearly neutral surface was evident over a pH range of 2–9 for starch-stabilized particles. In contrast, CMC-stabilized nanoparticles displayed a much more negative surface with a ζ value ranging from −120 to −150 mV at pH above the pK_a value of CMC. The pH of the point of zero charge (PZC) is estimated to be <pH 2 for CMC-stabilized particles. For bare magnetite particles, the ζ potential shifted from +15 mV at pH 3 to −31 mV at pH 11,

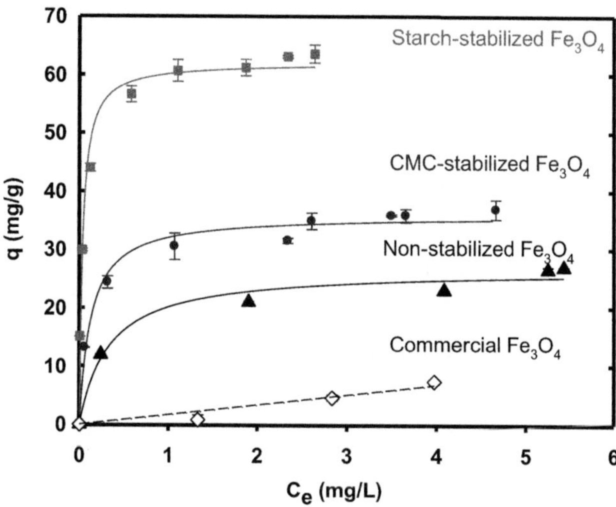

FIGURE 12-4 Arsenate sorption isotherms for various types of magnetite particles. Magnetite dosage = 0.1 g/L as Fe, starch, or CMC = 0.1 wt% for stabilized particles; initial As = 0.03–8.24 mg/L; final pH = 6.8 ± 0.4; equilibration time = 144 h. Data are given as the mean of duplicates, and errors refer to deviation from the mean [16]. (For color version of this figure, the reader is referred to the online version of this book.)

resulting in a pH_{pzc} value of 6.1. The PZC value for the stabilized nanoparticles differed significantly from that for bare magnetite particles, for example, 7.9–8.0 obtained by Illés and Tombácz [24] or 6.3–6.8 reported by Yean et al. [13] and Marmier et al. [25]. Based on Figure 12-3, it is expected that the nearly neutral starch-stabilized nanoparticles would offer a more favorable surface condition for arsenate uptake than the highly negatively charged CMC-stabilized counterparts.

12.2.3 Effects of Stabilizer on As Uptake

The simple surface modification can substantially enhance arsenate adsorption capacity. Figure 12-4 compares arsenate sorption isotherms at an equilibrium pH of 6.8 ± 0.4 for bare and stabilized nanoparticles. The classic Langmuir isotherm model was used to interpret the equilibrium data:

$$q = \frac{b * Q * C_e}{1 + b * C_e} \tag{12.5}$$

where b is the Langmuir affinity coefficient related to sorption energy (L/mg), C_e is the aqueous-phase concentration of As (mg/L), q is the solid-phase concentration of As (mg/g), and Q (mg/g) is the Langmuir capacity.

The starch-stabilized nanoparticles displayed a very favorable, nearly rectangular isotherm with a Q value of 62.1 mg/g, which is 72% greater than that for the CMC-stabilized counterparts. Both stabilized nanoparticles offered a much greater sorption capacity than did the bare magnetite (Q = 26.8 mg/g). The commercial magnetite

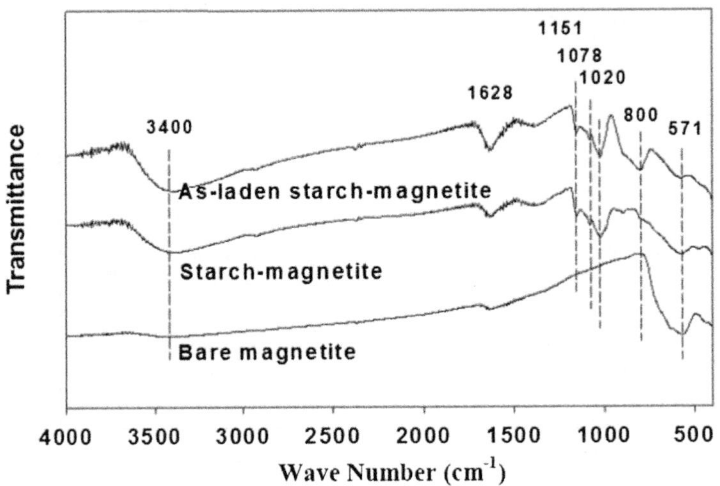

FIGURE 12-5 FTIR spectra of bare and starch-bridged magnetite nanoparticles with or without arsenate [16].

powder offered an even lower As sorption capacity than the laboratory-prepared bare magnetite.

Yean et al. [13] synthesized and tested a class of magnetite nanoparticles (11.72 nm) and observed a comparable Langmuir Q of 64 mg-As/g-Fe at pH $=$ 8 at As(V) concentrations $<$3.7 mg/L. However, compared to the particle preparation method, starch-stabilized magnetite nanoparticles are not only much more easy to prepare but are also likely to be more cost effective and more environmentally friendly as the technique eliminates the need for organic solvents and heating.

An et al. [16] investigated the nature of the chemical bonding between magnetite, starch, and arsenate through Fourier transform infrared (FTIR) spectroscopy. Figure 12-5 shows the FTIR spectra for bare magnetite particles and starch-bridged samples before and after sorption of arsenate. The absorption band at 571 per cm was obtained for all three types of magnetite samples tested, and it can be assigned to the Fe–O stretching vibration for the Fe_3O_4 particles [26]. The presence of starch coating reduced the absorption of the radiation by the core magnetite and, thus, weakened the peak intensity for the two starch-coated samples. On comparing the spectra of the starch-bridged magnetite before and after arsenate sorption, we clearly observed a new broad band between 750 and 850 per cm for the arsenate-laden magnetite, which was due to the Fe–O–As complexes. This is consistent with the results obtained by Pena et al. [27] who observed an FTIR band at 808 per cm for As(V) sorbed on TiO_2 and ascribed the band to the Ti–O–As groups. Pena et al. [27] compared FTIR bands for sorbed and dissolved arsenate species, and they noted a marked shift in the band positions of arsenate on sorption. For dissolved $H_2AsO_4^-$, two peaks were observed at 878 and 909 per cm corresponding to the symmetric and asymmetric stretching vibrations of As–O bonds. On adsorption, the peaks were shifted to 808 and 830 per cm, respectively. The shifts in band

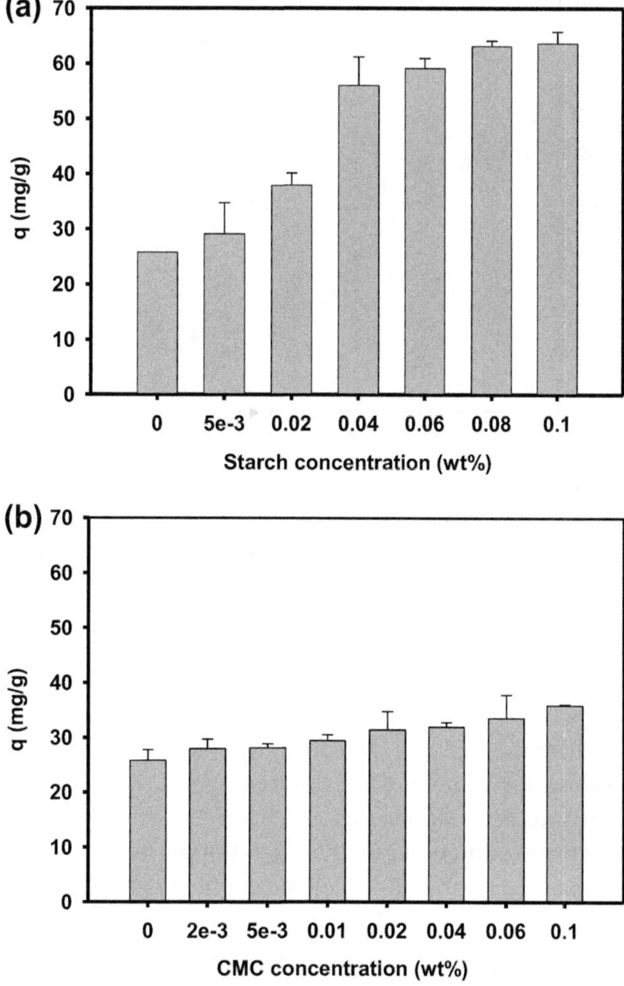

FIGURE 12-6 Effect of starch (a) or CMC (b) concentration on As(V) uptake (q). The nanoparticle dosage = 0.1 g/L as Fe; initial As(V) = 8.24 mg/L; equilibrium pH = 6.8 ± 0.4; equilibration time = 144 h. Data are given as the mean of duplicates, and errors refer to the standard deviation from the mean [16].

positions were attributed to symmetry reduction resulting from inner-sphere complex formation (i.e. the formation Fe–O–As complexes) [27,28]. Goldberg and Johnston [28] also reported that for arsenate sorbed on amorphous iron oxide, there existed two distinct bands corresponding to surface complexed and non-surface-complexed As–O groups, respectively. For the case of starch-bridged nanoparticles, only one single band was observed, and the wave number was shifted to approximately 800 per cm, indicating that surface complexation was the predominant mechanism for arsenate sorption to the starch-bridged nanoparticles. In addition, Zhang et al. [29] reported that the rise in

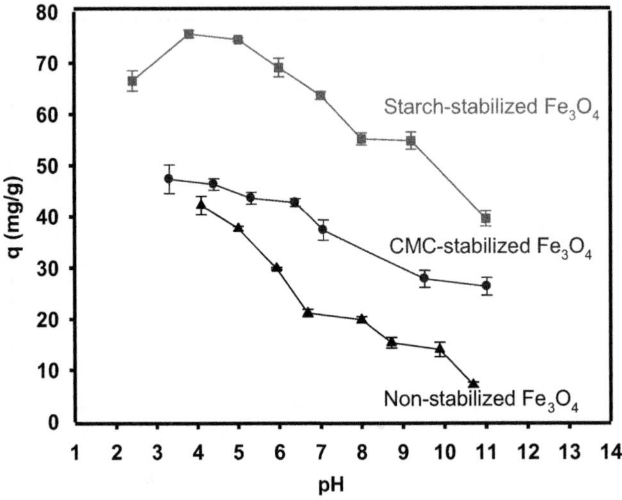

FIGURE 12-7 Equilibrium arsenate uptake as a function of the pH of the final solution. Experimental conditions: initial As = 8.0 mg/L; nanoparticles = 0.1 g/L (as Fe); CMC or starch (for stabilized nanoparticles) = 0.1 wt%; equilibration time = 144 h. Data given as the mean of duplicates, and errors refer to deviation from the mean [16]. (For color version of this figure, the reader is referred to the online version of this book.)

M–As–O (M: metal) peak was coupled with weakening of the M–OH peak when As(V) was adsorbed onto Fe–Ce oxide, and the researchers ascribed this phenomenon to the ion exchange reaction between –OH and arsenate anions. However, no M–OH bond was evident for the starch-bridged magnetite, which suggests that the sorption of As(V) did not necessarily conform to the standard ion exchange stoichiometry.

As the stabilizer concentration can greatly affect the particle stability and morphology, Liang et al. [17] also observed strong effects of the stabilizer concentration on the arsenate uptake capacity, especially when starch was used. Figure 12-6a shows the equilibrium uptake of As(V) with magnetite particles prepared in the presence of various starch concentrations. Overall, the As uptake increased with increasing stabilizer concentrations or with decreasing particle sizes. In accord with the observed physical stability, when the nanoparticles are fully stabilized (starch = 0.04 wt% or higher), a sorption plateau was evident. Compared to bare magnetite particles, the 0.04 wt% starch-stabilized particles offered a 2.2 times greater As(V) uptake. However, when the stabilizer concentration was further increased from 0.04 to 0.1 wt%, the capacity gain was only 14%. The capacity enhancement for stabilized magnetite particles can be attributed to the smaller size and, thus, to a greater specific surface area for more stable particles. However, a denser layer of starch results in elevated mass transfer resistance and increases the energy barrier for some of the sorption sites, and thermodynamically, the presorbed starch molecules diminish the effective sorption sites of the particles.

Figure 12-6b shows that CMC stabilization does not offer as much capacity enhancement for As(V) removal. At the experimental pH of 6.8, both $H_2AsO_4^-$ and $HAsO_4^{2-}$

are predominant arsenate species. As CMC-stabilized magnetite particles carry a highly negative surface, sorption of these anions would have to overcome a substantial energy barrier due to the electrostatic repulsion between like charges. Consequently, even though CMC stabilization gives rise to a smaller particle size and greater specific surface area, the arsenate uptake by CMC-stabilized magnetite was much lower than that by the starch-stabilized counterparts.

12.2.4 Effect of pH on Arsenate Sorption

The solution pH can affect the surface potential of the nanoparticles (as shown in Figure 12-3) and the speciation of arsenate in the aqueous phase, both of which can affect arsenate sorption. Figure 12-7 shows arsenate uptake as a function of solution pH for the bare, starch-, and CMC-stabilized magnetite particles. Starch-stabilized nanoparticles displayed a much greater As sorption capacity throughout the pH range tested. In general, the lower the pH, the greater the sorption capacity. This is attributed to the fact that the surface of the magnetite particles is less negatively charged at lower pHs, and the surface is positively charged at pH $<pH_{PZC}$ (Figure 12-3). Although the ζ potential for starch-stabilized particles seems to be nearly the same at $<$pH 7 (Figure 12-3) due to the starch coating, the more positive core particle surface becomes more favorable for arsenate anions at lower pH values. At an extremely low pH of 2, however, the As uptake was lowered primarily because of partial dissolution of the nanoparticles and partial decomposition of starch [17]. Arsenate uptake diminished with increasing pH due to elevated electrostatic repulsion and more fierce competition of hydroxyl anions.

12.2.5 Effects of Salinity and Organic Matter on Arsenate Sorption

One of the potential uses of the magnetite nanoparticles is to remove As from spent ion exchange brine [16]. To this end, the nanoparticles should be able to tolerate high salinity and competition from high concentrations of competing anions [16]. The presence of brine at concentrations as high as 10% did not show any appreciable suppression in arsenate sorption capacity. This observation violates the principle of selectivity reversal commonly cited in standard ion exchange regeneration [30], but this can be attributed to strong specific interactions such as strong inner-sphere complexation between As and the nanoparticles. To quantify the relative affinity, the arsenate and chloride binary separation factor $\alpha_{As/Cl}$ is calculated based on the equilibrium sorption data [30]:

$$\alpha_{As/Cl} = \frac{q_{As}C_{Cl}}{C_{As}q_{Cl}} \tag{12.6}$$

where q_{As} and q_{Cl} are the concentrations of arsenate and chloride, respectively, in the solid phase, whereas C_{As} and C_{Cl} are the concentrations of arsenate and sulfate in the solution, respectively. As a rule, a separation factor value of >1 indicates greater selectivity for arsenate than for chloride. The calculated $\alpha_{As/Cl}$ values were 102, 139, 147, and

290 at 2%, 4%, 6%, and 10% of NaCl, respectively, confirming the nanoparticles' strong affinity for arsenate over chloride.

The observed minimal salt effect contrasts with the results obtained with bare magnetite nanoparticles (20 nm) reported by Shipley et al. [31]. For bare particles, in the presence of high concentrations of cations, the electrostatic double layer is compressed, increasing particle aggregation and decreasing the available sorption sites. For starch-bridged magnetite particles, the particles were bridged, yet they were held apart due to the interparticle steric repulsion arising from the osmotic force of the starch on the particle surface, and increasing ionic strength did not affect the surface charge as the ionic strength exerts little effect on the steric repulsion.

Natural Organic Matter can affect particle stabilization and may compete with arsenate. Liang et al. [17] compared arsenate sorption isotherms for starch-stabilized magnetite nanoparticles in the presence of 0, 10 and 20 mg/L as TOC (Total Organic Carbon) of DOM (Dissolved Organic Matter). They observed that the presence of high concentrations of DOM can inhibit arsenate sorption capacity. Similar results were reported by others, who studied As uptake by amorphous iron oxides [31–34].

12.3 Potential Application for in situ Immobilization of As(V) in Soil and Groundwater

Although As has been detected widely in soil and groundwater, effective technologies have been lacking for in situ remediation of As contaminated soil and groundwater. It is even more challenging to remove As from groundwater in deep aquifers. Traditionally, As contaminated soils have been treated by capping, soil replacement, and solidification/stabilization and acid washing [35]. These methods are costly and only temporarily solve the contamination problem. The stabilized nanoparticles hold promise to develop an innovative in situ immobilization technology that may revolutionize conventional remediation practices. For example, the stabilized nanoparticles may be injected into deep aquifers to facilitate in situ adsorption and immobilization of As in groundwater. To this end, the nanoparticles must satisfy some of the key attributes, including the following: (1) The particles must be deliverable, that is, mobile under pressure so that they can be delivered into the contaminated source zones; (2) they must have a high As sorption capacity; and (3) the delivered nanoparticles will be immobile under the natural groundwater conditions, that is, once the external injection pressure is removed, the nanoparticles remain in position and serve as a permanent sink for As.

Liang et al. [36] studied breakthrough behaviors of starch-stabilized magnetite nanoparticles through 1-D column experiments. The results demonstrates that the water leachable As(V) from As(V)-laden sandy soil is greatly reduced on the nanoparticle treatment compared to simulated groundwater. Under various pore flow velocities ranging from 10^{-3} to 10^{-7} cm/s, the faster flow simulates the injecting flow condition, whereas the lower end velocity simulates the slow groundwater flow. The results indicate that the mobility of nanoparticles in soil could be manipulated by controlling the injection velocity. For example, at a magnitude of 10^{-3} cm/s, the nanoparticles are readily deliverable with the full breakthrough C_e/C_0 level remaining at >0.90. When we halved the pore velocity to 1.2×10^{-3} cm/s, the full breakthrough C_e/C_0 level was lowered to 0.80. When the flow rate was reduced to 10^{-7} cm/s, there was no breakthrough of the nanoparticles. Based on the breakthrough curves and by applying the classical filtration model, Liang et al. claimed that the maximum distance can be predicted as a function of pore flow velocity.

12.4 Conclusions

Stabilized magnetite nanoparticles have great potential for the enhanced removal of As from contaminated water and for soil remediation. Both starch and CMC can act as effective stabilizers in the preparation of highly stable magnetite nanoparticles of a much greater As sorption capacity than conventional iron oxide particles. The particle stability, degree of aggregation, and size may be controlled by manipulating the type and concentration of the stabilizer. The starch coating renders the ζ potential of particles nearly neutral over a broad pH range, whereas the use of CMC results in a highly negative surface. Consequently, CMC acts as a more effective stabilizer than starch does, whereas starch-stabilized magnetite particles offer a much greater arsenate uptake capacity. Acidic pH favors As(V) uptake, whereas As(III) sorption was observed over a wide pH range. Stabilization broadens the pH range of maximum sorption. Preliminary results indicated that stabilized magnetite nanoparticles can be used for in situ injection into contaminated soil, thereby facilitating the effective immobilization of As. Column breakthrough tests revealed that stabilized nanoparticles are highly transportable through a sandy soil under external pressure. Once the external pressure is removed, the nanoparticles remain immobile under natural conditions.

References

[1] Holm TR. J Am Water Works Assoc 2002;94.

[2] An B, Steinwinder TR, Zhao DY. Water Res 2005;39.

[3] N. R. C. (National Research Council). Arsenic in drinking water: 2001 update. Washington, DC: National Academy Press; 2001.

[4] E. P. A. Environmental Protection Agency. Technical fact sheet: final rule for arsenic in drinking water (EPA 815-F-00-016). Washington, DC: Office of Research and Development; 2001b.

[5] Frey M, Chwirka JD, Kommineni S, Chowdhury Z, Marasimhan R. Cost implications of a lower arsenic MCL. CO: Denver; 2000.

[6] Amini M, Abbaspour KC, Berg M, Winkel L, Hug SJ, Hoehn E, et al. Environ Sci Technol 2008;42.

[7] Jang M, Min SH, Kim TH, Park JK. Environ Sci Technol 2006;40:1636.

[8] Bissen M, Frimmel FH. Acta Hydrochim Hydrobiol 2003;31:97.

[9] Pierce ML, Moore CB. Water Res 1982;16:1247.

[10] Raven KP, Jain A, Loeppert RH. Environ Sci Technol 1998;32:344.

[11] Farquhar ML, Charnock JM, Livens FR, Vaughan DJ. Environ Sci Technol 2002;36:1757.

[12] Zhang GS, Qu JH, Liu HJ, Liu RP, Li GT. Environ Sci Technol 2007;41:4613.

[13] Yean S, Cong L, Yavuz CT, Mayo JT, Yu WW, Kan AT, et al. J Mater Res 2005;20:3255.

[14] He F, Zhao DY. Environ Sci Technol 2005;39:3314.

[15] Si S, Kotal A, Mandal TK, Giri S, Nakamura H, Kohara T. Chem Mater 2004;16.

[16] An B, Liang QQ, Zhao DY. Water Res 2011;45:1961.

[17] Liang QQ, Qian TW, Zhao DY, Freeland K, Feng YC. Ind Eng Chem Res 2012.

[18] Yavuz CT, Mayo JT, Yu WW, Prakash A, Falkner JC, Yean S, et al. Science 2006;314:964.

[19] Sun SH, Zeng H. J Am Chem Soc 2002;124:8204.

[20] Roonasi P, Holmgren A. A study on the mechanism of magnetite formation based on iron isotope fractionation. In: EPD Congress, TMS 2009 annual meeting and exhibition San Francisco, California; 2009.

[21] He F, Zhao DY, Liu JC, Roberts CB. Ind Eng Chem Res 2007;46:29.

[22] He F, Zhao DY. Environ Sci Technol 2007;41.

[23] Riddick TM. In: Control of colloid stability through zeta potential. With a closing chapter on its relationship to cardiovascular disease. New York: Livingston; 1968.

[24] Illes E, Tombacz E. Colloids Surf A Physicochem Eng Asp 2003;230.

[25] Marmier N, Delisee A, Fromage F. J Colloid Interface Sci 1999;211.

[26] Cornell RM, Schwertmann U. The iron oxides: structure, properties, reactions, occurrences, and uses. 2nd ed. Weinheim, Germany: Wiley-VCH; 2003.

[27] Pena M, Meng XG, Korfiatis GP, Jing CY. Environ Sci Technol 2006;40.

[28] Goldberg S, Johnston CT. J Colloid Interface Sci 2001;234:204.

[29] Zhang Y, Yang M, Dou XM, He H, Wang DS. Environ Sci Technol 2005;39.

[30] Clifford DS. Ion exchange and inorganic adsorption, water quality and treatment, a handbook of community water supplies, vol. 9.1–9.91. New York: McGraw Hill; 1999.

[31] Shipley HJ, Yean S, Kan AT, Tomson MB. Environ Toxicol Chem 2009;28:509.

[32] Redman AD, Macalady DL, Ahmann D. Environ Sci Technol 2002;36.

[33] Liu GJ, Zhang XR, Talley JW, Neal CR, Wang HY. Water Res 2008;42:2309.

[34] Grafe M, Eick MJ, Grossl PR, Saunders AM. J Environ Qual 2002;31.

[35] Tokunaga S, Hakuta T. Chemosphere 2002;46:31.

[36] Liang QQ, Zhao DY. Presentation at the 244th ACS National Meeting and Exposition. Philadelphia, 2012; August 19–23.

Transforming the Arsenic Crisis into an Economic Enterprise: Example from Indian Villages

Sudipta Sarkar*, Anirban Gupta[†], Arup K. SenGupta[‡]

*DEPARTMENT OF CIVIL ENGINEERING, INDIAN INSTITUTE OF TECHNOLOGY ROORKEE, ROORKEE, UTTARAKHAND, INDIA

[†]DEPARTMENT OF CIVIL ENGINEERING, BENGAL ENGINEERING AND SCIENCE UNIVERSITY, HOWRAH, WEST BENGAL, INDIA

[‡]DEPARTMENT OF CIVIL AND ENVIRONMENTAL ENGINEERING, LEHIGH UNIVERSITY, BETHLEHEM, PA, USA

CHAPTER OUTLINE

13.1 Introduction

During 1992–2012, arsenic contamination in the drinking water of millions of people living in different parts of the world has been the focus of attention of public health scientists and engineers [1–3]. Drinking of arsenic-contaminated water over a long period of time causes severe damages to the human body that culminate in various forms of cancers [4,5]. Apart from being fatal, health-related impairments are also known to cause a wide range of socioeconomic problems, especially in developing countries. The arsenic

Monitoring Water Quality, http://dx.doi.org/10.1016/B978-0-444-59395-5.00013-3

crisis prevailing over a large area of Bangladesh and India is arguably one of the worst calamities of the world in recent times [6–8]. The crisis is slowly unfolding in Southeast Asia also, affecting several other countries including Cambodia, Vietnam, Laos, and Myanmar [9,10].

The genesis of arsenic in groundwater is considered to be the geochemical leaching of arsenic from underground arsenic-bearing rocks caused by excessive extraction of the groundwater for agricultural purposes [11–14]. Surface water does not contain any arsenic, but its use for drinking purposes in many developing nations is restricted by the wide range of microbial contamination caused by the absence of proper sanitation practices. The use of groundwater in these regions is favored by its easy availability, microbial safety, and the absence of proper infrastructure for treatment and distribution of surface water. Although the best solution to the crisis is to switch over to treated surface water, development and maintenance of surface water-based drinking water systems are expensive, time consuming, and investment intensive. With all these difficulties, it is unlikely that a developing country would be able to switch the source of water from groundwater to surface water within a short period of time. To save lives before such a changeover is possible, it is imperative to build sustainable arsenic-removal systems on an urgent basis. During 1992–2012, several technologies have been developed for the removal of arsenic from contaminated drinking water. Many of them are capable of selective removal of arsenic from the contaminated water; however, only a few of them have gained wide-scale applications in the field [15]. Here, we present an example of one such arsenic-removal system that, apart from finding wide acceptance in the field, has also resulted in transforming the crisis into an economic enterprise. Between 1997 and 2010, >150 community-scale arsenic-removal units have been installed in the villages of West Bengal, a state of India, neighboring Bangladesh. The treatment units along with ancillaries and protocols were developed by Bengal Engineering and Science University, India, jointly with Lehigh University, USA. The project has been implemented by the Bengal Engineering and Science University with financial help mostly from Water For People, Denver, USA [16]. For all the treatment units, the villagers themselves take care of the related maintenance, upkeep, and management. Here, an account of the evolution of the treatment units is presented along with their performance for arsenic removal. This chapter also describes the extent and effect of community participation in this project and the socioeconomic changes it has brought in.

13.2 Arsenic Remediation—Choice of Scale of Operation

Success in field-level implementation of a treatment process developed in a laboratory depends on many factors. One such factor is the choice of the appropriate scale to which the technology is to be upgraded. For field-level implementation in the rural areas in developing countries, there may be two different types of applications: one is the point-of-use (PoU) type household treatment unit, and the other is a community-scale arsenic-removal device that can cater to a whole community or a group of neighboring villages. The household water treatment units have a modular design, portability, and ease of installation and use. They offer flexibility of operation and maintenance that can be dictated by the owner himself. One community-based water treatment unit, based on its size, can replace the installation of hundreds of individual household arsenic-removal units. Although the household treatment units offer several advantages over a community-based water treatment unit, for arsenic-removal purposes, the latter is chosen because of the following advantages:

- Analysis of treated water for arsenic at predetermined time intervals is necessary for any arsenic-treatment unit to ensure reliability in operation and quality control. Frequent measurement and detection of arsenic in treated water is difficult and cost prohibitive for an individual. For every single analysis of a water sample from a community-based unit, hundreds of water analyses would be required for individual household units, all other conditions remaining identical.
- Household PoU units that use adsorbent media or coagulants will always generate arsenic-laden sludge or solids. Coordination of collection and safe disposal of sludge or used media from individual households poses a level of complexity and requires enforcement effort that is difficult to sustain in remote villages. The wellhead type of community-based units often reduces such management problems.
- For an arsenic-removal unit that uses adsorbent media, replacement or regeneration of exhausted adsorbent can be cost prohibitive for a household unit user. But for a community-scale unit, collective effort from the users makes it affordable to replace or regenerate the exhausted adsorbent media.
- Even for household units, the villagers (mostly women) need to go every day to the existing wellhead units to collect water. Thus, a simple-to-operate arsenic-removal system mounted on the wellhead unit provides continued collective vigilance with regard to color (caused by iron), smell (because of biological activity), or other abnormal behavior that may otherwise go unnoticed for household units.
- Cost is one of the key parameters to be considered while choosing and adopting a technology by a community. According to our estimate, both fixed and operating costs of the community-based wellhead unit are significantly lower than the total sum of the same for the household units that it replaces. Also, it is relatively easy to introduce modifications and innovations in wellhead units for performance enhancement through the collective and direct involvement of the community through a water committee appointed by the villagers.

13.3 Design of the Treatment Unit

Traditionally, the sources of drinking water are groundwater wells. These wells, known as tube wells, were first introduced in India about 50 years ago and have hand pumps mounted on the top. These wells are popular for their ease of installation, maintenance, and use. According to a rough estimate, there are >10 million such tube wells in the arsenic-affected area of Bangladesh alone [6]. Over more than the last 50 years, villagers have been using such kinds of wells to meet their daily needs. Although microbiologically safe, many of these wells have been reported to contain arsenic in alarming concentrations. The immediate need was to develop a treatment unit that effectively removes arsenic from the contaminated drinking water. To qualify as an appropriate technology in a developing country, special considerations are to be made to conserve the traditional practices and values while ensuring that the local needs are effectively satisfied. Community-scale treatment units are installed as an attachment to the public tube wells that had alarming concentrations of arsenic. Figure 13-1A shows a photograph of the usual method of collection of water. Figure 13-1B shows a water treatment unit. It may be noted that even after the installation of the arsenic-removal devices, there have been no changes in the villagers' water collection practices, except for the fact that they now collect water from the exit of the treatment unit, instead of collecting water from the spout of the hand pump as practiced earlier. Thus, the technology does not call for

FIGURE 13-1 (A) Conventional method of water collection and (B) an arsenic-removal unit showing the method of water collection. (For color version of this figure, the reader is referred to the online version of this book.)

a change in the traditional ways of collection of drinking water, yet it delivers arsenic-safe and superior quality drinking water. Thus, this was immediately accepted by the end users.

Figure 13-2 shows a schematic of the treatment unit and describes the arsenic-removal mechanism. The arsenic-removal unit essentially consists of a stainless-steel column filled up with about 100 L of adsorbent. It may be observed from the schematic that the raw water enters from the top of the column, trickles down through the column, and treated water is collected at the bottom of the unit. The top part of the unit is deliberately kept open to the atmosphere through vent connections provided at the top of the unit. This arrangement serves two purposes: first, it allows the air to get dissolved in the droplets that are formed when the water is sprayed as fine droplets at the inlet of the column. The dissolved air helps oxidize the ferrous iron in the raw water to form fine precipitates of hydrated ferric oxide (HFO). Second, as the water sprayed is open to the atmosphere, hydraulically, the column acts like a gravity-based filter and not as an online filter where hydraulic transients can take place because of the irregular rate of manual pumping. The freshly formed HFO precipitates are responsible for the adsorption of a significant portion of the total arsenic present in the raw water. The HFO particles are trapped in the adsorbent bed with a size of 500–900 μ. The rest of the arsenic remaining in the water after the adsorption by HFO particles is removed by the adsorbent media.

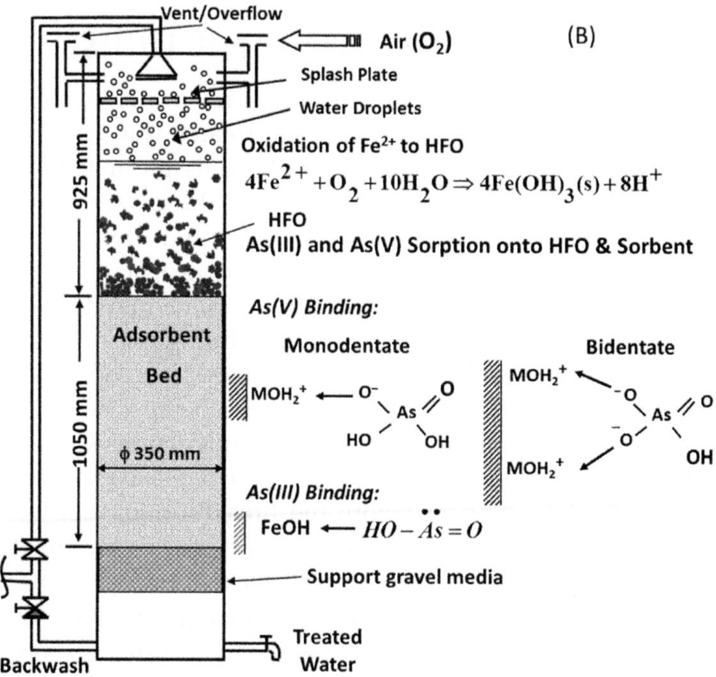

FIGURE 13-2 Schematic diagram of an arsenic-removal unit and salient reactions taking place inside the unit. (For color version of this figure, the reader is referred to the online version of this book.)

Traditionally, activated alumina, a locally available low-cost adsorbent, had been used. The more recent units (~15 in number) use hybrid anion exchange resins (HAIX) as adsorbent media [17].

The flow rate of treated water is 10–12 L/min. The operation of the unit is simple and user friendly. The flow rate of the treated water has been observed to be quite sufficient for the villagers. Because of the formation and precipitation of the HFO particles within the bed, the flow rate tends to get diminished over time. Backwashing of the column every other day to drive out the HFO particles is necessary to maintain a good flow rate through the column. The arsenic-laden HFO particles in the waste backwash are trapped on the top of a sand filter provided on the same premises.

The treatment unit is fabricated locally with materials available there. The units are robustly built and do not require regular maintenance. The adsorption column acts like a plug flow reactor. Therefore, it is forgiving toward any occasional spike in arsenic concentration in the raw water. Unlike batch-type reactors that are mostly used for coprecipitation methods of arsenic removal, these units consistently produce arsenic-safe water in a reliable manner. Reliability of operation and consistency in producing safe drinking water are necessary requirements, particularly in rural settings in a developing country where the frequent measurement and detection of arsenic in treated water is difficult and cost prohibitive.

13.4 Performance of the Treatment Units

Depending on the arsenic and iron concentrations in raw water, arsenic-removal units produce about 1,000,000 L or 10,000 bed volumes of treated water on an average before the concentration of arsenic in the treated water exceeds the maximum contaminant level (MCL) in India, which was set at 50 μg/L [18]. Once the arsenic concentration in the treated water exceeds the MCL, the exhausted adsorption medium is taken out for regeneration at a central regeneration facility located nearby. The operation of the arsenic-removal unit is resumed with another adsorption medium that has already been regenerated. Figure 13-3 represents the arsenic breakthrough history of one such treatment unit located at Habra, West Bengal, for over five cycles of operation. The unit has been in operation since 2001 and uses activated alumina as the adsorbent. The raw water had an average arsenic concentration of 140 μg/L. Although there was seasonal variation in the arsenic concentration in raw water, it may be observed that it consistently produced arsenic-safe drinking water before the breakthrough took place. It may be noted that in each cycle of operation subsequent to regeneration, there was an initial leakage of arsenic. Although, theoretically, it is possible to fully regenerate the adsorption media, using locally available chemicals, practically, the regeneration efficiency does not exceed 70%. The length of column runs have gradually shortened over the treatment cycles because of the loss of active sites of the adsorbent that is due to incompleteness of regeneration.

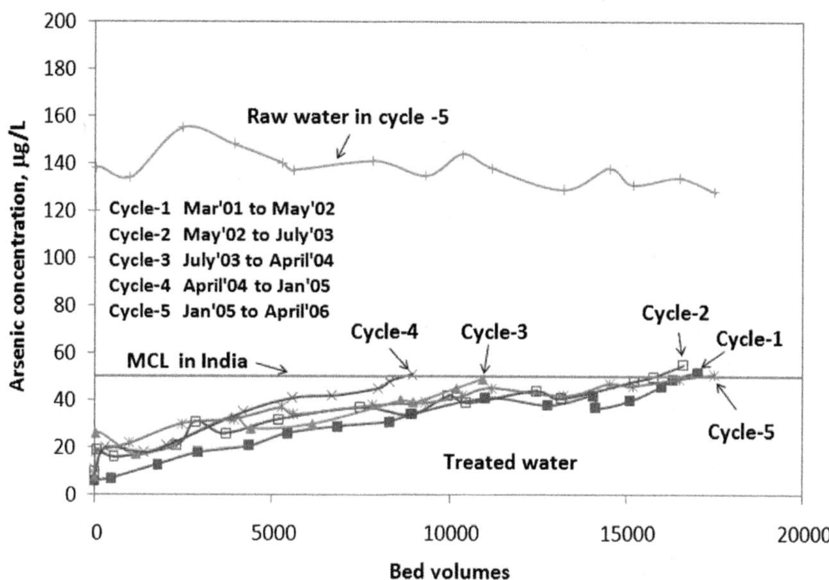

FIGURE 13-3 Arsenic breakthrough profile for the treatment unit at Deshbandhu Park, Habra, in West Bengal, India. (For color version of this figure, the reader is referred to the online version of this book.)

Because of their unique design, the arsenic-removal units are also capable of achieving a high degree of removal of dissolved iron present in the raw water. The raw water in this area has dissolved iron concentrations in the range of 2–10 mg/L. Iron is not a restricted element for safe drinking water, but it is esthetically displeasing and known to cause minor health problems. The units consistently produce water with an iron content of <0.5 mg/L. Figure 13-4 shows the iron breakthrough history of the same arsenic-removal unit.

Figure 13-5 provides the arsenic breakthrough history of another arsenic-removal unit, where an HAIX, developed in a Lehigh University laboratory, was used in place of activated alumina as the adsorbent. HAIX is known to be a better adsorbent compared to activated alumina. The length of the column run is much more compared to that of the units using activated alumina.

13.5 Regeneration of Exhausted Media

On exhaustion of the adsorption column, media from the unit are replaced by fresh or already-regenerated media. The exhausted medium is taken to a central facility where it is regenerated inside a stainless-steel batch reactor that can be manually rotated about its horizontal axis. Figure 13-6 shows the batch reactor being used for regeneration.

For regeneration of activated alumina, a solution containing 2% NaOH is used; the exhausted activated alumina is reacted with two bed volumes of regenerant solution in

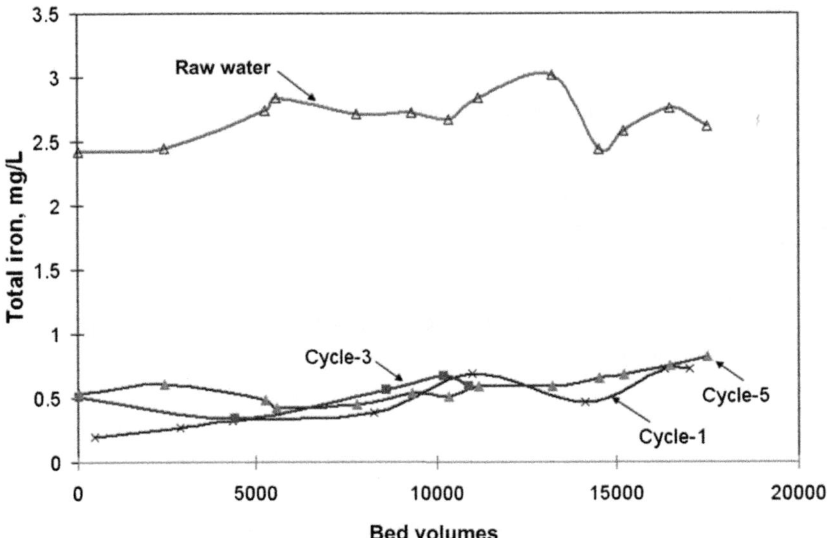

FIGURE 13-4 Iron breakthrough profiles for the arsenic-removal unit at Deshbandhu Park, Habra, in West Bengal, India. (For color version of this figure, the reader is referred to the online version of this book.)

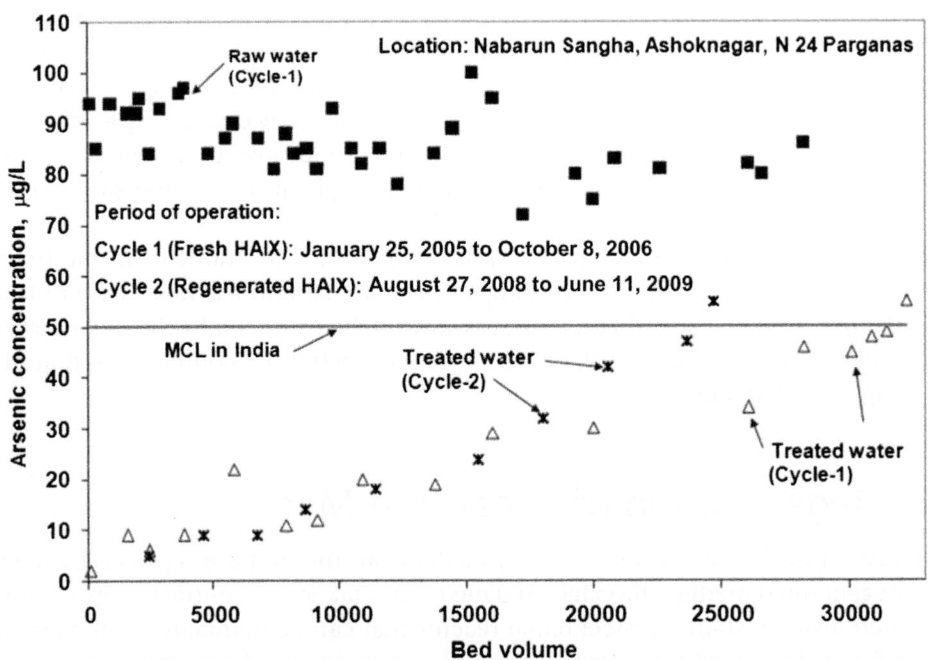

FIGURE 13-5 Arsenic breakthrough profile of a treatment unit using HAIX resin located at Nabarun Sangha, Ashoknagar, in West Bengal, India. (For color version of this figure, the reader is referred to the online version of this book.)

FIGURE 13-6 A regeneration reactor being operated at the central regeneration facility at N 24 Parganas, West Bengal. (For color version of this figure, the reader is referred to the online version of this book.)

the batch reactor for 45 min. The process is repeated again with a fresh regenerant solution. During regeneration, the pH is approximately 12.0, and the spent alkali is collected. After thoroughly rinsing with well water, the medium is subjected to two bed volumes of dilute HCl solution to neutralize and protonate the media so that the pH of the resultant solution is 5.5; subsequently, the spent acid is collected. The medium is then rinsed with well water. The general procedure for the regeneration of exhausted HAIX is similar, except that along with 2% NaOH, 2% NaCl is used in the regenerant. The regenerated medium is air dried and kept in a safe place for reuse. At the end of the regeneration, spent acid, alkali, and rinse water are mixed, and the pH is adjusted to 6.5 by adding 10% hydrochloric acid. A thick brown slurry forms immediately, and it settles down overnight before it is disposed of on top of a coarse sand filter kept at the same premises. Arsenic-laden solids and HFO particles are intercepted and retained at the top of the filter. The entire regeneration, including the spent regenerant treatment, is completed in 5 h and is performed by a group of trained villagers. The weight of the dried sludge obtained as treatment residual is never >2 kg. It may also be observed that the liquid waste remaining after the treatment is innocuous, and it contains only a neutral solution of sodium chloride with an insignificant concentration of arsenic (<50 μg/L) and some dissolved iron. Thus, the dried sludge having a mass of about 2 kg and containing a high concentration of arsenic in the solid phase is the treatment residual that needs proper disposal.

13.6 Regenerable Versus Throwaway Adsorbent

Economy in operation is the major reason behind the choice of an adsorbent that can be regenerated and reused over many cycles. This is especially true in developing countries where for a technology to be acceptable to the masses, it has to be easily affordable. A significant amount of savings in terms of the cost of treated water is possible when a regenerable adsorbent is used instead of a one-time-use or throwaway adsorbent. This is especially true in the developing countries where labor and chemicals, the two major resources required for regeneration, come at a low price. Moreover, as the regeneration procedure is simple and easy, even a group of trained villagers with little or no previous technical knowledge can effectively carry out the whole regeneration process. The cost of regeneration is just one-fourth of the price of freshly activated alumina. In the case of HAIX, the cost of regeneration is even less, compared to the price of fresh adsorbent.

Another important parameter is the management of treatment residuals. Treatment residuals are exceedingly rich in arsenic. Hence, they need to be suitably stored and disposed of so that any leaching of arsenic can be avoided. In the case of one-time use or throwaway adsorbents, the amount of the treatment residual generated is approximately 100 L, or about 100 kg. However, after regeneration, the treatment residual is obtained in the form of sludge containing mainly HFO and arsenic. The amount of the sludge is not >2 kg. It is easy to handle and manage this small amount of treatment residual, as opposed to 100 kg of treatment residual in the case of nonregenerable adsorbents. The cost of disposal of treatment residuals, if taken into account for the calculation of the overall cost of the project and the treated water, the one-time-use adsorbent will be an even worse choice. Moreover, in the countries where hazardous-waste regulations are not enforced properly, the mismanagement that is possible because of the handling of large amounts of treatment residuals would cause a menace that is comparable with the arsenic crisis in the groundwater itself, if not more. Thus, the use of regenerable adsorbent is a better choice for the project [19].

13.7 The Ultimate Fate of Arsenic: Ecological Sustainability

In a community-based wellhead arsenic-removal system, arsenic-laden wastes evolve from two separate processes: first, ferric hydroxide precipitates or HFO particulates that are formed because of the oxidation of ferrous iron to ferric species. These particles tend to clog the treatment unit and cause an excessive pressure drop, thereby reducing the flow rate. It is imperative to backwash the wellhead column once each day; Arsenic-laden HFO particles are collected on the top of a coarse sand filter located at the same premises. Second, the adsorbents used are regenerated periodically in the central regeneration facility. The spent regenerants obtained after treatment produce arsenic-laden solids. Chemically, these two wastes are similar, and both are rich in iron and in

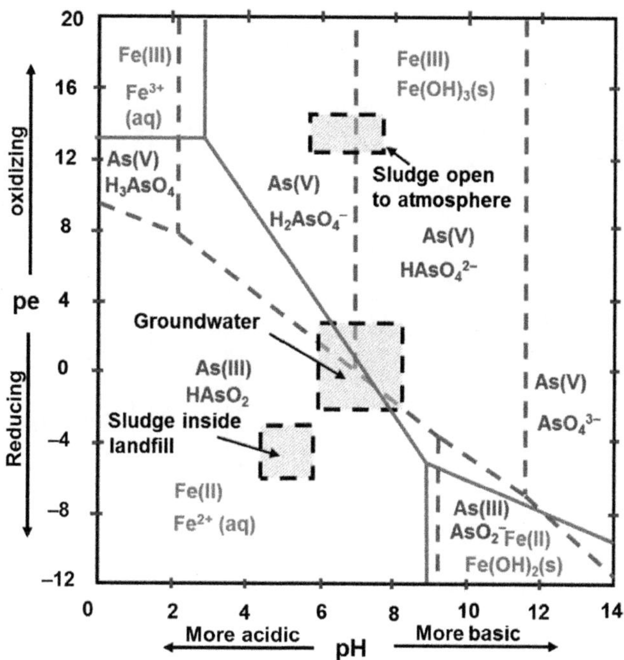

FIGURE 13-7 Predominance, or pe–pH, diagram for several arsenic and iron species. (For color version of this figure, the reader is referred to the online version of this book.)

arsenic. Local environmental laws/regulations with regard to the safe disposal of arsenic-containing sludge in remote villages in developing nations either do not exist or they are not enforceable. Thus, containing the arsenic removed from the groundwater with no adverse ecological impact and human health endangerment is as important as its removal to provide safe drinking water. Currently, in the developed western nations, arsenic-laden sludge or adsorbents are routinely disposed of in landfills. However, several recent investigations have revealed that leaching of arsenic is stimulated or enhanced in landfills or in hazardous-waste site environments [20–22]. Both pH and redox conditions uniquely determine speciation of arsenic and iron that, in turn, control arsenic leachability [23,24]. Figure 13-7 shows the composite predominance, or the pe–pH, diagram for various arsenic and iron species, using equilibrium relationships available in the literature [25,26]. The figure highlights (shaded rectangles) the three separate predominance zones of interest: neutralized HFO-laden sludge, groundwater, and landfill leachate. Note that Fe(III) and As(V) predominate in the aerated HFO-laden sludge, where Fe(III) is also insoluble. On the contrary, reduced Fe(II) and As(III) are practically the sole species present in the more-reducing landfill environments. Relatively high solubility of Fe(II) and the low sorption affinity of As(III) would always render the iron-laden sludge more susceptible to rapid leaching under the oxygen-starved

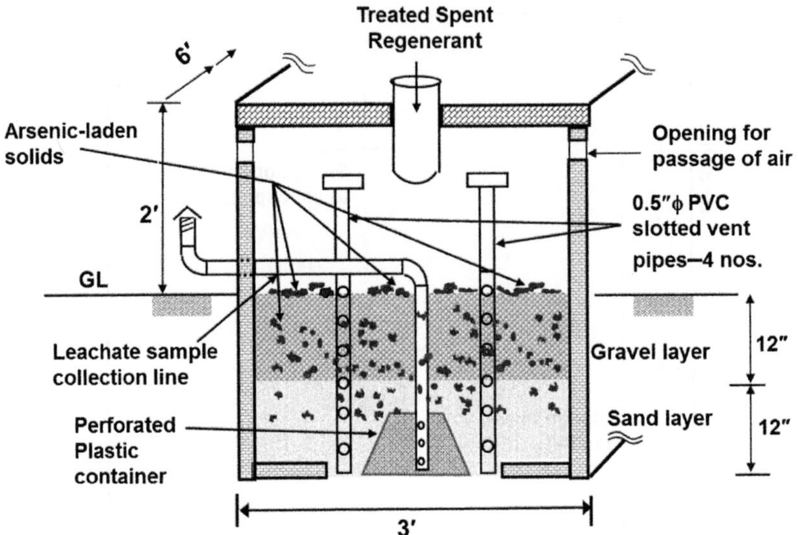

FIGURE 13-8 Schematic of an aerated coarse sand filter (GL, ground level). (For color version of this figure, the reader is referred to the online version of this book.)

environment of the landfill or underground waste site. This is why in a remote rural environment, arsenic-laden HFO sludge needs to be preserved in an aerated (i.e. oxidizing) environment to minimize arsenic leaching. Figure 13-8 provides a schematic of the aerated coarse sand filter.

The project not only involved arsenic removal from contaminated groundwater in the villages of West Bengal, India, but it also successfully considered the arsenic removed in low-volume sludge, ensured its safe storage and containment. Most projects surrounding the arsenic crisis have thus far ignored the issue of safe disposal of treatment residuals, which are obviously highly toxic and have the potential to affect the environment through possible leaching, thereby defeating the whole definition of sustainability. Figure 13-9 indicates the summary of the whole treatment process in terms of overall ecological sustainability.

13.8 Community Participation

Community participation has long been recognized as an effective means of helping rural and urban people focus their energy and mobilize resources to solve their health, environmental, and economic problems. Community participation at every stage of the installation, as well as operation of the treatment unit, was considered to be essential during the implementation of the project. Technology developed in a laboratory can become successful in the field when there is spontaneous and active participation of the end users in every aspect of the operation and upkeep of the unit.

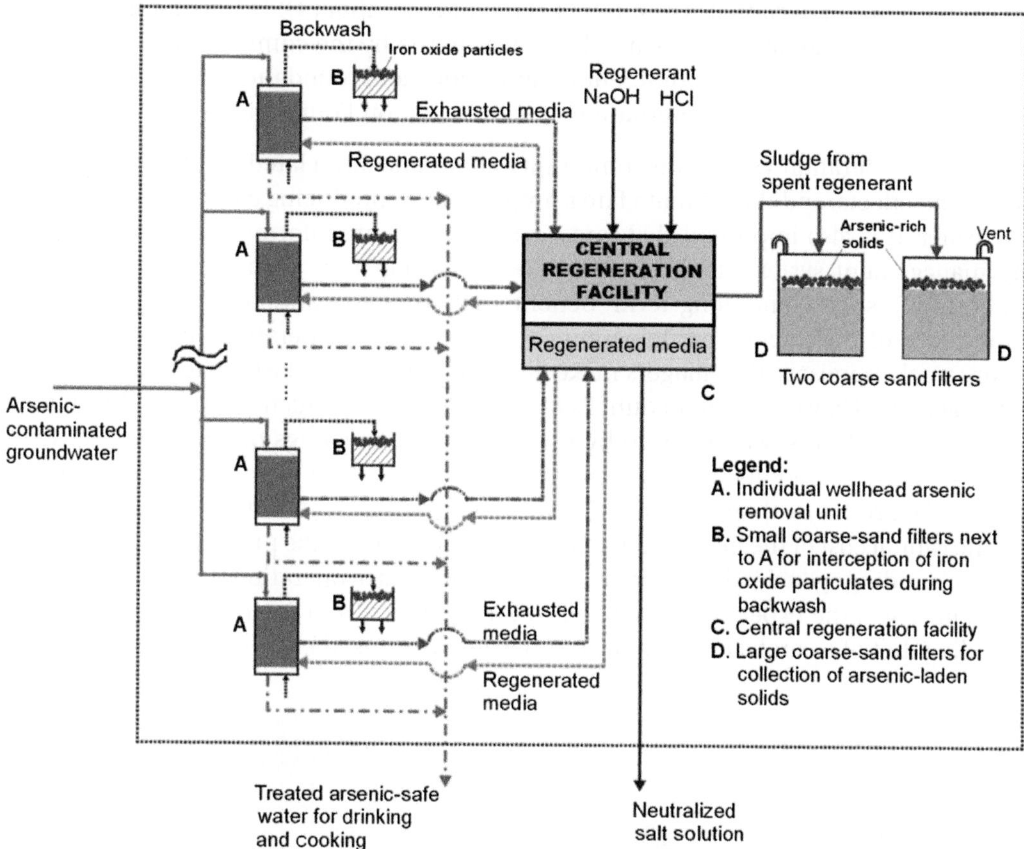

FIGURE 13-9 Schematic demonstrating the overall ecological sustainability of the arsenic-removal process. (For color version of this figure, the reader is referred to the online version of this book.)

The essential criterion for community participation is the acceptance of the technology by the community. Arsenic-removal units are highly acceptable to villagers for the following reasons:

- The units do not change the way the water is collected; they are culturally appropriate and take into account end-user lifestyles, preferences, etc.
- The cost of treated water produced is within the affordability limit of the villagers.
- The units are of a community level but are small enough to be owned, maintained, and operated by villagers.
- The units can be operated and maintained by using locally available resource material by only the villagers, with or without very little training.
- The design is simple and flexible. It can be adapted for use in various places and changing circumstances.

- A small amount of capital is required. Each unit costs about USD 1200 for installation. In most cases, no part of the capital cost was borne by the community, as initial funding was received from different organizations and foundations that included Water For People, Rotary International, and the Hilton Foundation.

Awareness generation within the community about arsenic-related health problems, its origin and effects, and the ultimate fate is important for producing a demand for arsenic-safe water. It is also important that target communities are informed of the technology, the management system, their duties, collective ownership, need for payment of the water tariff, short- and long-term benefits of arsenic-treatment systems, and the importance of their active participation for sustaining the operation. Through participatory rural appraisals, the villagers make decisions whether and where to install arsenic-removal units. They form water councils to manage the units. Such informal education to sensitize the villagers was achieved by means of group meetings organized by activists; paper pamphlets; and audiovisual communication, including movies and street-corner dramas. Figure 13-10 shows a street-corner drama being enacted in one of the villages.

The capital cost of the installation of the treatment units in the project was provided by external funding agencies; however, the subsequent costs of operation were not supported by them. Thus, the expenditures related to the operation and maintenance of the unit, including that of a caretaker, once-a-month water analysis, incidental maintenance, and

FIGURE 13-10 Photograph of a street-corner drama being enacted in a village in West Bengal, India. (For color version of this figure, the reader is referred to the online version of this book.)

once-a-year regeneration of the unit, are to be met by the villagers who benefit from the treated water. To attain long-term economic sustainability for the units, it was necessary to meet the cost of treated water by collection of water tariffs, a concept never practiced before in the villages of West Bengal and Bangladesh. The water tariff to be provided per family per month was only Rs. 15–20 (~30–40 US cents). The amount is low compared to that people spent for entertainment and even for cell phone use. Still, there was an initial resistance from the villagers to pay the water tariff. It was not easy to introduce this new concept of water tariff in the villages, where people used to get drinking water for free. However, through persistent communications, the villagers were motivated to understand that the water tariff was negligible, compared to the possible gain from the recovery of workdays that otherwise would have been lost because of illness. The villagers understood that owning and maintaining a community water treatment unit through collective payment in the form of water tariff was a good investment for the health and prosperity of present and future generations. This realization came after intense awareness-generation efforts were undertaken. Once started, the concept soon became popular. The return on the investment in the form of savings made on the cost of medicine could be immediately realized when the villagers switched over from contaminated water to treated water.

The installation process begins with a project staff visiting an affected village and screening the public wells for arsenic. The project group meets the villagers and discusses the possibility of installation of an arsenic-removal unit there. The villagers form a water committee that comprises local residents, with at least one-third of the members being women. They decide on the specific contaminated well where they want the arsenic-removal unit to be installed. After installation and initial monitoring for a few months, the ownership of the unit is transferred to the villagers. The duties of the members of the water committee are distributed, and they meet at least once a month to discuss the issues being faced. Normally, each committee appoints one or two local youths as caretakers who participate in daily backwashing, the operation, and upkeep of the unit. This water committee collects monthly water tariffs from the user's families, updates the tariff cards, pays the salaries of the caretakers, pays for monthly water tests, and deposits the money in a nearby bank for future expenditures. Figure 13-11 shows a bank deposit book and water tariff card used in two villages in West Bengal, India. The members of the water committees are elected every year from among the users. The water committee is responsible for reporting to the project office once a year about the qualities of water, experiences, difficulties, accumulated funds, etc. In the case of any emergency related to water quality, apart from contacting the local plumbers who are trained to undertake mechanical maintenance, the water committees can directly contact local representatives of the project team.

13.9 Toward Economic Prosperity

The project successfully showed that when people from the community organize, plan, share tasks with professionals, contribute financially to projects or programs, and help make decisions about activities that affect their lives, programs are more likely to achieve

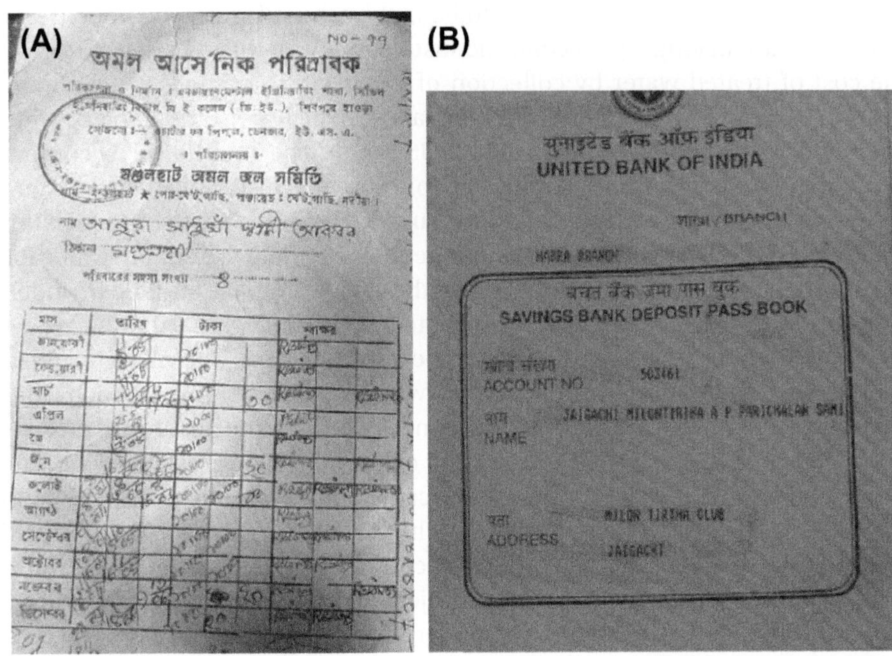

FIGURE 13-11 Photographs of (A) the water tariff card and (B) savings bank deposit book maintained by the village water councils in West Bengal, India. (For color version of this figure, the reader is referred to the online version of this book.)

their objectives. In this case, the people involved in the program have already shown the achievement of a successful community-managed arsenic-removal system, and now they are showing promise to take the development beyond the arsenic crisis. If guided properly, the communities can, in the long term, help promote practices of water-related hygiene and initiate sound sanitation practices.

The short-term benefits are well identified. The arsenic-related health hazards before becoming fatal affect the victim through various manifestations. The victims lose the ability to work and earn their livelihood, and they have to buy medicines. However, the first step to stop arsenic-related health hazards starts with switching over to arsenic-safe drinking water. Thus, provision of arsenic-safe drinking water not only saves lives directly, but it also saves families from being trodden down through the spirals of poverty. Moreover, the socioeconomic institution formed for the upkeep, operation, and maintenance of the arsenic-removal units provides direct and indirect employment. Jobs are created in every step of the system, starting from the fabrication of the arsenic-removal units and the manufacture of adsorbents. Installation of every unit in the village requires at least 10 labor workdays. Each unit employs one or two persons, on a part-time basis, for its operation and maintenance. Also, a group of six trained villagers work as entrepreneurs at the central regeneration facility. They earn their livelihoods from regeneration activities only. Apart from these direct employments, a host of secondary employment opportunities are

FIGURE 13-12 A water vendor carrying arsenic-safe water to the households in West Bengal, India. (For color version of this figure, the reader is referred to the online version of this book.)

generated for water vendors who earn their livelihood by transporting the treated water to householder who live at a far-off place or are unable to physically come to collect the treated water. In our estimate, each treatment unit has more than one water vendor associated with it. Figure 13-12 shows one such vendor associated with the treatment unit.

One example is the arsenic-treatment unit located in Nabarun Sangha, in the small town of Ashoknagar, in the North 24 Parganas district in West Bengal, India. There has been a continuous increase in the number of user families ever since it's commissioning in 2004. The number of user families ultimately reached 400 in 2009. The revenue earned each year has continuously increased over time from 2004 to 2009, as is evident from the curve shown in Figure 13-13A. In 2010, a second treatment unit was installed at the location. Although the first unit was funded by Water For People, the users contributed to the cost of the second treatment unit from the revenue that accumulated from the water tariffs. The number of users jumped to 450 families immediately after the second unit was installed. Figure 13-13B and C show the unit in 2004 and 2009, respectively. The improvements are evident. The users built an overhead storage tank, changed the operation from manual to electricity-driven pumps, and provided an enclosure and shed for all-weather operation. All these improvements were possible with the help of the revenue generated from the water tariffs.

The crisis is huge and so is the potential for development of widespread business activities related to the manufacturing of the treatment units, spare parts, and adsorbents. As the demand grows for arsenic-removal units, there is a possibility of widespread use of the adsorbents. To enjoy the advantage of locally manufactured adsorbents over the imported

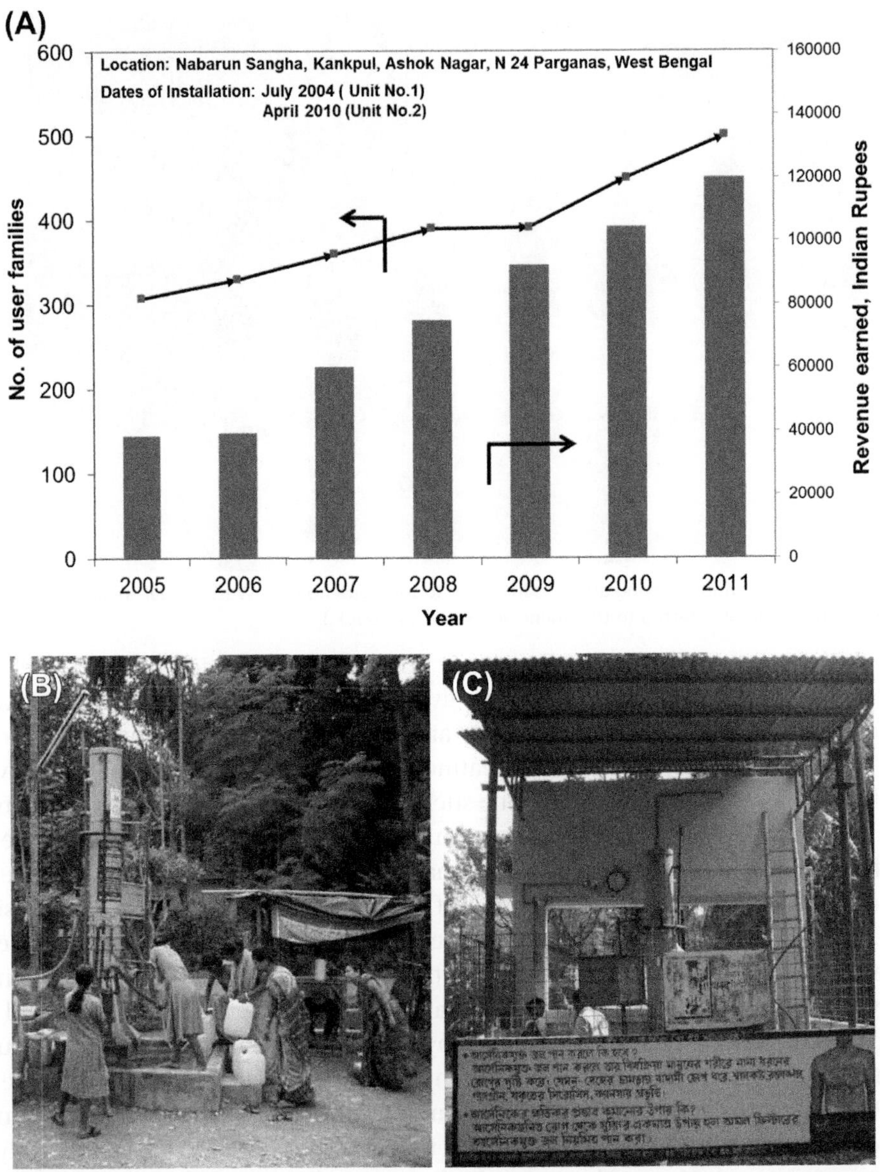

FIGURE 13-13 (A) Revenue earned from the collection of water tariff at the arsenic-removal unit in Ashoknagar, West Bengal, India. An arsenic-removal unit installed at Nabarun Sangha, Ashoknagar, in West Bengal, India, in 2004 (B) and also in 2009 (C). (For color version of this figure, the reader is referred to the online version of this book.)

ones, many of the arsenic-selective adsorbents are expected to be locally manufactured in the affected country itself. We have already observed such efforts being increasingly made in the local markets. If widespread, collectively, such efforts shall ultimately have a positive impact on the local economy and will create further new jobs in the region.

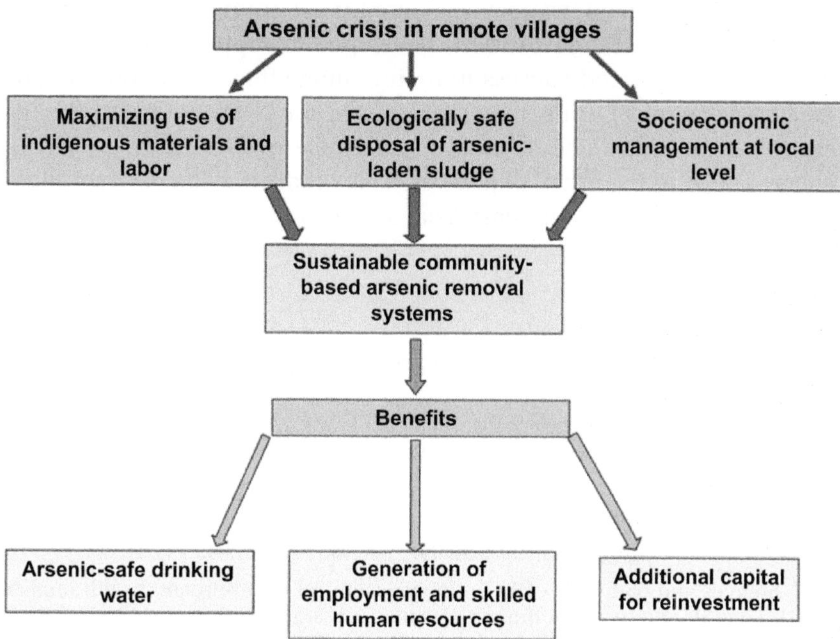

FIGURE 13-14 Overall business model of the arsenic-removal project, showing socioeconomic sustainability. (For color version of this figure, the reader is referred to the online version of this book.)

Thus, the small capital and technical know-how, when properly infused within the crisis-ridden areas, went a long way in benefiting arsenic-affected villagers who are also stakeholders of each unit. The entire system involving the treatment of arsenic-contaminated water has evolved as a viable business model where everybody enjoyed benefits. Figure 13-14 summarizes the relationship among the input conditions of an arsenic-affected community, development of technology, and the benefits accrued through the community-based project. The social institution acts like an economic enterprise owned indirectly by the villagers. The excess of the water tariff, saved after the expenditures, is viewed as small savings or a social capital that the villagers can use to bring about new improvements related to the water and hygiene-related infrastructure in their relative villages. There are even places where the users' committee has started to make use of the accumulated fund as the seed capital for microfinancing the village's small businesses. As these kinds of activities grow over time, the economic enterprise that grew out of the arsenic crisis will gain more momentum so that it can fuel bigger changes in society.

13.10 Conclusions

Like all other crises, the arsenic crisis also has an opportunity for newer development embedded into it. Once such opportunities are properly identified and carefully explored in a community-based mitigation program, not only does the mitigation program become a success, but it also opens up a new door for economic prosperity for the society as a whole. The socioeconomic institutions developed to support the arsenic mitigation program in India demonstrated success in maintaining effective operation and maintenance of the arsenic-removal units; they also showed the potential for further economic development for the society. The economic enterprise evolved may go a long way in achieving success beyond the initial objectives. This model may be used for mitigation of the arsenic crisis in South and Southeast Asia, and also in Latin American countries.

References

[1] Ng JC, Wang J, Shraim A. Global health problem caused by arsenic from natural sources. Chemosphere 2003;52:1353–9.

[2] Bhattacharya P, Welch AH, Stollenwerk KG, McLaughlin MJ, Bundschuh J, Panaullah G. Arsenic in the environment: biology and chemistry. Sci Total Environ 2007;379(2–3):109–20.

[3] Bundschuh J, editor. Natural arsenic in groundwaters of Latin America. Mexico City: Proc. Int. Cong. natural arsenic in groundwaters of Latin America; 2008. 20-24 June 2006, Taylor and Francis.

[4] Ratnaike RN. Acute and chronic arsenic toxicity. Postgrad Med J 2003;79:391–6.

[5] Thornton I. Sources and pathways of As in the geochemical environment: health implications. In: Appleton JD, Fuge R, editors. Environmental chemistry and health, vol. 113. GJH McCall; 1996. p. 153–61 [Geol Soc Spec Publ UK].

[6] Kinniburgh DG, Smedley PL, editors. Arsenic contamination of groundwater in Bangladesh. (British Geologic Survey Report WC/00/19, 2001).

[7] Chatterjee A, Das D, Mandal BK, Chowdhuri TR, Samanta G, Chakraborti D. Arsenic in groundwater in six districts of West Bengal, India: the biggest arsenic calamity in the world. Part 1—arsenic species in drinking water and urine of the affected people. Analyst 1995;120.

[8] Bearak D. New Bangladesh disaster: wells that pump poison. New York Times 1998. November 10.

[9] Berg M, Stengel C, Trang PTK, Viet PH, Sampson ML, Leng M, et al. Magnitude of arsenic pollution in the Mekong and Red river Deltas—Cambodia and Vietnam. Sci Total Environ 2007;372(2–3):413–25.

[10] Stanger G, Truong TV, Ngoc KSLTM, Luyen TV, Thanh TT. Arsenic in groundwaters of the lower Mekong. Environ Geochem Health 2005;27(4):341–57.

[11] Stueben D, Berner Z, Chandrasekharam D, Karmarkar J. Arsenic enrichment in groundwater of West Bengal, India: geochemical evidence for mobilization of As under reducing conditions. Appl Geochem 2003;18:1417–34.

[12] Nickson RT, McArthur JM, Ravenscroft P, Burgess WG, Ahmed KM. Mechanism of arsenic release to groundwater, Bangladesh and West Bengal. Appl Geochem 2000;15:403–13.

[13] Saunders JA, Lee MK, Uddin A, Mohammad SM, Dhakal P, Chowdhury MT, et al. Geochemistry and mineralogy of arsenic in (natural) anaerobic groundwaters. Appl Geochem 2008;23:3205–14.

[14] Nickson R, McArthur J, Burgess W, Ahmed KM, Ravenscroft P, Rahman M. Arsenic poisoning of Bangladesh groundwater. Nature 1998;395:338.

[15] Miller SM, Zimmerman JB. Novel, bio-based, photoactive arsenic sorbent: TiO_2-impregnated chitosan bead. Wat Res 2010;44:5722–9.

[16] Sarkar S, Greenleaf JE, Gupta A, Ghosh D, Blaney LM, Bandyopadhyay P, et al. Evolution of community-based arsenic removal systems in remote villages: assessment of decade-long operation. Water Res 2010;44:5813–22.

[17] Sarkar S, Blaney LM, Gupta A, Ghosh D, SenGupta AK. Use of ArsenXnp, a hybrid anion exchanger for arsenic removal in remote villages in the Indian subcontinent. React Funct Polym 2007;67(12): 1599–611.

[18] Bureau of Indian Standards. Drinking water-specification, IS–10500, 1993 (Amendment 2003) Bureau of Indian Standards, New Delhi.

[19] Sarkar S, Gupta A, Blaney LM, Greenleaf JE, Ghosh D, Biswas RK, et al. Community based well-head arsenic removal units in remote villages of West Bengal, India: a sustainable arsenic removal process. In: Ahuja S, editor. Arsenic contamination of groundwater: mechanism, analysis and remediation. Hoboken, New Jersey: John Wiley & Sons, Inc; 2008.

[20] Ghosh A, Mukiibi M, Ela W. TCLP underestimates leaching of arsenic from solid residuals under landfill conditions. Environ Sci Technol 2004;38:4677–82.

[21] Ghosh A, Mukiibi M, Saez A, Ela W. Leaching of arsenic from granular ferric hydroxide residuals under mature landfill conditions. Environ Sci Technol 2006;40:6070–5.

[22] Delemos JL, Bostick BC, Renshaw CE, Sturup S, Feng X. Landfill-stimulated iron reduction and arsenic release at the Coakly Superfund site (NH). Environ Sci Technol 2006;40:67–73.

[23] Meng XG, Korfiatis GP, Jing CY, Christodoulatos C. Redox transformations of arsenic and iron in water treatment sludge during aging and TCLP extraction. Environ Sci Technol 2001;35:3476–81.

[24] Sarkar S, Blaney LM, Gupta A, Ghosh D, SenGupta AK. Arsenic removal from groundwater and its safe containment in a rural environment: validation of a sustainable approach. Environ Sci Technol 2008;42(12):4268–73.

[25] Drever JI. The geochemistry of natural water. Englewood Cliffs, NJ: Prentice-Hall; 1988.

[26] Stumm W, Morgan JJ. Aquatic chemistry: chemical equilibria and rates in natural waters. 3rd ed. New York: John Wiley & Sons, Inc.; 1996.

Monitoring from Source to Tap: The New Paradigm for Ensuring Water Security and Quality

Dan Kroll

HACH HOMELAND SECURITY TECHNOLOGIES, LOVELAND, CO, USA

CHAPTER OUTLINE

Monitoring Water Quality, http://dx.doi.org/10.1016/B978-0-444-59395-5.00014-5

14.1 Introduction

Drinking water is one of the Nation's key infrastructure assets that have been deemed vulnerable to deliberate terrorist attacks; however, terrorism is not the only risk. Currently, our water supplies are vulnerable to a wide variety of potential hazards. These hazards can be intentional or accidental in nature. These hazards range from the threat of water quality deterioration because of our aging infrastructure, to accidental contamination of sources and treated water from pollution, to the potential of degrading our water quality as we make treatment changes to comply with new regulatory regimes. The rapid detection and characterization of breaches of integrity in the water supply network are crucial in initiating appropriate corrective action. The ability of a technology system to detect such incursions on a real-time basis and give indications as to their cause can dramatically reduce the impact of any such scenario.

Continuous monitoring of the entire water supply network is a goal that until recently seemed unachievable. Historically, most monitoring done outside of the water treatment plant has been relegated to the occasional snapshot provided by grab sampling for a few limited parameters or the infrequent regulatory testing required by mandates such as the Total Coliform Rule. The development of water security monitoring in the years since September 2011 has the potential to change this paradigm.

14.2 Where Do the System Vulnerabilities Lie?

The forms that an attack on the drinking water supply system could take are as many and varied as the components of the system that are vulnerable. The provision of drinking water to our homes is a complex process that involves many steps and components. All these steps are to some degree vulnerable to compromise by terrorist acts.

14.2.1 Source Water

The sources of our drinking water are diverse. Each municipality has its own source. These can include rivers, lakes, streams, groundwater from deep wells, groundwater from wells affected by surface water, desalinization of ocean water, or even recycled sewage water. Some places rely on a combination of different sources, for example, a river and groundwater wells.

Although source waters are usually not vulnerable to physical attack, they are vulnerable to contamination. As a general rule, source waters are easy to access, because it would be impossible to guard all the points along a river that a terrorist could use as a gateway to the water. Also, most source water supplies have a built-in hedge against intentional contamination. In the water industry, this is known as "The solution to pollution is dilution." Large sources are much less vulnerable than are small sources. Also, the fact that many of the potential contaminants are degraded over time by exposure to sunlight and the elements gives these sources a self-cleaning characteristic that makes them difficult, but not impossible, to contaminate.

14.2.2 Untreated Water Storage

Untreated water storage includes facilities such as dams, reservoirs, and holding tanks. They have many of the same attributes as source waters do. Like source water supplies, they vary dramatically in size, and their vulnerability to attack is linked directly to that size factor. They tend to be more closely watched than are source waters and do not present as large a footprint. This makes them a little harder to access, but overall, they are not difficult to approach. Many are of a dual-use nature, for example, reservoirs that are used for fishing and boating as well as for supply of water. Reservoirs have limited vulnerability to contamination because of their volumes. The same constraints apply to them as to source waters, and a mass casualty contamination event seems unlikely, although, nuisance attacks are very possible.

14.2.3 Treatment Plants

In most cases, the treatment plant represents the last barrier between natural, accidental, or deliberate contamination and the final end user. In many cases, it is also the final point where continual routine monitoring of the chemical and physical characteristics of the water occurs. Like the previous components of the water supply system, treatment plants can vary greatly in size and the amount of water treated per day. Different plants also use different methods of treatment and purification to make the source water potable. The method of treatment varies depending on the water quality of a source. Some sources require simple settling and disinfection, whereas others require a more elaborate process.

The treatment plant offers the would-be terrorist many opportunities to inflict damage. Al Qaeda and other terrorist organizations are well aware of this and have expressed an interest in recruiting workers in such facilities to orchestrate such an attack. An Federal Bureau Of Investigation bulletin issued on August 11, 2004, states "Information recently brought to the attention of Department of Homeland Security and FBI indicates that prior to the September 11, 2001 attacks; terrorists discussed possible attacks against United States facilities and systems to disrupt drinking water supplies serving major urban areas, which include large-capacity water reservoirs and water treatment facilities. Although no specific targets were selected, one specific site in the northeastern United States was mentioned as an example. While the original thought focused on large-capacity water supplies, terrorists thought it would be futile to attempt to directly poison a large water reservoir because of the dilution factor. Rather, they focused on the possibility of poisoning the water during the water treatment process. Terrorists mentioned inserting a poison (not further identified) into the chlorination section of the water treatment facility. To accomplish this objective, they discussed recruiting insiders to work with them" [1].

Another important design aspect to bear in mind regarding drinking water treatment plants is that as part of the drinking water treatment process, various chemicals are intentionally added to the water. These can include flocculating agents, caustics, acids, disinfectants, and fluoride among others. All of the dosing equipment is in place at the facility to feed massive quantities of solid, liquid, or gaseous chemicals into the finished water supply. If a terrorist could gain access to the plant, he could feed any of a large number of toxic substances into the system. There has also been some concern that a terrorist would not even have to gain access to the facility to mount such an attack. It may be much simpler, from the terrorist's viewpoint, to infiltrate one of the companies responsible for delivery of chemicals to the treatment plant. They could then replace or contaminate the usual shipment with a toxic compound and deliver it to the plant as normal. Then, when the plant operators added the replaced or adulterated treatment chemical to the water, they would in fact be poisoning the finished water.

Although the introduction of toxic chemicals from an outside source is of a grave concern, in fact, it would not even be necessary to bring the toxic material in from the outside. Many of the normal treatment chemicals already present at the site can be toxic. All that would be needed to cause harm would be for the terrorist to increase the dosages of these usually benign chemicals to toxic levels. An example is fluoride. It is usually fed into the system at a rate that gives a final concentration at the tap of between 0.8 and 1.2 mg/L. The United States Environmental Protection Agency (USEPA) maximum contaminant level for fluoride is 4 mg/L. Fluoride can be extremely toxic at higher doses. This is especially true for young children in whom the lethal dose can be as low as 16 mg/kg. Patients undergoing dialysis are even more susceptible. There have been many documented incidents in the past when equipment failure or operator error has resulted in the overfeeding of fluoride into drinking water supplies. Many of these incidents have resulted in illness and, a few deaths have been documented [2]. It would be very easy for a terrorist to simply turn up the dosage on a feeder and achieve similar results. It may seem that such an attack would be contrary to the purposes of a terrorist—to create terror, as this would seem to be just an unfortunate accident and could easily be covered up by authorities to make it seem so. However, it is very unlikely that the authorities (especially the local water authorities) could or would want to cover-up any incident that resulted in injuries or casualties. The local water utility would be held fiscally and criminally liable for any accident, whereas the liability burden shifts for a terrorist act, especially, if the utility uses systems and methods approved under "the Safety Act." Although federal authorities may have a vested interest in masking a successful attack, such a cover-up would be difficult as any successful attack will be claimed by the responsible terrorist group. This claim would be even more credible if multiple sites were involved or a public warning, sans location, were given immediately before the attack took place.

As was noted previously, in many cases, the treatment plant is the last location where routine continuous monitoring of water quality parameters is carried out. Drinking water metrics, such as pH, turbidity, disinfectant levels, fluoride, are commonly examined by the use of online continuous reading monitors at larger plants and by regular grab sampling and laboratory testing at smaller facilities. Other tests, such as bacterial testing, metal testing, occur off-line and at regular intervals. When the same person is responsible for both testing and dosing of treatment chemicals, it would be quite easy to elude detection of an attack until investigators look into it, and then, it becomes obvious as to who did it. This becomes more difficult if separate personnel are responsible or if the data are automatically collected and downloaded to a third party. Although disruption to many parts of the system is feasible, the treatment plant is the first location in the system where the possibility of deliberate contamination in an attempt to cause mass casualties actually becomes realistically workable because of the ease of introducing chemicals and the decreased levels of dilution that are afforded at this point in the water delivery system.

14.2.4 Finished Water-Holding Tanks

These sites represented by water towers and standpipes are also locations with a limited geographic footprint so they can also benefit from some degree of physical security. They do tend to be located in remote and widely dispersed areas, so in some locations, physical monitoring can be burdensome.

The volume of water involved is also greatly reduced from that in previous stages of the water supply chain, but there is still some potential for dilution. Most of these tanks are covered, which thus reduces the chances for natural attenuation. However, most systems maintain a disinfectant residual that offers some barrier to contamination. These tanks are post treatment plants, and if the residual disinfectant barrier is breached, there is little opportunity for contaminant neutralization before the water reaches the end user.

14.2.5 The Distribution System

The distribution system comprises the network of pipes, valves, pumps stations, and other accoutrements that help supply the water from the treatment plant to the end user. The distribution system is widely recognized as the most vulnerable component of the water supply network. At this point, the potential for dilution is vastly reduced as is the time available for attenuation.

The most likely scenario for such an attack, in which the goal is to inflict mass casualties, is to orchestrate a simple backflow contamination event. A backflow attack occurs when a pump is used to overcome the pressure gradient that is present in the pipes of a distribution system. This is usually approximately $\leq$80 lbs/in^2 and can be easily achieved by using pumps available for rent or purchase at most home improvement stores. After the pressure gradient present in the system has been overcome and a contaminant is introduced, siphoning effects act to pull the contaminant into the flowing system. Once the contaminant is present in the pipes, the normal movement of water in the system acts to disseminate the contaminant throughout the network-effecting areas surrounding the introduction point. The introduction point can be anywhere in the system and can be a fire hydrant, commercial building, or residence.

Backflows occur via accident on a regular basis and are of great concern to the water industry. Accidental backflow events have been found to be responsible for many incidents of waterborne illness and even death in the United States. According to the USEPA, backflow events caused 57 disease outbreaks and 9734 cases of waterborne disease between 1981 and 1998 [3].

To prevent such accidental backflows, many systems have been equipped with backflow-prevention devices. These means of preventing backflow are very useful in preventing the all too common accidental event, but it should be noted that these devices are installed to prevent accidental backflows. They are all physical devices that can be removed or disabled quite easily by a would-be terrorist, thus rendering them ineffective in preventing deliberate attempts at contamination by all but the most amateurish perpetrators.

Intentional dissemination of contaminants through a backflow event is in fact a very critical vulnerability. Studies conducted by the US Air Force and Colorado State University have shown this to be a highly effective means of contaminating a system [4]. These studies show that a few gallons of highly toxic material was enough, if injected properly at a strategic location, to contaminate an entire system supplying a population of 100,000 in a matter of a few hours. Material and significant contamination was not relegated to only the areas surrounding the introduction point. Material flowed through each neighborhood and then reentered main trunk lines and thus made its way to the next area until the contaminant had permeated the entire system. By using computer simulations, when a military nerve agent material was injected, >20% of the population was determined to have received a dose adequate to result in death. When a common chemical was used instead of the warfare agent, the result was a casualty rate of >10% [4]. Thousands of deaths could result from this very inexpensive and low-tech mode of attack. There is no doubt that this form of assault meets all the terrorist's criteria for planning an attack. It would cause mass casualties, be inexpensive, and actually offer the terrorists a good chance of avoiding apprehension. Unfortunately, because monitoring for contamination in the distribution system is typically limited to infrequent grab samples, the first indications of such an attack are likely to be casualties showing up at local hospitals. A terrorist could launch such an attack and be on a plane out of the country before the first casualty is reported.

These sorts of attacks can occur from any access point to the water system. Wherever water can be drawn out, material can be forced back into the system. Some areas, however, are more vulnerable than are others. Access points near high flow areas and larger pipes would be favored because they would disseminate the material to a wider area more quickly; however, any point except for those at the very end of long deadhead lines could be used to effectively access the system.

It is obvious from the large number of accidental backflows that occur and the fact that terrorist organizations have shown an interest in contaminating water that the distribution system is a prime candidate for such an attack. The fact is that a bona fide terrorist is virtually inundated by possible candidate substances and locations that would be very effective in such a situation. The possibilities are virtually endless. Protection against or detection of such an attack is difficult to achieve.

Recent breakthroughs in the online detection of contaminants have made the deployment of a cost-effective early warning system capable of detecting and categorizing such events a reality. The concepts behind the development and the use of such a system are described in the remainder of this chapter.

14.3 Bulk Parameter Monitoring

The ability to detect and act on changes in water quality is a critical component in the drive to protect our drinking water supplies from intentional or accidental contamination. The twin motivators of the terrorist threat to water along with consumer demands for safe and potable supplies has led to a sea change in the drinking water industry. Several studies conducted since September 2011 have shown that bulk monitoring of basic water quality parameters has the potential to indicate the presence of many harmful agents in water at the levels of interest [5–7]. The realization of the potential of bulk parameter monitoring as a practical tool to detect terrorism-related events has led to the development of many sensor packages designed for deployment in the distribution system.

Since September 2011, numerous communities have installed multiparameter monitoring stations in various locations throughout their distribution network as early warning systems to detect potential water security threats and to provide operational data. These continuous online systems have recorded large streams of data (at some sites for several years) relevant to water quality in the distribution systems in which they have been deployed.

These data streams are quite complex, and it becomes difficult to differentiate normal background noise and fluctuations because of normal everyday events from changes that indicate a deviation in water quality and deserve further attention. Intelligent algorithms are a needed addition to bulk parameter monitoring if signals useful in the decision-making process are to be obtained. Intelligent algorithms are necessary to streamline the process of data interpretation. These algorithms should be capable of detecting the subtle changes in bulk parameter readings that indicate an incursion into the system without burdening the operators with a constant stream of false or trivial alarms. They should also be capable of differentiating the unique pattern of responses that are elicited by different classes of agents. These differences may be enough to narrow down the cause of events and, possibly, to fingerprint the class of disturbance that caused the event. Over the past several years, many of such algorithms designed for this use have been under development by private industry, government programs, universities, and national laboratories.

14.4 An Early-Warning System

Since late 2001, scientists at Hach Homeland Security Technologies (HST) have been actively engaged in the development and testing of an early warning system for detecting water quality problems, including those related to an intentional terrorist attack in the drinking water distribution system. Monitoring in the distribution system, as in source water, is a difficult proposition. The sheer number and diversity of potential threat agents that could be used in an attack against the system makes monitoring them on an individual basis an effort that is doomed to failure from the start. This does not even take into account the even larger number and diversity of compounds that may accidentally find their way into our drinking water. To counter and detect the unprecedented number and types of compounds that may be encountered, what is needed is a broad-spectrum analyzer that can respond to any likely threat and even to unknown or unanticipated events.

Rather than attempting to develop individual sensors to detect contaminants or classes of contaminants, the Hach HST approach was to use a sensor suite of commonly available off-the-shelf water quality monitors such as pH, electrolytic conductivity, turbidity, chlorine residual, and total organic carbon (TOC) linked in an intelligent network. The logic behind this is that these are proven technologies that have been extensively deployed in the water supply industry for several years and have been shown to be stable in such situations. One of the difficulties encountered when designing such a device is that the normal fluctuations in these parameters found within the water can be quite pronounced.

The problem then becomes whether we can differentiate between the changes that are seen as a result of the introduction of a contaminant and those that are a result of everyday system perturbation? The secret to success, in a situation such as this, is to have a robust and workable baseline estimator. Extracting the deviation signals in the presence of noise is absolutely necessary for achieving good sensitivity. Several methods of baseline estimation were investigated. Finally, a proprietary, patented, nonclassical method was derived and found to be effective.

In the system as it is designed, signals from five separate orthogonal measurements of water quality (pH, conductivity, turbidity, chlorine residual, and TOC) are processed from five-parameter signals into a single scalar trigger signal in an event monitor computer system that contains the algorithms. The signal then goes through the crucial proprietary baseline estimator. A deviation of the signal from the established baseline is then derived. A gain matrix is then applied that weights the various parameters based on experimental data for the presence of a wide variety of possible threat agents. The magnitude of the deviation signal is then compared to a preset threshold level. If the signal exceeds the threshold value, the trigger is activated (Figure 14-1).

The deviation vector that is derived from the trigger algorithm is then used for further classification of the cause of the trigger. The direction of the deviation vector relates to the

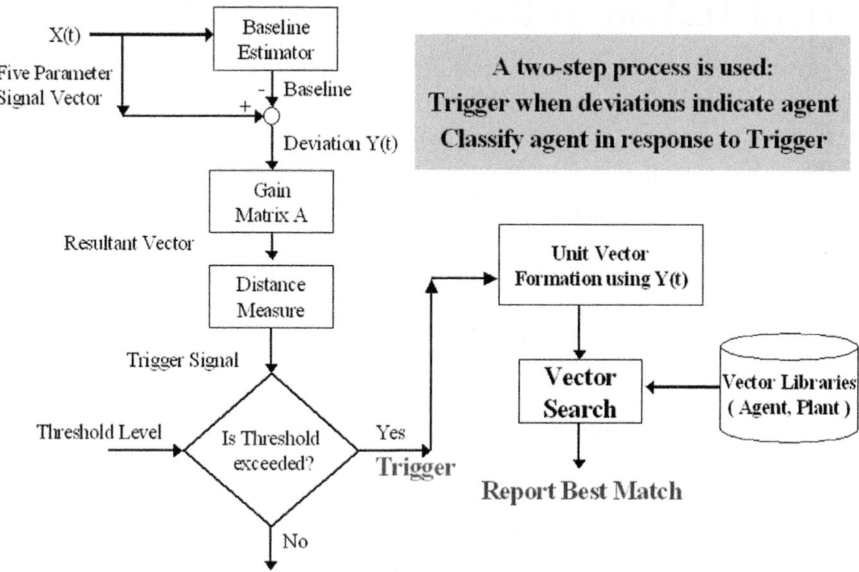

FIGURE 14-1 Flowchart depicting algorithm functionality.

agent's characteristics. On seeing that this is the case, one can use laboratory agent data to build a threat agent library of deviation vectors. A deviation vector from the monitor can be compared to agent vectors in the threat agent library to see if there is a match within a tolerance. This system can be used to classify what caused the trigger event. This system can also be very useful in developing a heuristic system for classifying normal operational events that may be significant enough in magnitude to activate the trigger. When such an event occurs, the profile for the vector causing it is stored in a plant library that is named and categorized by the system operator. When the event trigger is exceeded, the library search begins.

The agent library is given priority and is searched first. If a match is made, the agent is classified. If no match is found, then the plant library is searched, and the event is identified if it matches one of the vectors in the plant library. If no match is found, the event is classified as being unknown and can be named if an investigation determines its cause. This is very significant, because no profile for a given event need be present in the libraries for the system to trigger. This gives the system the unique ability to trigger on unknown threats. Also, the existence of the plant library, with its ability to store plant events, results in a substantial and rapid decrease in unknown alarms over time. The agent library is generalized and is transferable form site to site. The plant library is site specific and requires some modest user input to identify and classify events so that they can be recognized on recurrence. The developed system has been subjected to strenuous testing in both laboratory and field scenarios.

14.5 Field Testing of the Developed System

Before the onset of this project, there was a definite lack of data concerning conditions in the distribution system. Since the initiation of this program, real-time data have been collected across a wide variety of different distribution systems to exhibit different water matrix profiles that reveal many interesting attributes of the distribution systems. The following are a few examples of incidents that have been recorded during these real-world deployments. Although not actual terrorist events, these incidents help to demonstrate the system's ability to learn and to become a useful tool, not just for security but also for everyday operational improvements.

14.5.1 Caustic Overfeed Event

In this deployment scenario, the plant uses caustic feed to control water pH. The system experienced a trigger that when investigated was identified as an operational problem that resulted in the feed of excess caustic. The result was that the overfeed affected the pH and the conductivity of the water, which caused the alarm (Figure 14-2). The reason behind this was that the vendor from whom the caustic was being purchased had delivered the wrong concentration of the solution. No one had checked to see if the concentration was correct before feeding in the material. New procedures were put in place to verify incoming raw materials. The event monitor learned this plant event and can now identify a recurrence of the event in the future if there is another failure in the system.

14.5.2 Roadwork Event

In this event, roadwork (jackhammers) near a distribution line dislodged biomass and other particulate matter from the lining of the pipe. There was a massive increase in turbidity, which not only showed up on the turbidimeter but also showed up as an interference in the chlorine measurement (optical). As expected, the conductivity and pH also showed minor changes. The increase in biomass in the water was indicated by the TOC analyzer (Figure 14-3). This event illustrates the ability of the event monitor to detect

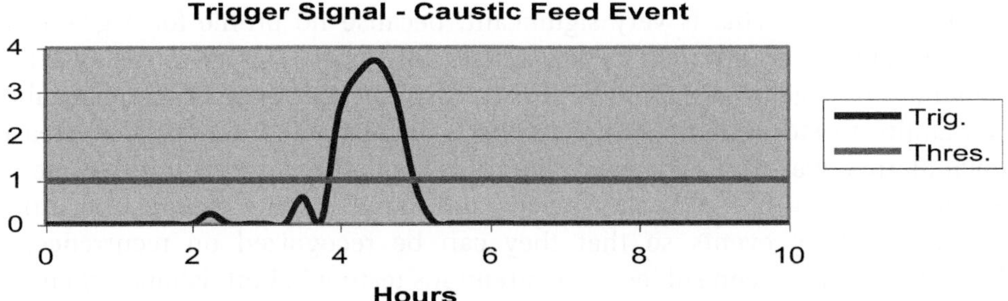

FIGURE 14-2 Caustic event. (For color version of this figure, the reader is referred to the online version of this book.)

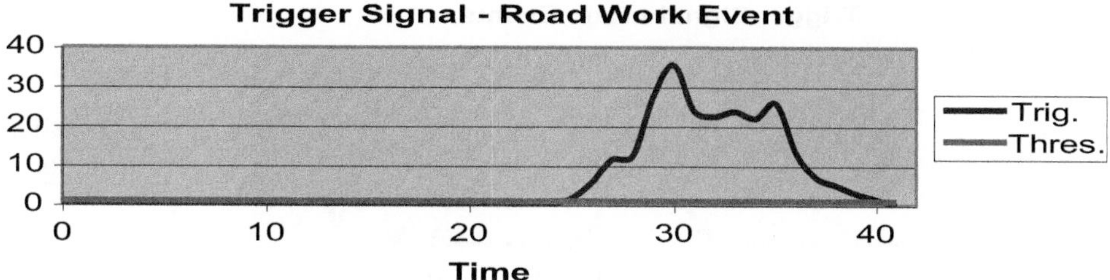

FIGURE 14-3 Roadwork event. (For color version of this figure, the reader is referred to the online version of this book.)

and trigger an alarm on unanticipated events. This event also provides a signature for the materials adhering to the walls of the pipes in this location and should recognize any future excursions of this type.

14.5.3 Pressure Event

In this scenario, the system was located in a building that experiences a daily variation in water pressure. The sample variation is associated with a turbidity increase that causes a trigger (Figure 14-4). There is also a small pH decrease at that time, possibly because of increased solubility of CO_2 in the water, which decreases the pH slightly. After recognition of the cause and proper naming of this pattern, it is recognized by the event monitor as a "normal" event, rather than as an alarm condition, and appropriately classified and named as such.

14.5.4 Rain Events

Heavy rainfall in the area of a reservoir raised turbidity and affected other water quality parameters. These events were large enough to cause a trigger (Figure 14-5). The system was able to store this pattern and recognize a recurrence.

14.5.5 Effect of Variable Demand

In this deployment, daily events influencing turbidity, chlorine, pH, and conductivity are not completely understood but are suspected to be caused by water demand fluctuations

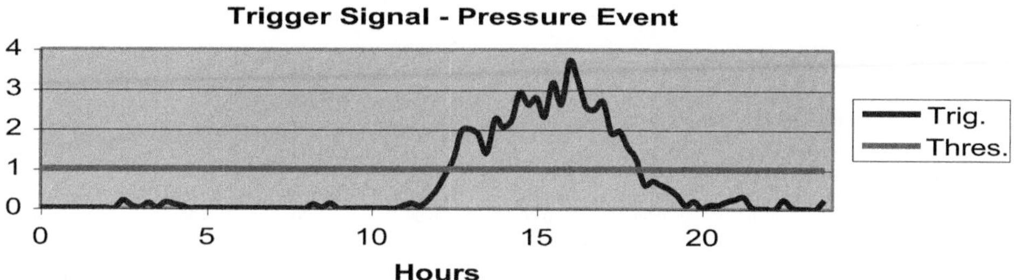

FIGURE 14-4 Pressure event. (For color version of this figure, the reader is referred to the online version of this book.)

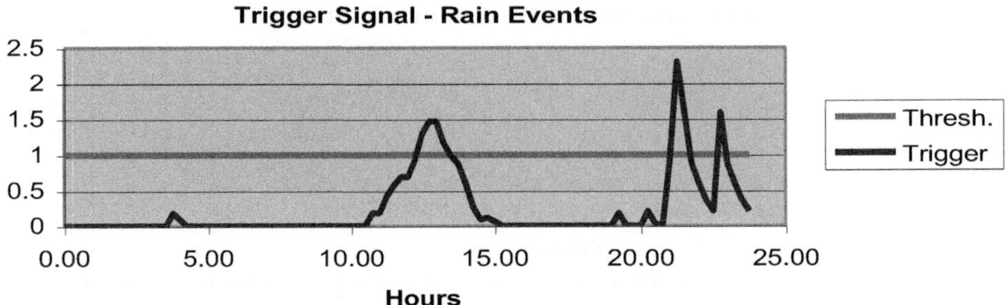

FIGURE 14-5 Rain event. (For color version of this figure, the reader is referred to the online version of this book.)

in the area. This may indicate a need for more routine flushing of the pipes in the area and the instillation of a chlorine booster station (Figure 14-6).

14.5.6 Ammonia Overfeed Events

On March 26, 2007, maintenance was performed at the plant supplying water to the distribution system being monitored. After the maintenance was completed, the plant was restarted, and the system that feeds ammonia to produce monochloramine as a residual disinfectant overfed the chemical. The person in charge of the online monitors immediately noticed the increase in pH and notified plant operations. Operations reported a problem with ammonia-feed pumps. The problem was temporarily fixed, but a slug of ammonia was sent into the distribution system. Several customers called up,

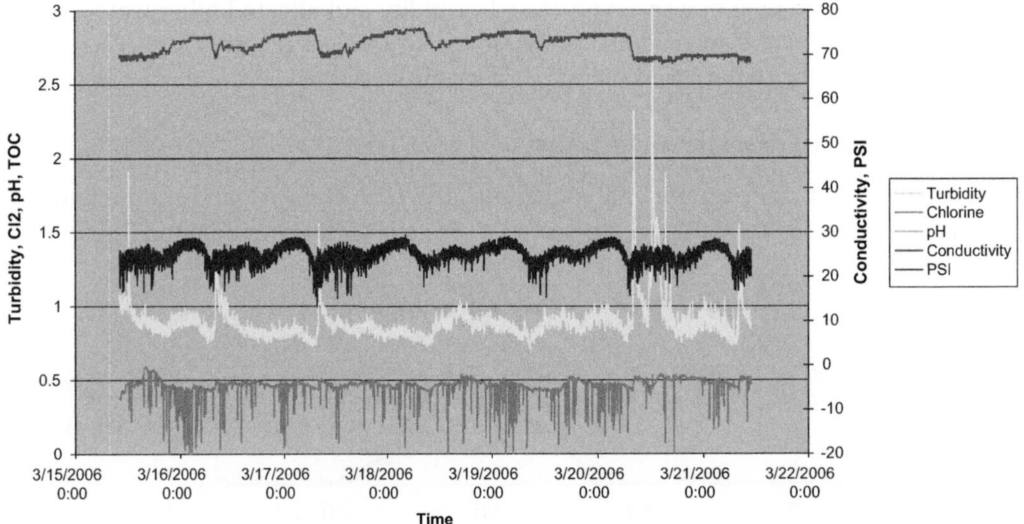

FIGURE 14-6 Variable demand. (For color version of this figure, the reader is referred to the online version of this book.)

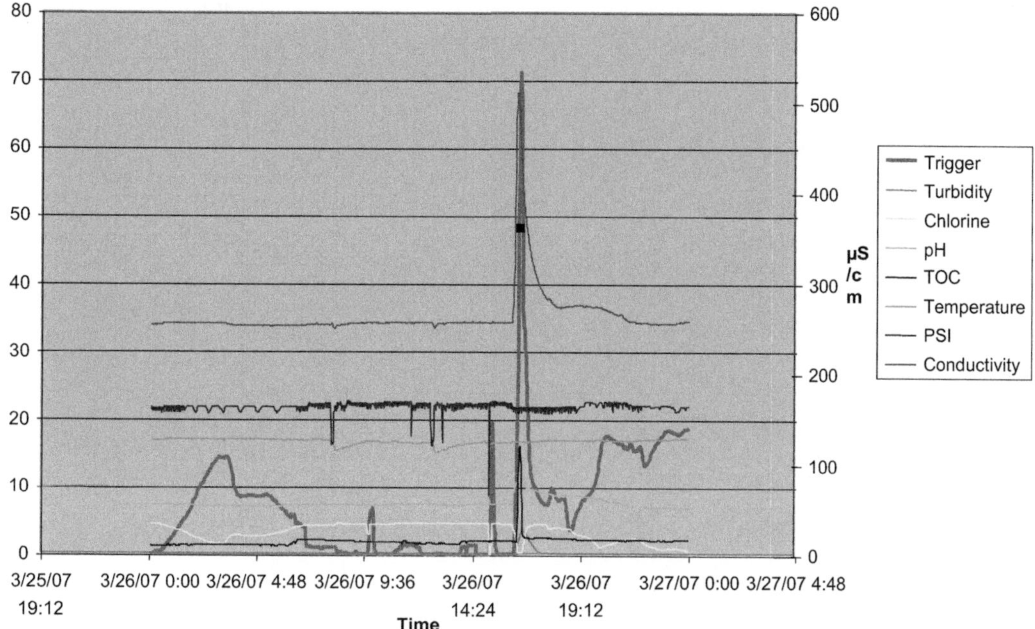

FIGURE 14-7 March 26: ammonia overfeed. (For color version of this figure, the reader is referred to the online version of this book.)

complaining of an ammonia smell and taste emanating from the tap. The exact amount of ammonia released was unknown but was believed to be <10 ppm. The facility continued operations but temporarily switched over to free chlorine as a disinfectant until July 2, 2007 (Figure 14-7).

On October 3, 2007, the same treatment plant experienced a brief ammonia overfeed. In this case, a pump was turned on and not switched off at the proper interval. There was a decrease in the chlorine concentration and a decrease in pH (Figure 14-8).

During August 22–23, 2007, three similar events occurred. Increases in turbidity pH and possibly TOC with a decrease in chlorine caused triggers. These changes resemble the large-scale ammonia event that occurred in March. The operator believes that these could be ammonia-feed-related events but could not confirm this or determine the cause (Figure 14-9).

14.5.7 Chlorine-Feed-Related Event

On April 3, 2007, there were turbidity and pH increases and a decrease in chlorine and conductivity. The operator suspects that there was a problem with the chlorine feed at the plant. However, this cannot be confirmed. The plant was using free chlorine at the time, which rules out the possibility of an ammonia-feed-related problem (Figure 14-10).

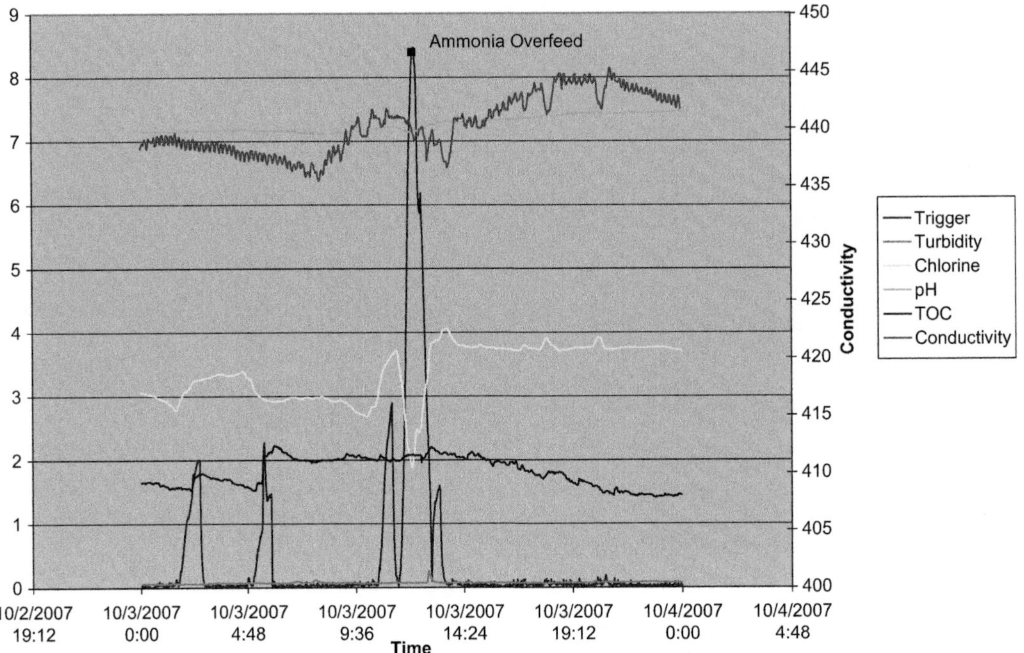

FIGURE 14-8 October 3: ammonia overfeed. (For color version of this figure, the reader is referred to the online version of this book.)

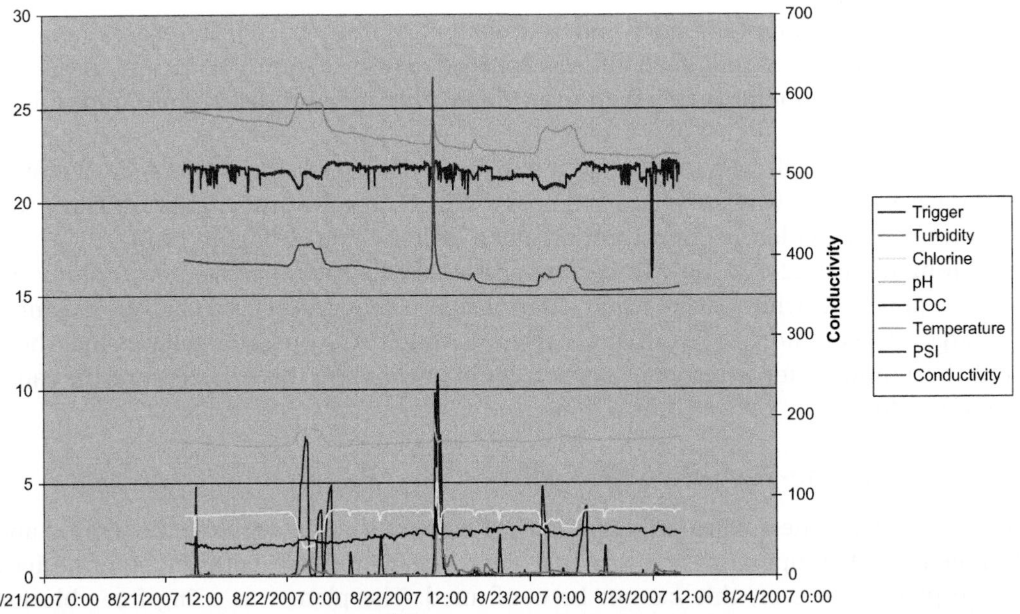

FIGURE 14-9 Possible August ammonia-related events. (For color version of this figure, the reader is referred to the online version of this book.)

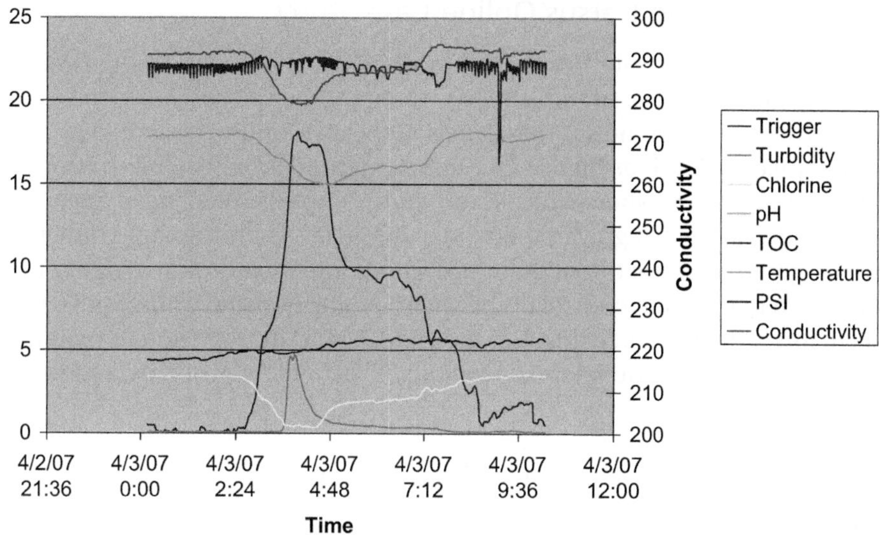

FIGURE 14-10 Possible chlorine-feed-related event. (For color version of this figure, the reader is referred to the online version of this book.)

14.5.8 Air Bubble Event

In the distribution system of one northern midwestern city, every Friday, the deployed sensors would behave extremely erratically. This resulted in multiple alarm signals being generated. Investigation led to the discovery of extreme amounts of entrained air bubbles being present in the system's water on Friday afternoons and evenings. Further investigation into the root cause of the air bubbles being present revealed that school buildings in the area that were to be vacant over the weekend had a policy of using air to blow out their water lines to prevent freezing so that the heat could be turned off over the weekend and holidays. A faulty check valve at one of the schools allowed the air to bleed into the distribution system. The valve was replaced, thus closing a possible backflow route into the system. After this maintenance was performed, the erratic readings ceased to exist.

14.5.9 Distribution Flushing Case Study

In this case, the water utility customers were experiencing poor water quality in certain areas of the distribution system. The systems operations team could not pinpoint the problem. A series of distribution system monitoring devices were installed at various locations throughout the distribution network. The devices were able to help in locating several transient dead ends and low flow areas based on the monitoring results. By using these data, the team revised the normal flushing schedule for these areas. This resulted in improved water quality and fewer customer complaints.

14.5.10 Grab Sample Versus Online Case Study

In this situation, the local water utility had in place an extensive system of water monitoring through grab samples. All indications were that the water quality was good. After installation of several water monitoring panels in the distribution system, they found that the turbidity spiked to levels as high as 20 Nephelometric Turbidity Units during the night and early morning hours when typically no grab samples would have been collected. They also found extremely high variability in chlorine levels during these time periods. A series of changes to their treatment plant operations and distribution system procedures allowed the system to regain control of the water quality in the distribution system. They were able to lower the turbidity spikes to around 1.5 NTU at night and to maintain more consistency in chlorine residual levels, resulting in better water quality and consistency for the end consumers.

14.5.11 The Strange Case of the Chlorine Spikes

In one field deployment scenario, the system was very quiescent and rarely came anywhere close to causing a trigger alarm to go off. Every night at around midnight, the chlorine level would spike dramatically and cause an alarm. This was deemed very strange, and extensive troubleshooting of the instruments and power supply revealed no abnormal conditions that could be causing the problem. After a thorough investigation was done, the night operator for the treatment plant was queried about the strange chlorine response. His reply was, "Of course the system's chlorine level spikes every night at midnight that is when he super chlorinates the system just like he had been told to do." It seems that several years ago, when there was a pipe rupture in the system that may have allowed contamination to seep in, the night operator had been told to superchlorinate the system. Unfortunately, the operator was new at the time, and the instructions were not explicit enough to make him understand that the superchlorination should take place that night only. The operator had continued to perform the operation every night for years, resulting in a huge unnecessary cost in procuring chlorine. This situation was remedied and should result in substantial chemical cost savings in the future and in better training.

14.5.12 Aviation Fuel Incident

In this situation, water from a well abandoned because of contamination with aviation fuel was being used to flush out some lines that had been physically isolated from the remainder of the distribution system. Somehow, the water got by a double-valve isolation setup and contaminated a water storage unit that was being monitored. The system was isolated, and customers were not exposed (Figure 14-11).

14.5.13 Fluoride Incident

In this scenario, the water utility was forced to revert to the use of an older water treatment plant while the new plant was being maintained. A pump that was responsible for

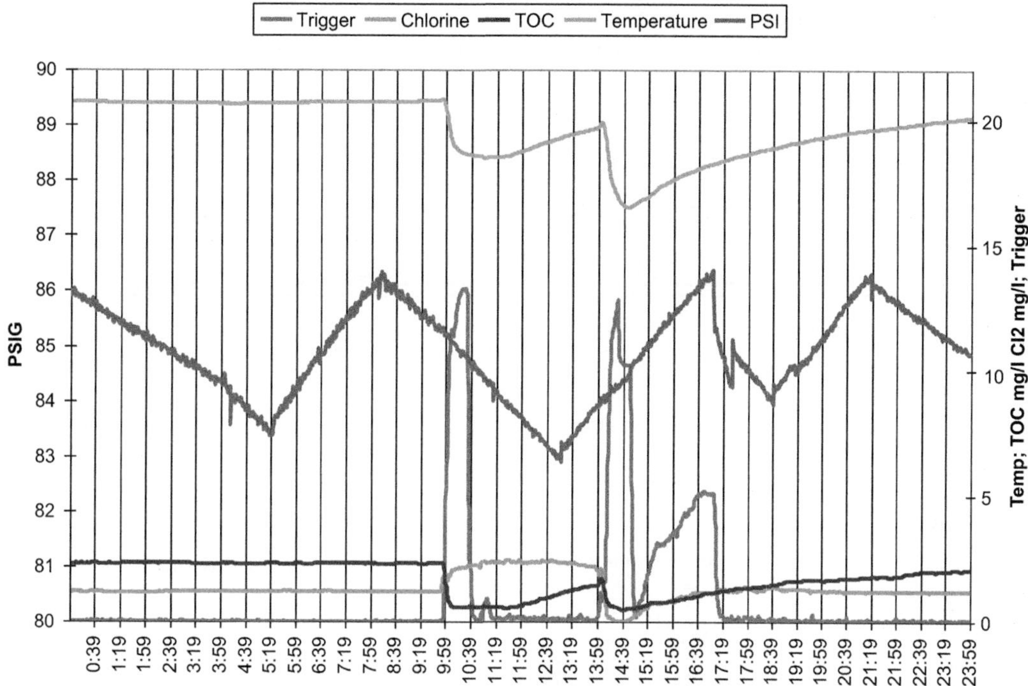

FIGURE 14-11 Aviation fuel event. (For color version of this figure, the reader is referred to the online version of this book.)

dosing fluoride into the treated water malfunctioned, causing the dose to increase over time. In this case, the monitor not only triggered an alarm but it also classified the likely cause of the problem to be a fluoride overfeed as that fingerprint was in the agent library. This allowed a rapid response before consumers were exposed to potentially dangerous levels of fluoride.

14.5.14 Main-Break Events

In this situation, the system had only just been installed a few days previously. Hach HST personnel were informed that the instruments were behaving abnormally and were giving strange readings (Figure 14-12). An investigation of the sensors found no problems. Soon, a major main ruptured catastrophically. The system was able to detect the perturbation in water quality parameters that were precursors to the main-break and trigger upon them. Unfortunately, the system was newly installed, and the event was not recognized until it was too late (Figure 14-13).

Another incident occurred on June 29, 2006. The pipe was an 8-inch line and was located 1.8 miles downstream of the monitoring sensors. There was an increase in turbidity and chlorine concentration. The turbidity seemed to have increased the day before the break had occurred (Figure 14-14).

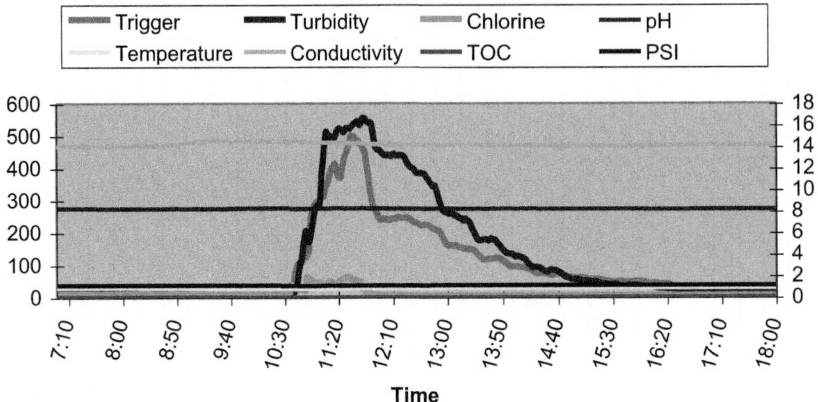

FIGURE 14-12 Erratic instrument readings soon after installation. (For color version of this figure, the reader is referred to the online version of this book.)

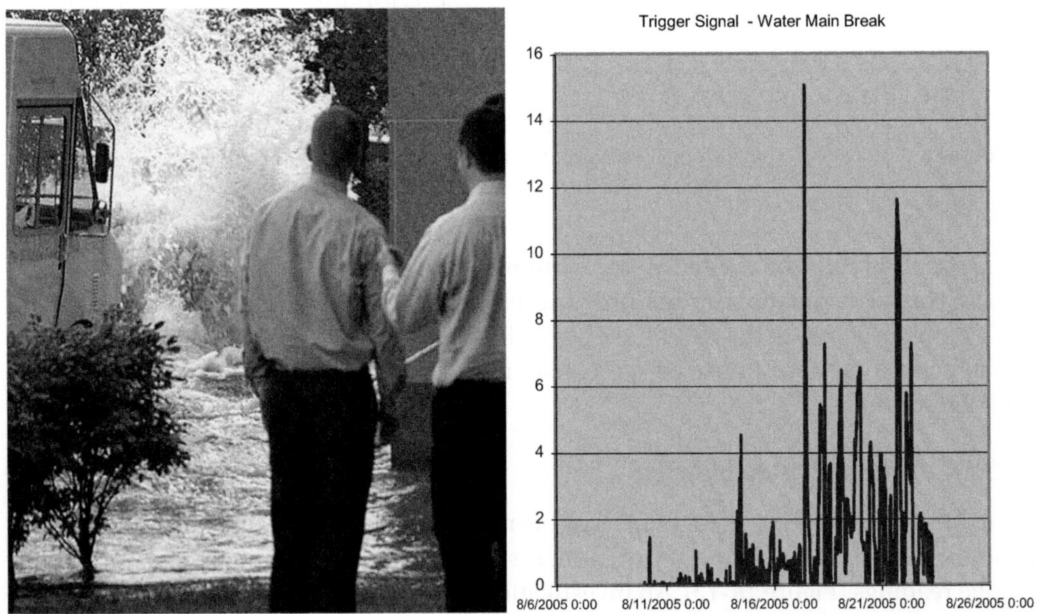

FIGURE 14-13 Photograph of the main-break event along with graph of trigger signal. (For color version of this figure, the reader is referred to the online version of this book.)

A third break occurred on July 30, 2006. The line was a 36-inch water main located 1 mile from the monitor. Conductivity, turbidity, and chlorine levels spiked. There seems to have been two water flow interruptions to the Water Distribution Monitoring Panel the night before, but it is unclear if they were related to the break on the 30th (Figure 14-15).

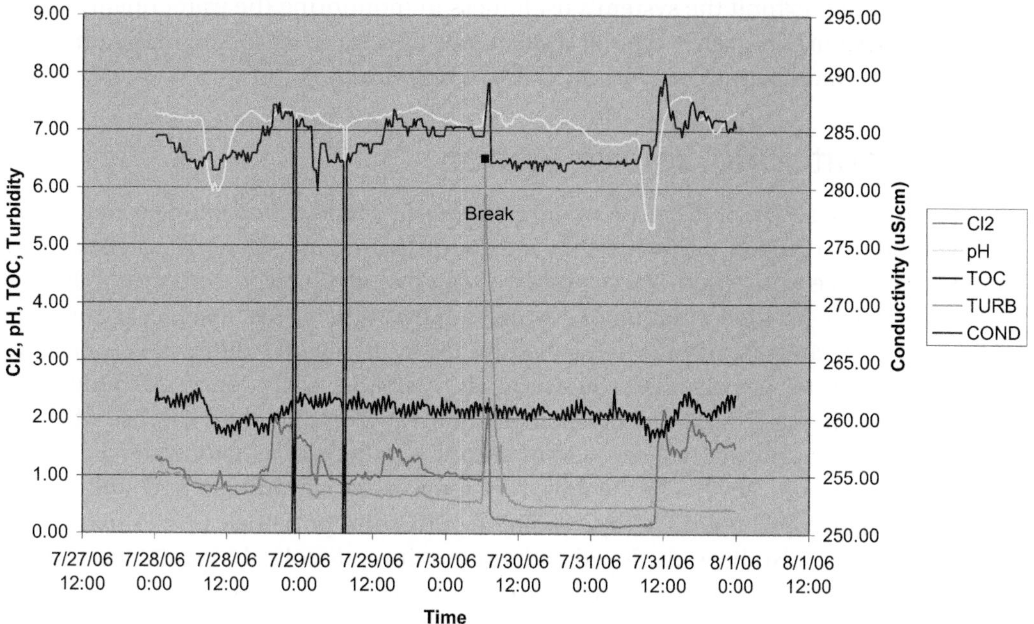

FIGURE 14-14 The 8-inch pipe break. (For color version of this figure, the reader is referred to the online version of this book.)

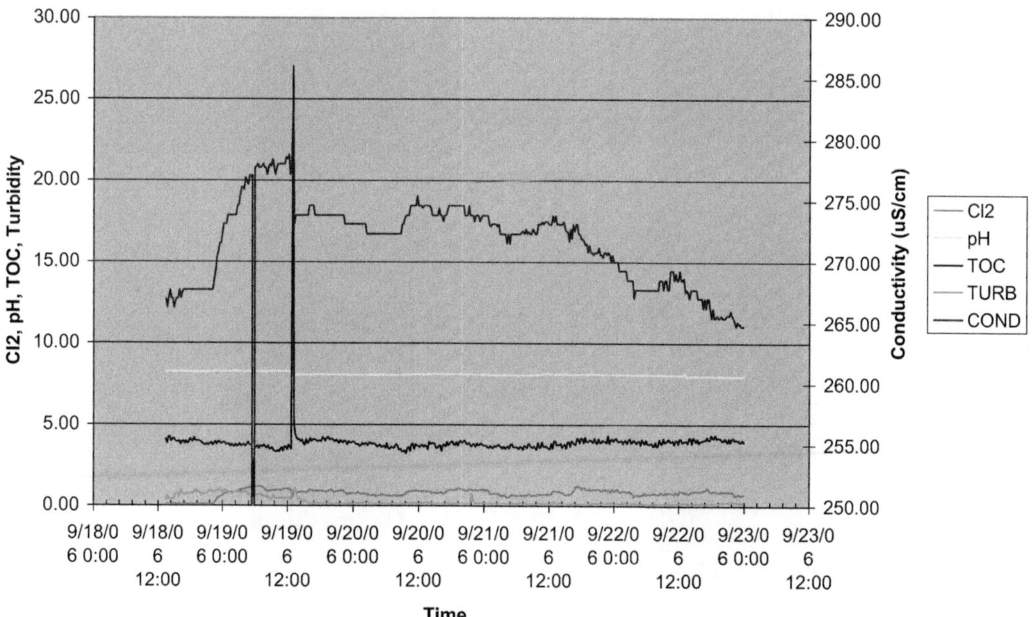

FIGURE 14-15 The 36-inch main-break. (For color version of this figure, the reader is referred to the online version of this book.)

These examples exhibit the system's usefulness in monitoring the water quality of the distribution system. The same type of system can also be used to keep tabs on other portions of the water supply network such as source water.

14.6 Monitoring Source Water

Besides reasons of security, there are many good reasons to monitor source water. There are a multitude of regulated parameters that are found in drinking water that can have their origin in the source water used. These problems can include fluoride, turbidity, TOC that can be responsible for the formation of disinfection byproducts, pH, arsenic, and nitrate.

The use of the same algorithms in the distribution system with a different sensor set to monitor source water has also been shown to be feasible. A trial program conducted in coordination with the United States Geological Survey demonstrated the use of these algorithms for this application. A series of monitoring stations in a major watershed were deployed in 2006 and 2007. The deployed systems were continuous and online and measured turbidity, pH, conductivity, oxidation–reduction potential, UV254 (organics); some stations have also been deployed with ammonium, dissolved oxygen, and nitrate. The systems were also equipped with automatic samplers that would kick in when triggered to take a sample for further analysis (Figure 14-16).

FIGURE 14-16 The instruments were deployed in a weir-type arrangement. (For color version of this figure, the reader is referred to the online version of this book.)

Data were collected for several months and run through the algorithms to see if events could be recognized and alarm threshold levels could be set (Figure 14-17). Although the algorithms were successful in detecting and raising an alarm when major events occurred, there were many more alarms than in the drinking water deployments. Much of this was due to the nonhomogeneous nature of the sample stream, which caused more noise in the sensor package. This resulted in more trivial alarms than are commonly present in drinking water samples because of the increased noise function.

To adjust for this factor, a new tuning function was added to the software package to compensate for the noise. By using the new algorithm, the software for the detection system can continually adjust its sensitivity based on the recent noise level in the multiple parameter readings. This adjustment helps to compensate for changes in baseline conditions over time. If the parameter readings become noisier, the system reduces its sensitivity, and increases its sensitivity should readings become less noisy. This method gives the system the best sensitivity it can have at any given time regardless of conditions. If the system did not automatically adjust its sensitivity, increased noise could generate alarms based on the increased noise in addition to alarms based on water events. With the new tuning function in place, the occurrence of alarms was dramatically reduced while still allowing the system to maintain a level sensitive enough to detect real events with a high degree of confidence (Figure 14-18).

Such systems have been widely deployed throughout the world to keep an eye on the quality of the water in many watersheds. One application in China uses profiles of the pollutants emitted by various industrial plants in the area. These profiles are stored in

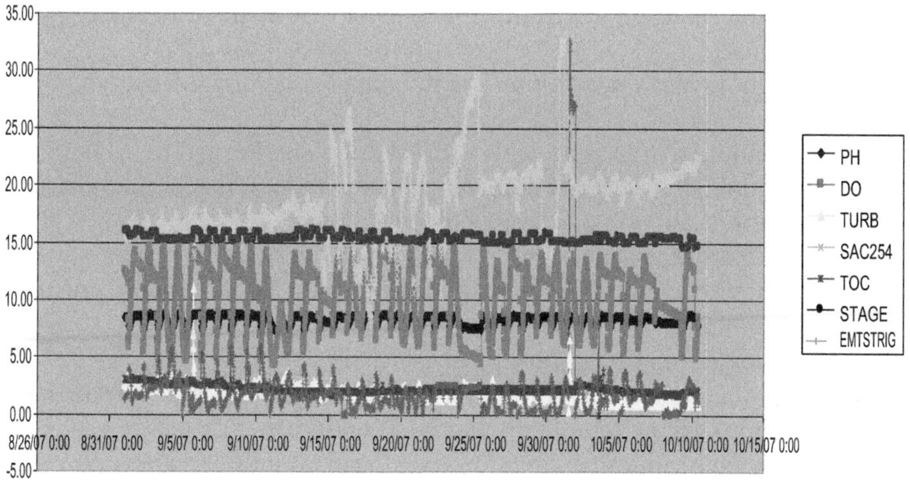

FIGURE 14-17 Six-weeks' worth of data from one site. (For color version of this figure, the reader is referred to the online version of this book.)

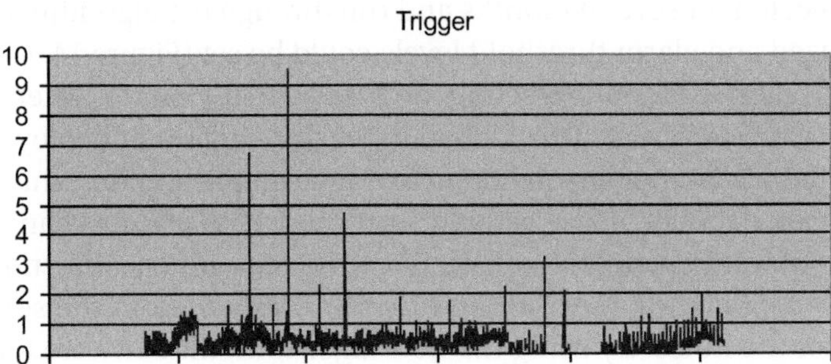

FIGURE 14-18 Graph of trigger signal of combined parameters. (For color version of this figure, the reader is referred to the online version of this book.)

the plant library of the event monitor used at the site. When a significant change in water quality occurs, the early warning system cannot only indicate that a change has begun but it also pinpoints the most likely sources of the industrial pollution so that rapid remedial action can be taken.

14.7 Bringing It All Together

There are several criteria that a successful early warning system for water should address [8]. These include the following:

(1) The system should provide a rapid response.
(2) The system should be capable of detecting a sufficiently wide range of potential contaminants.
(3) The system should exhibit a significant degree of automation, including automatic sample archiving.
(4) The system should allow acquisition, maintenance, and upgrades at an affordable cost.
(5) The systems should require staff with a low skill set and limited training to operate.
(6) The system should demonstrate sufficient sensitivity to detect contaminants at the levels of interest.
(7) The system should experience minimal false positives/false negatives.
(8) The system should exhibit robustness and ruggedness in continually operating in a water environment.
(9) The system should function continuously.
(10) The systems should allow for third party testing, evaluation, and verification.
(11) The system should allow remote operation and adjustment.
(12) The systems should identify the source of the contaminant and allow accurate prediction of the location and concentration downstream of the detection point.

The algorithm sensor packages currently available do an excellent job of addressing the first 10 criteria, but they fall short on points 11 and 12. Further software deployment can help to address the remaining criteria.

14.8 Remote Operation and Adjustment

The widespread geographical dispersion of sensor packages in an early-warning system network often becomes an enormous problem for those responsible for the operation and maintenance of a system. Travel to remote and widespread areas for performing routine functions and troubleshooting can be an enormous drain on money and personnel. If the system is not easy to maintain and operate, it is likely to be neglected and fall into disrepair and inoperability.

A new software communication package (City Guard) allows the user to visualize the instrument readouts as if the operator were on site (Figure 14-19). End users can

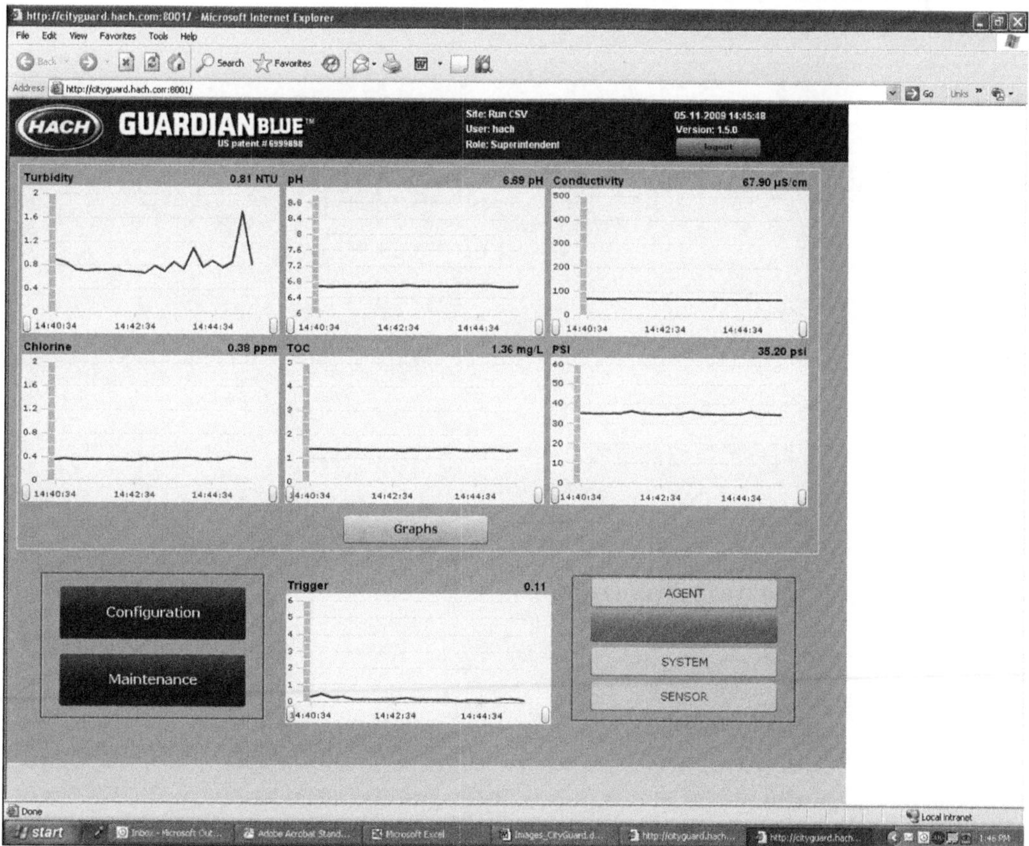

FIGURE 14-19 Individual parameters can be visualized as if you were on the site. (For color version of this figure, the reader is referred to the online version of this book.)

drill down and visualize individual aspects of the data and compare them to historical trends or rapidly compare them to other sites in the network. Sensor faults are accessible, which allows maintenance teams to be dispatched when needed. Users are able to access the event monitor software from remote locations of their choice rather than going to the site at which the instruments are deployed. Once the event monitor has been accessed, operators can exert bidirectional control of numerous event monitors from one or several remote locations. This will enable the end user to clear alarms, name events, reconfigure event monitor settings and alarm thresholds, and troubleshoot without traveling to the site. While travel to individual sites for functions such as calibration and routine maintenance will still be required, the decrease in travel for other functions now capable of remote manipulation can be substantial.

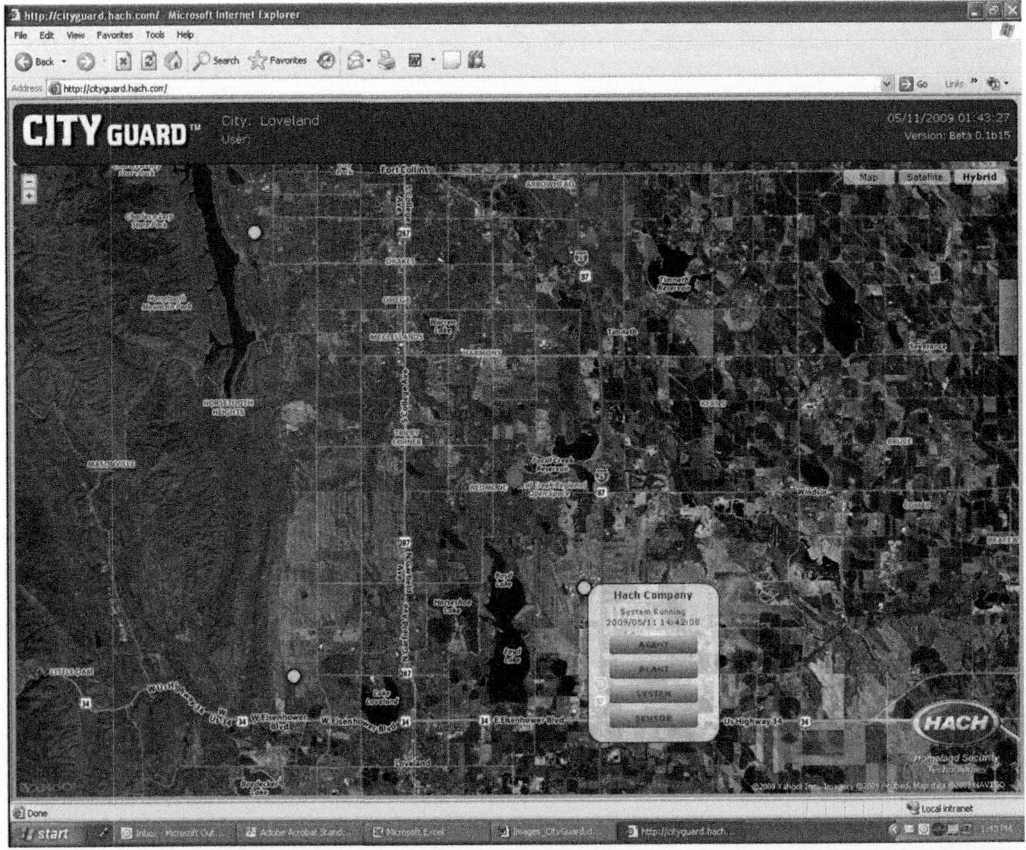

FIGURE 14-20 A simple color-coded icon scheme allows for rapid visualization of which nodes in a system are alarming and which are not. Circles denote monitoring nodes. Green nodes are not alarming. Red nodes are alarming. (For interpretation of the references to color in this figure legend, the reader is referred to the online version of this book.)

14.9 Identification of Contaminant Sources and Prediction

This system allows for the overlay of early warning system monitoring sites with street maps and aerial photographs so that a detailed spatial temporal snapshot of the early warning system network can be obtained. This allows easy visualization of monitoring platforms and their alarm status. A simple color-coded (red, yellow, green) icon scheme is used to denote alarm status. This allows for a quick visualization of which nodes in the system are alarming and which are not (Figure 14-20).

This combined with flow information can allow for prediction of which nodes in the system are likely to alarm in the future if an actual event is occurring, and the alarm is not due to other causes. This visualization can be very useful in accelerating public health response, determining isolation schemes to prevent spread of contamination, and reducing damage to infrastructure and cost of remediation. When prompt and appropriate action is taken water service can be restored on a timely basis.

Simply clicking on a node will bring up a window with the online data for that site. Multiple sites can be brought up at once to compare instrument readings (Figure 14-21).

FIGURE 14-21 Data from multiple sites can be simultaneously accessed by simply clicking on their icons. (For color version of this figure, the reader is referred to the online version of this book.)

This allows evaluation of whether predictions about plume movement within the system are correct without visiting the sites.

The benefits of such a software package as the one described here are that it allows you to derive enhanced value from remote monitoring while making operation and maintenance less onerous. The software allows the viewing of multiple monitoring points at a glance. This combined with the mapping function allows users to better visualize and understand patterns of sensor response to events, which in turn can be used to streamline response.

The system gives the user bidirectional control of multiple monitoring stations at the same time from any location with Internet access. This allows a savings in time and costs associated with travel to remote monitoring sites. End users can also drill down, view and analyze water quality data from multiple sites without leaving the office.

14.10 Conclusions

Extensive laboratory and pipe loop tests indicate that multiparameter systems are a good choice for detecting water quality excursions that could be linked to water security events. The deployment incidents detailed in this paper confirm this and also demonstrate the applicability of utilizing these everyday parameters by linking them with advanced algorithms. It is foreseeable that these devices will become a much more valuable tool than the purpose for which they were originally designed. While their antiterrorism function is important, their day to day application is the key to adoption and successful implementation by end users. They are rapidly becoming a critical tool for improving everyday operations.

For example, through many years of experience, seasoned treatment plant operators have developed "a sense" for knowing something in the treatment system is amiss. Generally, similar experience-based knowledge is not present as it concerns the distribution system because there has typically been little measurement done upon which to gain these senses and, therefore; bulk parameter monitoring in the distribution system with interpretive algorithms has the potential to become the artificial sense able to quickly "learn" the quirks of the distribution system and have those quirks labeled by those with extensive experience so that less experienced employees have the benefit of that knowledge. With the aging of the workforce and rapid employee turnover experience has the chance of quickly dying out. Above and beyond their obvious security benefits, algorithms could be a way to circumvent this loss of knowledge and to build a knowledge base where none has previously existed. This in turn could allow improvements in system operation that may result in cost savings and definitely will result in higher quality water being delivered to the consumer. The use of these technology innovations are not relegated to the distribution system only, but they have wide applicability for ensuring water quality integrity throughout the water supply network from source to tap.

References

[1] FBI Information Bulletin. Potential vulnerabilities of U.S. drinking water and wastewater treatment facilities to insider terrorism. Issued August 11, 2004.

[2] Fluoride Action Network Website. Health effects; examples of acute poisoning from drinking water fluoridation. Available from: http://www.fluoridealert.org/health/accidents/fluoridation.html [accessed 12.07.11].

[3] USEPA 2002. Potential contamination due to cross-connections and backflow and the associated health risks: an issue paper. Available from: www.eap.gov/ogwdw/tcr/pdf/ccrwhite.pdf.

[4] Allman T, Carlson K. Modeling intentional distribution system contamination and detection. J Am Water Works Assoc; January 2005 [Note: The executive summary of this article is still available online but the full text has been pulled from the AWWA website for security reasons].

[5] ETV. Multi-parameter water monitors for distribution systems. A series of EPA environmental technology verification reports performed by Batelle. Available from: http://www.battelle.org/PRODUCTSCONTRACTS/etv/verifications.aspx#W12; 2005.

[6] Hall J, Zaffiro AD, Radha KE, Haught RC, Herman JG. On-line water quality parameters as indicators of distribution system contamination. J Am Water Works Assoc 2007;99(1).

[7] Kroll D. Safeguarding the distribution system: on-line monitoring for security and enhancing operational performance. J New England Water Works Assoc; June 2006.

[8] EPA. Technologies and techniques for early warning systems to monitor and evaluate drinking water quality: a state-of-the-art review. US Environmental Protection Agency Office of Research and Development; 2005. National Homeland Security Research Center Research Report.

Evaluation of Sustainability Strategies

Taha Marhaba*, Ashish Borgaonkar
NEW JERSEY INSTITUTE OF TECHNOLOGY, NEWARK, NJ, USA
**CORRESPONDING AUTHOR*

CHAPTER OUTLINE

15.1 Introduction

15.1.1 Sustainability

The word "sustainability" finds its root in the Latin word "sustinere" (to hold). The word "sustain"—the root word in sustainable or sustainability—has been defined in various ways in several dictionaries. The important definitions include the words

Monitoring Water Quality, http://dx.doi.org/10.1016/B978-0-444-59395-5.00015-7

support, endure, and maintain. However, in the recent past, sustainability has been referred to for some of its specific definitions hinting toward human sustainability on planet Earth.

One of the most widely quoted definitions of sustainability includes that of the Burndtland Commission of the United Nations: "Meeting our needs while not compromising the ability of future generations to meet theirs [1,2]," by Marthis Wackernagel: "Living well within limits of nature [3,4]," and one of the simplest definitions by former UK Minister John Gummer: "Not cheating on our children." The United States Environmental Protection Agency's (USEPA) definition of sustainability is based on a simple principle: We depend totally on our natural environment for everything that we need for our survival and well-being. Sustainability helps to create and maintain conditions under which humans can coexist in harmony with nature, while fulfilling economic, social, and other requirements of present and future generations. Sustainability is important for ensuring that we will continue to have all the resources, especially water and materials, to protect human existence, growth, and the environment [5].

Regardless of the definition, sustainability has three realms or pillars: economic, social, and environmental. All the three realms are highly important toward achieving sustainability, as you cannot have one without the others, especially in the long term. For example, many countries—before the emergence of the term sustainability—have been pushing only for economic development, only to realize now that the environmental costs (resource depletion, pollution, health problems, and frequent occurrences of calamities such as flooding) of such single-minded growth nullify all the gains. To understand the importance of sustainability, it is important to realize that the three pillars of sustainability are intimately intertwined. A vibrant community is absolutely essential for healthy business growth and to have the luxury to care about environmental problems. A healthy economy offers ample revenues and resources to minimize social and environmental problems. Finally, a healthy environment means a healthy supply of resources to sustain economic and social growth for a very long period of time. When we understand these interdependencies, we are in a much better position to make sound decisions and to set long-term goals. A lack of this understanding will surely result in making poor decisions that may give short-term profits with any of the three realms but will result in long-term damages that may not be easily repaired or recovered. Sustainability avoids the trade-off between any two of these three realms and aims to optimize all three. Figure 15-1 indicates that both society and economy are constrained by the limits of the environment.

The Venn diagram (Figure 15-1) of sustainable development shown has been presented several times and in many versions, but the first person to use it was economist Edward Barbier [6]. However, this Venn diagram approach has often been criticized for failing to capture the integration between the three pillars: economic, social, and environmental. Pearce et al. [7] suggested that the Venn approach is inconsistent with the Brundtland Commission Report [1], which emphasized the interrelationship among environmental degradation, economic development, and social pressure. Figure 15-2 presents attempts to enhance the Venn diagram approach

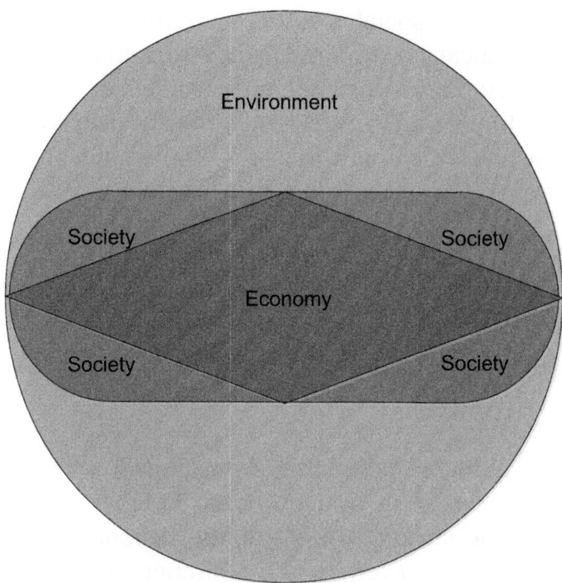

FIGURE 15-1 Three realms of sustainability: economy, society, and environment. (For color version of this figure, the reader is referred to the online version of this book.)

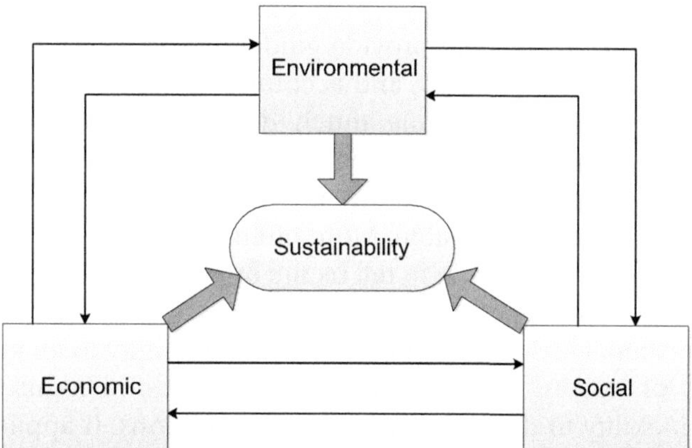

FIGURE 15-2 Relationship among the three realms of sustainability. (For color version of this figure, the reader is referred to the online version of this book.)

to incorporate the interdependencies among the three pillars: economic, social, and environmental.

15.1.2 Sustainable Development

Sustainable development is that which meets our present needs without compromising the needs of future generations. It represents a multifaceted approach to managing our environmental, economic, and social resources for the long term. In this approach, federal, state, tribal, and local governments work together to achieve environmental protection goals and set standards for cooperation between communities, businesses, and governments. The ultimate aim of this collaborative decision-making process is to promote more sensible use of human, natural, and financial resources by creating a widely held ethic of environmental stewardship among individuals, institutions, and corporations [8].

15.2 Summary of Methods for Evaluation of Sustainability Strategies

15.2.1 General

Numerous literature and guidance manuals are available on sustainability. Most of them agree that the biggest challenge in evaluating sustainability strategies is the lack of sufficient understanding and clear vision of the interdependencies of economic, social, and environmental aspects. However, most researchers also agree that evaluation of sustainability strategies within plans, legislations, policies, and projects is essential for a sound, all-inclusive, and informed decision-making process. Evaluation of sustainable strategies needs to broaden its scope to incorporate interrelations and interdependencies of the three realms. It should also attempt to look at the synergistic and cumulative impacts of management practices to provide guidelines on a comprehensive level that will ensure better clarity, transparency, and accountability in decision making. In 2007, a report for the USEPA stated: "While much discussion and effort has gone into sustainability indicators, none of the resulting systems clearly tells us whether our society is sustainable. At best, they can tell us that we are heading in the wrong direction, or that our current activities are not sustainable. More often, they simply draw our attention to the existence of problems, doing little to tell us the origin of those problems and nothing to tell us how to solve them" [9].

Evaluation of sustainable strategies or simply sustainability assessment is an assessment of proposed or existing policies, plans, and projects in terms of achieving and maintaining sustainability to determine optimized conditions. It applies a flexible set of techniques and procedures for the assessment of interdependencies, impacts, and implications of the three realms, on policies, plans, and projects in an integrated manner to facilitate net long-term benefits. Lawrence [10] defines sustainability assessment as applying the broad principles of sustainability "to the assessment of whether, and to what extent, various actions might advance the cause of sustainability."

Robert Gibson conducted a study for the government of Canada, Specification of Sustainability-Based Environmental Assessment Decision Criteria and Implications for Determining "Significance" in Environmental Assessment [11]. In his study, he suggested that "Sustainability-based environmental assessment is certainly different from the more common, narrower exercises that typically consider only some aspects of the environment and focus chiefly on negative effects. It is more ambitious, more demanding, and much more positive. But it is also in important ways, not a huge step from present practice and present capabilities." In an ideal-case scenario, evaluation of sustainability strategies should integrate all the three sustainability pillars at every stage in the decision-making process. However, it is oftentimes very difficult and impractical to achieve such integration. Most entities that attempt sustainability assessment, in fact, conduct a "triple bottom line" assessment that lists economic, social, and environmental issues and concerns to be analyzed and considered in decision making. Although these entities provide the groundwork for sustainability-based assessment, in reality, a widely accepted and truly integrative sustainability-based framework/mechanism to evaluate sustainability strategies is yet to be established. A framework for sustainability assessment should apply on all levels and across the range of the project [12]. In their report "Sustainability Appraisal," Smith and Sheate [13,14] suggest that a true sustainability assessment process would involve and document what changes would be needed to build existing sustainability appraisal practice up to preset standards. Devuyst [15], in How Green Is the City?, suggests that the framework to evaluate sustainability strategies should be flexible enough to accommodate several different applications while operating within the limits of the environment's carrying capacity in which the activity takes place. Rees [16] provides four initial steps to adapting sustainability:

- **(a)** extend the scope of the activity subject to cover the full range of ecologically and socially relevant public and private sector proposals and actions,
- **(b)** create a variety of institutional frameworks for sustainability assessment adapted to the increased diversity of initiatives and activities to be assessed,
- **(c)** develop methods for environmental assessment that reflect the discontinuous temporal and spatial dynamics, and the resilience properties of ecosystems, and
- **(d)** implement the preceding as part of a broader planning and decision-making framework that effectively recognizes ecological functions as limiting factors.

Integration of qualitative and quantitative information has always been a critical issue for the evaluation of sustainability strategies. Only a few techniques are able to combine quantitative and qualitative data in an effective and accurate manner to aid in decision making. Most methods are based on a qualitative rather than on a quantitative approach, mainly because of the complexity involved in fully analyzing a variety of quantitative data. This is not to say that well-defined quantitative analysis tools are not available to evaluate sustainability alternatives. In fact, many of the standard statistical analysis techniques can be modified effectively to integrate economic, social, and environmental factors in the sustainability assessment.

15.2.2 Strategic Environmental Assessment

Strategic environmental assessment (SEA) has been practiced for quite some time in the US, especially in the state of California, and in countries such as Australia, Canada, and

New Zealand [17–23], as well as in the European Union (EU), particularly for urban planning in the Netherlands [24]. SEA is also used for strategic planning and project appraisal in Hong Kong [25]. Most Organization for Economic Cooperation and Development (OECD) countries now have formal provisions for SEA, while those that do not (e.g., Austria, Ireland, Luxembourg, Portugal, and the accession countries of Central and Eastern Europe) are already in the process of adopting the SEA (they will be required to adopt it as a condition of the EU membership) [26]. Sadler [26] further comments that even though an increasing number of countries and international organizations now undertake some form of SEA, general approaches, institutional arrangements, requirements, scope of application and procedures vary considerably. Also, the extent to which SEA accommodates sustainability and the extent to which intended SEA goals are met depend on a variety of things including a skill set available within the organization, availability of dialog, financial and human resources, flexibility afforded by decision makers, and the degree to which overall strategies of sustainability are established [12,26–28].

15.2.3 Integrated Assessment

Integrated assessment is being practiced in several countries, such as Canada, Denmark, Hong Kong, New Zealand, and UK [18,29]. Some of the prominent areas attracting an integrated assessment approach include project preparation and appraisal, and water management projects. Kolhoff et al. [30] in "Towards Integration of Assessments, with Reference to Integrated Water Management Projects in Third World Countries," present the inadequacies of current or "aspect-by-aspect" approaches and the need for the development of a conceptual framework for integration of assessments. They further suggest that a lack of the capacity to encompass crosslinks between aspects, "aspect-by-aspect" approaches carry "the risk of misjudgment of impacts which may compromise the quality of the project proposal and its appraisal" [30].

Drawing on Dalal-Clayton's theory of the world as a system with three subsystems (social, economic, and environmental) working together and facilitating human existence, Kolhoff et al. assert that an integrated impact assessment framework is a must "to improve the dependability and reliability of the information covering the three subsystems on which decision making for development projects is based" [12,27,30]. This framework would analyze "the effects of proposed and alternative interventions on the three subsystems and their intersystem relations" and attempt "to identify beneficial interventions and to explicate unavoidable trade-offs"; it would use existing assessment methods and analytical tools [12,27,30]. Higher-level integrated assessment would have to focus primarily on working routines to enable sufficient management of the multidisciplinary process [30]. Kolhoff et al. conclude: "The function evaluation method may be an acceptable basis for analysis for all disciplines, and may thus play an instrumental role in process management. Work has to be done to complement the function evaluation method; for instance, cultural aspects will have to be incorporated. The water sector is an excellent area to focus on in the process of methodology development [30].” Kolhoff et al. also list techniques used in integrated assessment as presented in Table 15-1.

The success of sustainability assessment depends on the effectiveness of integration of information available. Eggenberger and Partidário have identified five different possible forms of integration [29], as presented in Table 15-2.

15.2.4 Decision-Aiding Techniques for Evaluation of Sustainability Strategies

As discussed earlier, integration of qualitative and quantitative information has always been a critical issue for the evaluation of sustainability strategies. Only a few techniques are able to combine quantitative and qualitative data in an effective and accurate manner to aid in decision making. Most methods are based on a qualitative approach rather than on quantitative approach, mainly because of the complexity involved in fully analyzing a variety of quantitative data. This is not to say that well-defined quantitative analysis tools are unavailable to evaluate sustainability alternatives.

“Decision-aiding techniques” involve stakeholders coming together to analyze alternatives when a decision is to be made. These techniques are significant to the evaluation of sustainability alternatives [18]. Decision-aiding techniques generally involve a set of alternatives, a set of criteria for comparing the given alternatives, and a method for ranking the alternatives based on how well they satisfy the criteria.

Annandale and Lantzke [18] list seven steps in the process:

(1) Specifying alternatives
(2) Specifying criteria
(3) Scoring alternatives

Table 15-1 Techniques Used in Integrated Assessment

Assessments	Environmental impact assessment (EIA)
	Health impact assessment
	Social impact assessment
	Poverty assessment
	Gender assessment
Analytical tools	Function evaluation
	Problem in context analysis
	Participatory appraisal techniques
Economic valuation methods	Using market values
	Using surrogate or estimated values
	Macroeconomic models
	Contingent valuation methods
Decision support	MCA
	Objective-oriented project planning

MCA=Multicomponent analysis
Adopted from Ref. [30].

Table 15-2 Forms of Integration

Form of Integration	Description
Substantive integration	Physical and biological issues with social and economic issues Health risk, biodiversity, climate change, and other rising issues Global and local issues
Methodological integration	Environmental, economic, and social (impact) assessment approaches such as cumulative assessment, risk assessment, technological assessment, CBA, and multicriteria analysis. Different applications and experiences with the use of particular tools such as a geographical information system The integration and clarification of terminologies
Procedural integration	Environmental, social, economic planning/assessment, spatial planning, and EIA Approval/licensing processes, spatial planning, and EIA The adoption of coordination, cooperation, and subsidiary as guiding principles for (governmental) planning at different levels of decision making Affected stakeholders (public, private, nongovernmental organizations in the decision-making process) Professionals in a truly interdisciplinary team
Institutional integration	The provision of capacities to cope with emerging issues and duties The definition of a governmental organization to ensure integration The exchange of information and possibilities of interventions between different sectors The definition of leading and participating agencies and their respective duties and responsibilities
Policy integration	Sustainable development as overall guiding principle in planning and EIA The integration of regulations The integration of strategies The timing and provisions for political interventions Accountability of government

Adopted from Ref. [29].

(4) Assigning weights to criteria
(5) Undertaking the computation (using a computer spreadsheet or a similar program)
(6) Dealing with uncertainty (reviewing)
(7) Presenting results

Consider the following basic decision-aiding model. In a waste management policy choice for a local council, there might be three viable alternatives: (a) traditional landfill, (b) incineration, and (c) composting combined with landfilling of residual waste. This might be analyzed based on criteria to compare alternatives:

- Capital cost
- Employment potential
- Area of land required
- Possibility of groundwater pollution

There is no guarantee that the best choice of alternatives will be obvious. One alternative may rank the highest under one criterion but may rank the lowest under another. Clearly,

Table 15-3 Example of Impact Matrix

	Options		
Criteria	**Conventional Landfill**	**Incineration**	**Composting and Residual Landfill**
Capital cost ($ million)	25	40	12
Area of land required (ha)	150	15	40
Employment (thousands)	25	12	60
Groundwater pollution potential	High	Minimal	Low

Adopted from Refs. [17,18].

the decision is not straightforward and will only increase in complexity with the availability of the number of options and of criteria. Annandale and Lantzke [18] concluded that "decision-aiding techniques can help in situations such as this by comparing the advantages and disadvantages of each alternative, one against the other." An "effects table" or "impact matrix" can be used (Table 15-3) where "scores" are entered in cells or "expert judgment" (or a panel of community members) sets values (as in the last criterion).

Table 15-4 presents strengths and weaknesses of decision-aiding techniques.

The following section provides a brief overview of a selection of multicomponent analysis (MCA) techniques (also called multicriteria analysis techniques) that could be used for the evaluation of sustainability strategies [31,32]. Different existing decision-aiding techniques could be applied to the evaluation of sustainability strategies, mainly drawn from applied statistics [31,32]. The following section discusses a few such techniques, namely, weighted summation, concordance/discordance analysis, planning a balance sheet (PBS), and a goals-achievement matrix (GAM).

Table 15-4 Strengths and Weaknesses of Decision-Aiding Techniques

Strengths	**Weaknesses**
Structured decision making with allowances when flexibility is needed. Arranges information clearly in complex problems. Allows decision makers to use available data, whether qualitative or quantitative. Weights make opinions/values explicit. Does not depend on assigning monetary values to all factors.	Does not overcome the fundamental problem associated with comparing quantities that some would argue are not comparable but does provide more flexibility than is available with, say, benefit–cost analysis. Does not provide guidance on which evaluation methods to use. Since many of the methods are complex and remain a "black box" to the decision maker, they can lead to either distrust or excessive faith in the results. Decision makers sometimes neglect to introduce implicit weights at all stages of the analysis because they focus too much on the definition of explicit weights. Methods for incorporating uncertainty explicitly into the analysis are not yet well-developed.

Adopted from Refs. [17,18].

15.2.5 Multicomponent Analysis

15.2.5.1 Weighted Summation

Weighted summation simply pertains to additive weighting and is one of the important forms of MCA. Weighing is based on many alternatives, several criteria to judge them, and a ranking system to differentiate among them. However, when it comes to alternatives that may not be ranked or quantitatively rated (especially social alternatives), weighted summation is of little use.

15.2.5.2 Concordance/Discordance Analysis

Concordance/discordance analysis uses an approach to calculate concordance index and discordance index for pairwise comparisons between alternatives (let us say A and B). The sum of the weights for criteria for which alternative A scores better than alternative B is the concordance index. Concordance analysis is useful only when a mixture of quantitative and qualitative data is available. Similarly, the discordance index suggests whether alternative A is worse than alternative B. It is the largest difference in the score for all the criteria for which alternative A does worse than alternative B. Concordance index and discordance index together produce a ranking of alternatives.

The major disadvantage of this technique is the dependence on ratio-scaled data.

15.2.5.3 PBS and GAM

These methods came into existence mainly because of the limitations of cost–benefit analysis (CBA) in evaluating sustainability strategies. PBS was developed in the UK and, along with the GAM approach, is used frequently in the US and the UK.

PBS is an extension of CBA in two ways: It has the ability to record and process detailed information on cost distribution, and it accounts even for intangible impacts.

GAM was developed to overcome deficiencies of both CBA and PBS. GAM is a step beyond PBS, such that impacts are organized and categorized more efficiently under specifically stated goals and according to the community groups that are affected. PBS does not overcome difficulties in handling impacts that may not be quantified, whereas GAM is only as good as the scaling or ranking system adopted.

In conclusion, it can be said that the above decision-aiding techniques depend on the availability of effective and accurate information. Only a few are able to analyze both qualitative and quantitative data. The choice of the technique based on available data and the type of decision to be made (application) is highly critical.

15.3 Towards a Sustainable Future

There are many processes around the world that present good examples of sustainable development that give equal consideration to economic, social, and environmental factors. Many national and international organizations have adopted sustainability in some or the other form and at one or more levels. Some of these organizations include the World Trade Organization, OECD the Dow Jones Sustainability Indexes, and the Global Mining Initiative, Mining Minerals Sustainable Development assessment process, and the World Economic Forum. Further, different levels of government, to varying degrees, are adopting sustainability and using sustainability assessments. Table 15-5 presents some other examples of various organizations that are sincerely pursuing sustainability.

15.3.1 Benefits and Risks

It is a common misconception that if a sustainable approach is adopted, it may result in added production costs for most products and companies. However, there are several benefits of pursuing sustainability. Hitchcock and Willard, in their book, "The Business Guide to Sustainability: Practical Guides and Tools for Organizations," list several benefits of pursuing sustainability, disadvantages of not pursuing sustainability, and risks involved in pursuing sustainability [33]. The benefits include the following:

(1) Reduction in energy and waste costs: the primary goal for many organizations is to achieve zero waste status, which is not possible without a solid sustainable approach. This results not only in the elimination of disposal costs but it also offers the potential to make money by selling residual products that were formerly discarded as waste. The Ford Motor Company at Detroit, Michigan, has adopted a firm sustainable approach with many innovative techniques, including solar panels, green roofs, and porous parking decks to minimize point source wastewater discharge.

(2) Standing out in competition: until such a time when sustainability is a standard practice, companies can stand out firmly by differentiating themselves from the competition with a sustainable approach. Scandic Hotels in Sweden adopted a sustainable approach and used it as a story to be told in their advertising campaign. This resulted in a significant increase in the number of guests, who not only found a bed to sleep in but also became part of an exciting trend.

(3) Leaping over of future regulations: the only certainty about regulations is that they continuously change. In the environmental sector, more and more stringent regulations are adopted and become a standard practice. The pursuit of a sustainable approach helps cope with these ever-changing regulations even before they are issued. This is a perfect opportunity to get ahead of the curve and be better prepared for future regulations. A Swiss textile giant switched over to benign chemicals, which resulted in their product becoming biodegradable, and their discard, which was formerly a hazardous waste, is now becoming a new product.

(4) Opening of new markets and customers: economic sustainability opens up many new customers and new markets. Amul, a dairy giant in India, realized that their main cost pertained to refrigeration of milk products. The company came up with a newer and cheaper way to store all their products, which resulted in significant cost reductions. Amul was now able to sell an ice-cream scoop for INR 1 and, in turn, was able to cater

Table 15-5 Examples of Various Organizations That are Sincerely Pursuing Sustainability

Government Agencies						
US	**ROW**	**Energy Sector**	**Academic Institutions**	**Manufacturing Industries**	**General Services**	**Food Industry**
Cities:	*Countries*:	BP	Cornell	Coca Cola	*Finance*:	Bon Appetit
Chicago, IL	Australia	Conoco-Philips	George Washington	Dell	Bank of America	Chiquita
Madison, WI	China	Florida Power and	Harvard	DuPont	Calvert	Fetzer Winery
San Francisco, CA	European Union	Light	Imperial College –	Electrolux	Goldman Sachs	Frito Lay
Santa Monica, CA	Japan	Public Service Electric	London	Epson	Munich Re	Heinz
Department of	New Zealand	and Gas Company	Iowa State	General Electric	Swiss Re	Unilever
Defense	Sweden	(PSE&G)	Massachusetts Institute	Herman Miller	*Other*:	
States Departments	UK	Royal Dutch Shell	of Technology	Hewlett-Packard	Aspen Skiing	
of Environmental	*Cities and States*:		Oregon State	IKEA	Company	
Protection (DEP)	Bogota, Columbia		Portland State	Intel	Kaiser Permanente	
USEPA	Curitiba, Brazil		Stanford	Interface	Price-Waterhouse-	
National Aeronautics	Kerala		Yale	Johnson Controls	Coopers	
and Space	Whistler		University of British	Mattel	Starbucks	
Administration			Columbia	Nike	Wal-Mart	
States:			University of California	Philips		
Arizona			(Berkeley, San Diego,	RR Donnelley & Sons		
Massachusetts			Santa Barbara)	*Vehicles*:		
Minnesota			University of East	Ford		
New Jersey			Anglia, UK	General Motors		
North Carolina			University of Victoria	Toyota		
Oregon			University of Virginia –	Volkswagen		
Washington			Darden			

Adopted from Ref. [33].

to 300 million + new customers, who could not afford their products earlier. This not only increased Amul's revenues significantly but it also gave them a sound competitive advantage.

(5) Creation of innovative new products and processes: sustainability presents a clear picture of the world's present and future needs. This gives anyone an advantage of being able to develop new products and processes that not only will generate more money but can also become part of the overall solution. ShoreBank Pacific from Ilwaco, Washington, started offering funds to companies adopting a sustainable approach. This attracted many clients from throughout the US and also encouraged many to pursue sustainability in the hope of obtaining a firm funding source. Toyota was one of the first to develop hybrid vehicles. This not only gave them a good name but it also generated side revenue for selling this technology to other car manufacturers.

(6) Attracting and retaining of the best employees: many qualified employees nowadays have a good sense of environment and sustainability. They want to work for environmentally friendly companies that share their ideas of a sustainable future. The work seems more meaningful and exciting. Bend, Oregon, is housing an entire cluster of renewable energy companies and is attracting the best employees even in this relatively remote area.

(7) Provision of a higher quality of life: sustainability helps communities adopt smart growth design principles. The whole city of Curitiba, Brazil, by integrating a design of urban planning, social programs into their transportation system, is designed to provide a much better quality of life to all the citizens, while reducing air pollution.

(8) Improvement of public image: green companies are favored in most markets. People show faith in a sustainable organization with the belief that an organization that cares for the future generations will also provide the best products for its customers. It will not only provide good publicity but in many cases free mouth-to-mouth publicity.

(9) Reduction of insurance costs and legal risks: while adopting sustainability, all organizations must keep an eye on social and environmental practices of present and future generations—this is exactly what is also necessary to manage and minimize risk. OKI, a semiconductor company, eliminated certain hazardous chemicals from their process chain, and this resulted in a significant reduction in their insurance costs.

In addition to the aforementioned benefits, organizations practicing sustainability will also minimize threats from changing regulations by staying ahead of worldwide trends. Several artificial disasters could have been avoided with a sustainable approach. Sustainability will also aid in reducing the strong effects of climate change around the world. Hitchcock and Willard [33] list the following threats one should watch out for before ignoring sustainability.

(1) Bad mouthing of products

(2) Getting closed out of certain markets

(3) Exceeding legal risks
(4) Damaging the image of your firm
(5) Increasing the liability for pollutants
(6) Shortage of raw materials and energy

Risks involved in adopting a sustainable approach: Sustainability is not all green—all the time. It is a very serious matter and should be treated as such. Most of the risks involved in adopting a sustainable approach have to do with half-hearted attempts toward attaining a sustainable development. Hitchcock and Willard, in their aforementioned book [33], discuss various risks in adopting a sustainable approach.

(1) Greenwashing: organizations often start publicizing of sustainability without backing their efforts with a sound action plan. Such organizations then get accused of greenwashing. The larger the organization, the bigger is the risk involved. Of course, one can easily avoid this by adopting a modest approach and a sound action plan rather than boasting about it without much stuff in the basket.
(2) Cannibalizing your business: more often than not, sustainability presents alternatives that completely oppose the core business idea of the organization. For example, some transportation companies are working on reducing the transport of material. Many organizations are forced to risk such decisions, but this risk is small compared to that of someone else doing this before you and presenting you with major revenue problems.
(3) Unrealistic expectations: sustainability is an excellent tool to possess, but it may not solve all the problems. Sustainability offers a sense of passion, purpose, and urgency, but it may still not be enough. There will always be someone who would argue that you can do more. It is very important to be realistic when it comes to benefits of sustainability, while having a good understanding of the risks involved.

15.3.2 Economic and Social Sustainability

Economic sustainability is a set of actions taken by the present generation that allows future generations to enjoy wealth, utility, and welfare. Economics plays a key role in sustainable development. Social and environmental consequences are valued or ranked based on their economic impact. Sustainability is much bigger than mere resource management, welfare, and profit margins. Economic integration into sustainability enables firms to grow steadily for longer periods of time. A shortsighted approach of ignoring sustainability in view of immediate economic gains often results in long-term and sometimes in damage to steady growth.

United Nations Environment Program defines green economy as "improvement in human well-being and social equity, while significantly reducing environmental risks and ecological scarcities." Green economics generates increases in wealth (especially a gain in ecological commons or natural capital), and offers a higher rate of growth in the gross domestic product (over a period of six years). Green economy creates new jobs and in turn minimize losses. Economic analysis and reform should target environmental effects of economic growth.

Sustainability often requires implementation of changes to well-established methods. This is more often than not a social challenge. Several key relationships, such as peace, security, social justice, sustainability, and poverty and human relationship to nature, must be considered while taking account of the social dimension of sustainability.

15.3.3 Environmental Sustainability

Environmental sustainability calls for an idea of keeping the environment as pristine as humanly possible in all processes of interaction with the environment. If the sum total of nature's resources is used at a faster rate than they can be replenished naturally, an unsustainable situation is the result. Sustainability calls for the means to contain the rates of consumption or to develop ways to replenish resources at a faster rate. Environmental degradation over prolonged periods could result in the extinction of humanity. Sustainable development demands a firm understanding of the concept of carrying capacity.

A sustainable water management system is one that meets the needs of the environment and the population and can continue to supply those needs, undiminished indefinitely.

The US uses an average of 410 billion gallons of water per day. Two-hundred and one billion gallons are used to produce 89% of the country's energy needs. One-hundred twenty-seven billion gallons are used for irrigation. Forty-five billion gallons are used for public consumption. Thirty-seven billion gallons are used for livestock, industry, aquaculture, and mining. To accomplish sustainable water management, it is necessary to determine the following:

- the quality and quantity of water needed to meet all stakeholder needs,
- the quality and quantity of water remaining and flows/volumes available to meet those needs,
- factors impacting the quality and quantity of water and possible remedies to minimize the impact,
- conflicts between the stakeholders' needs and implementable solutions to solve these conflicts,
- periodic assessments of water management effectiveness.

The list of impaired water sources in the US has grown significantly in the past several decades. As shown in Table 15-6, the majority of water sources are lakes and reservoirs, and they are impaired by 66%. The principal causes of impairment in water quality include sediments, pathogens, nutrients, metals, dissolved oxygen, temperature, pH, pesticides, mercury, organics, ammonia.

Table 15-6 Water Sources in the US

	Rivers & Streams (miles)	Lakes, Reservoirs & Ponds (acres)	Bays & Estuaries (sq. miles)	Costal Shorelines (miles)	Wetlands (acres)	Great Lakes Shoreline (miles)
Good waters	464,764	5,926,646	6687	1287	1,304,892	75
Threatened waters	6355	47,330	17	–	805	–
Impaired waters	463,698	11,602,447	11,740	791	746,163	1110
Total assessed waters	934,799	17,576,423	18,443	2078	2,051,861	1184
% Impaired (assessed)	50%	66%	64%	38%	36%	94%

Adopted from various USEPA sources.

Some important facts that need to be considered about water availability in the US include the following:

- rapid growth of the nation's population, urban expansion, and development, demanding clean fresh water
- increased salinity caused by inefficient irrigation and overextraction in coastal developed areas
- water shortages and degradation in water quality resources
- decline in groundwater levels and aging water infrastructure.

Water availability impacts power supply, the largest stakeholder, and societal and economic infrastructure sustainability.

15.3.4 Sustainable Solutions to Water Quality Problems

Sustainable technologies and strategies are needed for combined sewer overflow treatment, stormwater treatment, wastewater treatment, and water treatment. In addition, the development and application of rapid, efficient techniques for detecting water quality problems and sources, uniform application of analytical parameters to monitor water quality, and effective management of water resources and allocations to all stakeholders. Some examples of sustainable technology solutions are as follows:
Stormwater treatment:

- Bioretention and infiltration drain field runoff is conveyed as sheet flow to the treatment area, which consists of a grass buffer strip, sand bed, ponding area, organic layer or mulch layer, planting soil, and plants.
- Water quality inlets, oil/grit separators, or oil/water separators entail sustainable technologies that consist of a series of chambers that promote sedimentation of coarse materials and separation of free oil (as opposed to emulsified or dissolved oil) from stormwater.
- Sand filters—able to achieve high removal efficiencies for sediment, biochemical oxygen demand (BOD), and fecal coliform bacteria.
- Hydrodynamic separators—settling or separation unit to remove sediments and other pollutants. Hydrodynamic separators are ideal for use in potential storm-water "hot

spots"—such as areas near gas stations, where higher concentrations of pollutants are more likely to occur. These can be installed via manholes.

Wastewater treatment:

- The living machine—an emerging sustainable wastewater treatment technology that uses a series of tanks, which support vegetation and a variety of other organisms (microorganisms, protozoa, snails, and plants). Ref. *Advanced Ecologically Engineered System* or AEES.
- Slow rate land treatment is the sustainable controlled application of primary or secondary wastewater to a vegetated land surface.
- Rapid infiltration land treatment—remedy, including filtration, adsorption, ion exchange, precipitation, and microbial action, occurs as the wastewater moves through the soil matrix. Phosphorus and most metals are retained in the soil, whereas toxic organics are degraded or adsorbed. It is rapid and requires little space.
- Wetlands: subsurface flow (SF)—very effective and reliable for removal of BOD, chemical oxygen demand, total suspended solids, metals, and some persistent organics in municipal wastewaters. SF wetland is also capable of removing nitrogen and phosphorus to low levels, and produces no residuals.

Water treatment:

- Advanced Oxidation Processes
  - Ultraviolet
  - Peroxide
  - Ozone
- Membrane Processes

15.4 Conclusions

The following summarizes the points discussed in this chapter:

- common goals, public and environmental health protection
- participation in federal rule making
- interaction with the states
- participation in stakeholder processes at the local watershed level
- participation and interaction with water utilities
- sharing of information
  - helps balance interconnected resources
    - maintains passing flows
    - maintains water quality
    - maximizes the water to be stored in upstream reservoirs
  - helps build a consensus on actions to be taken

- helps determine the presence and control of nonregulated contaminants, antibiotics, pesticides, etc.
- common use of parameters and cost-effective techniques to monitor spatial and temporal variations of water quality

References

[1] Nations U. Annex to report of the world commission on environment and development (WCED); 1987.

[2] Brundtland GH. Our common future: the world commission on environment and development. UN; 1987.

[3] Wackernagel M, Rees W. Our ecological footprint: reducing human impact on the earth; 1996.

[4] Niccolucci V, Bastianoni S, Tiezzi EBP, Wackernagei M, Marchettini N. How deep is the footprint? A 3D representation. Ecol Modell 2009;220(20):2819–23.

[5] USEPA. Sustainability: basic information. Available from: http://www.epa.gov/sustainability/basicinfo.htm; 2011 [accessed 12.22.11].

[6] Barbier EB. The concept of sustainable economic development. Environ Conserv 1987;14(2):101–10.

[7] Pearce D. Blueprint 3: measuring sustainable development; 1993.

[8] Novotny V, Ahern J, Brown P. Water centric sustainable communities: planning, retrofitting and constructing the next urban environments. Hoboken, NJ: Wiley & Sons; 2010.

[9] USEPA. Can indicators and accounts really measure sustainability? Considerations for the U.S. Environmental Protection Agency. USEPA; 2007.

[10] Lawrence DP. Integrating sustainability and environmental impact assessment. Environ Manage 1997;21(1):23–42.

[11] Gibson R. Specification of sustainability-based environmental assessment decision criteria and implications for determining "significance" in environmental assessment. Available from: http://www.sustreport.org/downloads/SustainabilityEA.doc; 2001 [accessed 12.22.11].

[12] Dalal-Clayton B. Modified EIA and indicators of sustainability: first steps towards sustainability analysis. In: Proceedings of the twelfth annual meeting of the international association for impact assessment, Washington DC; 1992. p. 19–22.

[13] Smith SP, Sheate WR. Sustainability appraisal of English regional plans: incorporating the requirements of the EU strategic environmental assessment directive. Impact Assess Proj Apprais 2001;19(4):263–76.

[14] Smith SP, Sheate WR. Sustainability appraisals of regional planning guidance and regional economic strategies in England: an assessment. J Environ Plan Manage 2001;44(5):735–55.

[15] Devuyst D, editor. How green is the city? Sustainability assessment and the management of urban environments. New York: Columbia University Press; 2001.

[16] Rees WE. A role for environmental assessment in achieving sustainable development. Environ Impact Assess Rev 1988;8(4):273–91.

[17] Annandale D, Bailey. Asian Development Bank strategic environmental assessment report. Perth, Western Australia: Institute for Environmental Science, Murdoch University; 1999.

[18] Annandale D, Lantzke R. Making good decisions: a guide to using decision-aiding techniques in waste facility siting; 2001.

[19] Arce R, Gullon N. The application of strategic environmental assessment to sustainability assessment of infrastructure development. Environ Impact Assess Rev 2000;20(3):393–402.

[20] Bailey JM, Saunders AN. Ongoing environmental impact assessment as a force for change. Proj Apprais 1988;3(1):37–47.

[21] Morrison-Saunders A. Environmental impact assessment as a tool for ongoing environmental management. Proj Apprais 1996;11(2):95–104.

[22] Morrison-Saunders A, Annandale D, Cappelluti J. Practitioner perspectives on what influences EIA quality. Impact Assess Proj Apprais 2001;19(4):321–5.

[23] Weaver A, Pope J, Morrison-Saunders A, Lochner P. Contributing to sustainability as an environmental impact assessment practitioner. Impact Assess Proj Apprais 2008;26(2):91–8.

[24] Nooteboom S. Environmental assessments of strategic decisions and project decisions: interactions and benefits. The Netherlands: Ministry of Housing, Spatial Planning and the Environment; 1999. DHV Environment and Infrastructure.

[25] Devuyst D. Linking impact assessment and sustainable development at the local level: the introduction of sustainability assessment systems. Sustain Dev 2000;8(2):67–78.

[26] Sadler B. Strategic environmental assessment: institutional arrangements, practical experience and future directions. International workshop on strategic environmental assessment. Tokyo, Japan: Japan Environment Agency; 1998.

[27] Dalal-Clayton B, Sadler B. Strategic environmental assessment: a rapidly evolving approach. Available from: http://www.nssd.net/References/KeyDocs/IIED02.pdf; 2002. [accessed 12.22.11].

[28] Baxter T. The sustainability assessment model (SAM). In: SPE international conference on health and safety and environment in oil and gas exploration and production; 2002. Kuala Lumpur, Malaysia.

[29] Eggenberger M, Partidario MDR. Development of a framework to assist the integration of environmental, social and economic issues in spatial planning. Impact Assess Proj Apprais 2000;18(3):201–7.

[30] Post RAM, Kolhoff AJ, Velthuyse BJAM. Towards integration of assessments, with reference to integrated water management projects in third world countries. Impact Assess Proj Apprais 1998;16(1):49–53.

[31] Garfi M, Ferrer-Marti L, Bonoli A, Tondelli S. Multi-criteria analysis for improving strategic environmental assessment of water programmes. A case study in semi-arid region of Brazil. J Environ Manage 2011;92(3):665–75.

[32] Avram O, Stroud I, Xirouchakis P. A multi-criteria decision method for sustainability assessment of the use phase of machine tool systems. Int J Adv Manufact Technol 2011;53(5–8):811–28.

[33] Hitchcock DE, Willard ML. The business guide to sustainability: practical strategies and tools for organizations. Earthscan/James & James; 2009.

Index

Note: Page numbers with "f" denote figures; "t" denote tables.

A

Printed and bound by CPI Group (UK) Ltd, Croydon, CR0 4YY
08/05/2025
01864822-0004